CHARIOTS OF FIREFIGHTERS

THE HISTORY AND PRACTICE OF FIREMATIC COMPETITION
IN
NEW YORK STATE

ISBN #978-0-578-02358-8

Acknowledgements

The author would like to extend his sincere thanks and gratitude to all the museums, historical societies, libraries, fire service organizations, officials and individuals listed below, without whose help the completion of this book would not have been possible:

The Museums

The New York State Fire Museum at Hudson
The F.D.N.Y. Fire Museum
The Brookhaven Town Volunteer Firefighters Museum
The Firemen's Memorial Exhibit Center of Western New York
The Sag Harbor Fire Museum
The Aurora Regional Fire Museum
The Thousand Island Museum

The Historical Societies

The Babylon Historical Society
The Historical Association of South Jefferson County
The Niagara County Historical Society
The Suffolk County Historical Society

The Libraries

The Amherst Public Library, Williamsville, N.Y.
The Brockport Public Library
The Brooklyn Public Library
The Dansville Public Library
The East Hampton Public Library
The F.D.N.Y. Mann Library
The Flower Memorial Library, Watertown, N.Y.
The Hawn Memorial Library, Clayton, N.Y.
The John Jermain Public Library, Sag Harbor, N.Y.
The Jervis Public Library, Rome, N.Y.
The Lockport Public Library
The Mascherry Library, Alexandria Bay, N.Y.
The New York Historical Society Library
The New York Public Library
The Niagara Falls Public Library
The North Tonawanda Public Library

The Libraries (Cont'd)

The Ogdensburg Public Library
The Onondaga Public Library, Syracuse, N.Y.
The Patchogue-Medford Library
The Potsdam Public Library
The Queens Borough Public Library
The Richmond Public Library, Batavia, N.Y.
The Riverhead Public Library
The Rochester Public Library
The Swan Public Library, Albion, N.Y.
The Warsaw Public Library

The Fire Service Organizations

The Bath, Maine Fire Department
The Firemen's Association of the State of New York
The Greenport Fire Department
The Huntington Fire Department
The New England States Veteran Firemen's League
The Postdam Fire Department
The Suffolk County Volunteer Firemen's Association
The Theresa Fire Department

The Officials

Jack Barry, Mark Jaffie, Tom Leggio, Rich Ligon, Robert McConville, Bill Rowse
...and all the other NYSVFP&DTCA officials

The Teams

The Bay Shore "Redskins"
The Central Islip "Hoboes"
The Copenhagen "Cubs"
The Copiague "Yellowbirds"
The Coram "Hellraisers"
The Elsmere "Wanderers"
The Islip "Wolves"
The Lindenhurst "Snails"
The Main-Transit "Slowpokes"
The New Hyde Park "Termites"

The Teams (Cont'd)

The North Lindenhurst "Piston Knockers"
The Patchogue "Forty Thieves"
The Riverhead "Ironmen"
The Selden "Slowpokes"
The West Carthage "Hornets"
The Hempstead "Yellow Hornets"
The West Islip "Raiders"
The West Sayville "Flying Dutchmen"

...and all the other teams and their members

The Individuals

Miguel Bermuda, John Bigotti, Tim Bigotti, Steve Boerner, Vince Bogert, Neil Brogan, Eugene Bryson, Jeff Champion, Mark Champion, Tom Champion, Floyd Chivvis, Richard Chudy, Kenny Conn, Kevin Connell, Dan Cornell, Kevin Davis, Stan Dixon, Stan Dybus, Jr., George Eysser, Dennis Eickhoff, Dave Falconi, Charlie Fitzpatrick, Jeff French, Kevin Hayes, Bob Henry, Jim Henry, Tom Horn, Sr., Erik Huber, John Ippolito, Jr., Pat Jeffares, Sarah Judd, Jim Kavanaugh, Winiford Kelsey, David Kerkhof, Thomas Kienle, Hugh King, Rob Kryger, Jules Krzenski, Jim Lawida, Jack Lerch, David Lewis, Sue Lund, Don Mauersberger, Tom McGrail, Ray McGuire, Mike McNamara, Chris Murphy, Ralph Nichols, Megan Oleksa, Patrick Patterson, Harry Penny, Herb & Marie Peterson, Matthew Podolski, Lou Reboli, Scott Rice, Walter Sammis, Philip Sartoli, Gary Schaum, Lou Scrima, Don Sherman, Linda Schleher, Dave Smith, Jamie Smith, Miranda Spencer, Paul Suskind, Edith Thompson, Marci Vail, Mike Ward, Rick Webster, Mike Welton, Kevin Wheelock, Mike Whelan, Don Wood, Jay Zinger

...and everyone else who chipped in and was so helpful

The author would like to extend a Special Thanks to his proofreaders, as well as Kevin Hayes, Gary Schaum, Vincent Bogert and the "Historian" – all of you went far above and beyond the call of duty to help me with this project, and for that I owe you a debt of gratitude. This one's for you guys!

TABLE OF CONTENTS

Preface

Firefighters are a strange breed.

I know this, since I have been closely related to the fire service – as both a firefighter and a photojournalist – for twenty years, and if there is one thing I've learned, it's that we are enigmatic: we are dedicated to the point of being obstinate, caring to the point of being overbearing, and loyal to our craft to the point of being narcissistic. But, paradoxically, we are often also at the same time sensitive about what we do, almost to the point of becoming paranoid: as public servants we *know* there are people out there watching us, watching everything we do and every move we make. Part of this can be attributed to simple fascination with firefighters and the fire service – after all, what we do is pretty exciting – but there is also a feeling of a Social Standard to which we are being held, an ideal, of sorts, of behavior that we are constantly being judged against and are constantly expected to meet, seeing as we are, by definition, Public Servants.

On the fireground this is not as much of an issue, because our behavior is dictated by the task at hand; our training, and tactics derived from the particular problem at hand, require us to "perform" properly or the job simply won't get done.

But these same standards – either real or imagined – exist for members of the fire service as much off duty as they do on. Whether at home or at play, we are at all times role models to the community, and, being that we are only human, it's not always easy; the pressure is always on.

This phenomenon, however, has come into being only fairly recently. As you will see, for close to one hundred and fifty years firefighters were considered men that could do no wrong, either at work or at play. Drunkenness, mischief or unruly behavior was forgiven, or folks would simply look the other way; after all, firefighters are "heroes", are they not?

As I began to do my initial research for this book, it therefore came as quite a surprise to me to learn just how *little* people knew about the modern-day sport of firematic drill team competition – even those people whose own fire departments have teams that compete on a regular basis. I learned that it has sometimes been called "the secret sport", simply because so few people have heard of it. I learned also that this was, perhaps, directly or indirectly intentional: it is a sport that has been surrounded in controversy as long as it has been in existence, and in these modern times controversy and firefighters do not mix…at least in any way that they are comfortable with.

But what I found most significant was that this was a sport that has a history that spans close to 200 years, and that – incredibly – it was a history that had, for the most part, never been purposefully documented at all, or even put down on paper in any comprehensive manner. There was information to be found – newspaper articles, single paragraphs in books, or a program from a drill years ago that was rescued from someone's attic – but just finding it at all was a challenge. By way of an analogy, researching this book was like having to search the entirety of New York State for pieces of a jigsaw puzzle that had been dropped from an airplane flying at 40,000 feet: it was necessary to travel far and wide to even find the pieces…and then try to fit them together, without even knowing what the finished picture was supposed to look like. As you will see from the Acknowledgements page (please do take a moment to read it), sources of information came from all over, and fitting them together took time, patience and perseverance. Some of the puzzle pieces are missing, having been lost to time; documents were lost, and many of those who lived during earlier eras of the sport had by this time passed on, taking their stories with them. Others, fortunately, are still with us, and it was from those competitors – both current and former – that I was able to learn a great deal about what there was to know.

I do have one disclaimer: there is no way this book, or any single history of drill team competition, will ever be complete. Like mathematical fractals, there are endless levels beneath levels beneath levels of stories and information about the sport that could be told: the specific history of each of the over 350 teams, or a particular truck, or a funny anecdote about a particular drill on a particular day. To do so would take volumes, and I regret that both time and practicality did not permit me to do so.

But I do say this: if you do find it at all interesting, *do* take the time to learn more. Attend a local drill – you can see where they are being held at any time by visiting www.nysdrillteams.com - and you may be amazed to find out how close they are to you, and how frequently they are held. Take some time to talk to any one of the team members. They may be reticent at first, since as an "outsider" they may not be sure if you are approaching them out of an honest interest in the sport or a desire to hold them up to the torch of public scrutiny, but if you gain their trust you will find them to be engaging, honest, forthright and even funny, as well as filled with a wealth of knowledge about the sport that you – and sadly, almost everyone else – never knew.

I know of no drill team member who doesn't wish for the sport to survive and grow, if perhaps not to the degree of fame it reached in its "glory days" at the turn of the twentieth century, but at least to the point where he or she feels it is once again accepted by the populace as a legitimate pastime for firefighters, of which they can be proud to be a part of. However, they are forced in part to depend upon people like you, the reader, to make that happen. With knowledge comes understanding, and it is my hope, by providing that knowledge with this book, that new heights in the history of the sport of firematic competition may someday become realized.

Chapter I: An Introduction to Firematic Competition

On August 16th, 2008, at roughly one thirty in the afternoon in a small town in western New York, four volunteer firefighters dressed in long pants, sneakers, helmets and knee pads stood behind an odd looking, squat, colorfully-painted competition vehicle and gazed down the length of a paved, arrow-straight, 800-foot track. At its far end stood a 20-foot high archway made of steel. A few feet in front of that arch was a strange, triangular-shaped object constructed out of aluminum pipe, with an 8-inch fluorescent orange aluminum square in the middle of it – a target. About half the distance up the track back towards them, on the right side of the course, stood a fire hydrant. In a few moments, at a signal from a nearby official, they would perform a short, time-honored ritual of handshakes, helmet slaps and high-fives, then take their places on a low, narrow tailboard situated at the back of the vehicle, arranging themselves next to each other and grabbing on tightly to a horizontal metal bar situated just in front of them. Their fifth member, the driver, would give two "revs" of the engine, drop the truck into gear, and together they would accelerate down the track, reaching a speed of close to 70 miles per hour. If all went according to plan, just before they reached the hydrant the driver would hit the brakes and two of the four on firemen on the back would jump off, carrying with them one end of a lightweight fire hose, which they would then connect to the hydrant with a lightning-fast flick of the wrist. At the same time, a ¼-turn valve on the hydrant would be slammed open, sending water through the hose as the truck, with its remaining two passengers, accelerated once again towards the arch, all the while laying hose onto the track from the miniature hose bed at its rear. As it reached a point roughly 50 feet from the target the driver would once again hit the brakes hard, and the remaining two firemen would jump off, focused solely on a particular hose coupling that came off the truck at about the same time they did: if their practice (and luck) paid off, they would unscrew that coupling and attach a small brass nozzle to the male end, just as the water from the hydrant was reaching them. With a little more skill and luck, one of the two firemen would immediately point the resulting spray of water so that it hit – and tripped – the 8" aluminum square target.

All this, in just over 12 seconds!

An aerial view of the New York State Volunteer Firemen's Parade and Drill Team Captain's Association annual Motorized State Drill in Ridge in August of 2007

The 3-Man Ladder contest

The event at which this contest took place was the New York State Volunteer Firemen's Parade and Drill Team Captain's Association annual Motorized State Drill, hosted that year by the Main-Transit Fire Department in Amherst. That contest, called "Motor Hose Class C", was the fourth that day of eight contests overall – dubbed the "Big 8" – that make up a standard modern-day firematic tournament. In the order that they are usually held, the other seven contests are:

The 3-MAN LADDER contest, which is usually the first contest of the day. In this contest, a wooden service ladder that has been secured to the above-mentioned arch is climbed by three firefighters running on foot from points 50, 75 and 100 feet away, respectively. Time is taken electronically from the moment the first of any of the three men trips a timing beam to the moment the third man (the one starting from the 100-foot mark) grasps and holds the top rung of the ladder. This contest is usually followed by

The MOTOR HOOK & LADDER CLASS B contest, wherein up

The Motor Hook & Ladder Class B contest

to nine firefighters (five is the norm, including the driver) carry a custom-made aluminum competition ladder on a fire apparatus or modified factory-built pickup truck ("B-Truck") from a point 150 feet from the arch to a point just in front of the arch, at which point the ladder is removed from the truck and placed against the arch. While three firemen anchor the ladder, a fourth climbs to the top and grasps the top rung. Time is taken electronically from the point the front wheels of the truck cross the 100-foot mark to the moment the top rung is grasped and held. Usually, this contest is followed by

The Motor Hook & Ladder Class C contest

The MOTOR HOOK & LADDER CLASS C contest, which is essentially the same as the Class B version, except for the type of competition truck that is used (a "C-Truck", which is basically a custom-built

The Motor Hose Class C contest

vehicle with certain restrictions), and the point on the track from which it starts (475 feet instead of 150 feet.) Time is taken the same way, except it is from the 450-foot mark, respectively. (After this comes the Motor Hose Class C contest, as described at the beginning of this chapter.)

The Motor Hose Class B contest

The MOTOR HOSE CLASS B contest, which is essentially the same contest as the Motor Hose Class C contest, but with a Class-B vehicle, and the timing starts once the front wheels of the truck break a timing beam 50 feet from the hydrant. This contest is usually followed by

The EFFICIENCY contest, wherein up to eight firefighters hand-carry three lengths of fire hose down the track on foot, in

The Efficiency contest

a contest that is essentially the same as the Motor Hose Class B contest described above but without the truck. Beginning at a line 50 feet from the hydrant, contestants must attach the hose couplings together as they run, connecting one end to a hydrant and tripping a target with a stream of water from the other. Time is taken from the moment the first firefighter breaks the beam at the start until the target is tripped. Normally following this contest is

The MOTOR PUMP contest, which uses the Class B truck. Another characteristic that makes the Class B vehicles different from the Class C vehicles is the requirement that they have an operational pump onboard, and this contest makes use of it. Up to eleven

The Motor Pump contest

The Bucket Brigade contest

contestants start onboard the truck at a mark 75 feet from the hydrant. As the truck moves forward, time begins at a mark 25 feet from the hydrant, at which point several things happen at once:

Firefighters stationed on the back step of the truck jump off and manually pull lengths of fire hose down the track, towards a target, while

As the truck draws even with the hydrant, firefighters stationed on the hydrant side of the truck connect a section of pre-attached hard-suction fire hose to the hydrant and turn on the valve, while

Firefighters in or near the hose bed of the truck disconnect the couplings of two lengths of the same hose that the first set of firefighters are deploying down the track towards the target, and re-connect the now-open coupling of same to the pump discharge pipe of the truck, while

The driver (or his assistant), after seeing all hoses are connected, transfers power from the truck's motor away from the rear axle to the pump, thus pumping water from the hydrant through the truck to the nozzle at the far end of the track.

As with the other water contests, time is taken when the water from the nozzle strikes – and trips – the target. Following this contest is usually the last contest of the day:

The BUCKET BRIGADE contest, or "Buckets." Up to 12 contestants begin at a line 75 feet from the base of the arch, to which a wooden service ladder has been affixed. Each contestant carries a collapsible bucket that may carry no more than 5 gallons of water when filled to the brim. As the contest begins, several contestants run to an 8-foot water trough that has been placed 21 feet from the arch, fill their buckets, and then pass them to other firefighters who have in the meantime stationed themselves at places up the length of the service ladder, who in turn pass them to a firefighter stationed on top of the arch, who pours the water from each bucket into a 45-gallon barrel. Time is taken from the point that the first firefighter breaks the beam at the starting line until the water level reaches the top of the barrel, at which time the clock is stopped either electronically via sensors built into the barrel or by hand, when water flows from a spout at the top.

There are many other types of contests that are sanctioned by the Drill Team Captain's Association as well – contests with names like Individual Ladder, Running Hook and Ladder, Motor Hose Replacement, Motor Hook and Ladder Dummy, Running Hose Replacement and Eight Man Hose, to name a few, but these "Big 8" just described are the standard contests most usually offered at modern motorized tournaments. Three of these "Big 8" – the Three Man Ladder, Efficiency and Bucket Brigade contests – are what are referred to as "Old Fashioned" contests, i.e. those that are run on foot and that do not involve any kind of motorized apparatus. The more modern "motorized" contests obviously date back only to the invention of the gasoline engine and the automobile, but the Old Fashioned contests have a much longer and interesting history, each having moved through numerous manifestations and permutations since their earliest beginnings in the early 1800's. Just exactly why these contests are run is a whole other question; one that is still asked to this day. On the surface it may seem obvious: they are a test of firefighting skills, in a non-emergency environment. But there is much more to it than this, as the reader will see. The answer to that question wasn't always so clear, and in fact changed several times throughout the course of history.

Chapter 2: The Beginnings

As long as humans have been living in dwellings that could catch on fire, there has been, understandably, a keen urge among us to put those fires out, to protect our health and property, and of that and our loved ones, which make up such an important part of our lives. We take this for granted; it's been that way ever since humans lived in caves. If a fire broke out, there was no debate about what to do: someone somewhere would need to extinguish it.

At first fire suppression was everyone's responsibility, but after a time (and a few burned up villages) folks realized that some people were better – or at least more trained – at it than they were, and the first fire service came into being.

One could define a "fire service" as "any group of people formally gathered together with the specific responsibility of protecting against and suppressing fire in their community." There are records stating that the first fire apparatus in New York City arrived in 1648 in the form of ladders, hooks and buckets that were "...procured from Holland for the purpose of extinguishing fires,"[i] but at that time, when the city was in its infancy, protection from fire was left up to each individual citizen or family, who would gather in the streets at the alarm of fire and do their untrained best to extinguish the blaze. However, as the city grew in leaps and bounds, this method of fighting fire became less and less effective, and it became apparent that something had do be done:

"Old measures such as requiring citizens to have leather fire buckets at the ready and placing hooks and ladders – used to tear down walls and roofs to prevent fire from spreading – in strategic places throughout the town would no longer suffice for a city of New York's size and ambitions. Night patrols charged with sounding the alarm at the first sight of fire were not particularly well regarded – citizens referred to them derisively as 'the prowlers.'"[ii]

So, on December 3rd, 1731, New York City took delivery of its first two new "machines" – Engine Co. 1 and Engine Co. 2 – and six years later, in December of 1737, the colony's General Assembly authorized the formation of a volunteer fire department, so it could be said that the first "formal" fire service in that city was organized on that date. They were probably not the first in New York State, however: historical records indicate that on June 17th, 1726, New York State Governor Burnett signed a measure creating "the first Long Island Fire Brigade at Hempstead Town, naming 22 residents to enforce it."[iii] And, it is difficult to say whether or not similar fire brigades in other settled areas of the state were formed prior to that time as well; many records of that time are lost to history.

Aside from the obvious gains in fire protection to a community, what is important about the creation of these "formal" fire services is that as soon as they were formed, a basis for *competition* between their members was formed as well.

There are several reasons for this. The first, most obvious, reason is self-evident: fire – then, and still – ranks as one of the most destructive elements known to mankind, and it is simple human nature to want to extinguish it as quickly as possible, so as to minimize its damage as much as possible. In the late 1700's, through the early- to mid-1800's, even a small fire could potentially wipe out many blocks of buildings, or even an entire city, so it goes without saying that anyone – volunteer firefighter or not – would want to hurry to the scene without any delay to put the fire out. If one did so, it also goes without saying that he would be the recipient of a great deal of gratitude from that community for saving what could potentially be the entire town from destruction. Terry Golway, in his history of the New York City Fire Department entitled So Others Might Live, writes:

"The volunteers quickly became much-admired figures in New York and in other American cities that were also developing fire departments. Observing the firefighters of Philadelphia, Benjamin Franklin wrote: '[He]re are brave Men, Men of Spirit and Humanity, good Citizens, or Neighbours, capable and worthy of civil Society, and the Enjoyment of a happy Government.' Franklin was not the most objective source on the subject, for he himself was a volunteer fireman, having helped organize Philadelphia's first fire company. Still, most citizens in Philadelphia, New York, and elsewhere seemed to agree with his assertion that the volunteer firemen were 'Heroes and effective Men.'"[iv]

Human nature – and human vanity – being what it is, it follows naturally that those members of the community who were members of the first fire company didn't mind all this attention and glory, and – like celebrities throughout the ages who revel in such attentions – it stands to reason that once one was in the spotlight he would do everything in his power to keep it that way. It was a pretty simple goal: if one was to maintain his celebrity for being "the best," one would have to make sure that he – or his fire company – was always first to arrive at – and extinguish – the fire. In his book The Romance of Firefighting, Robert Holzman writes, "The firemen were proudly conscious of their importance in the community. Yet no man considered himself to be a member of the community's fire department, for there *was* none. There was merely a collection of fire companies. Each man strove mightily to make *his* company the best one in town. That meant humbling the other companies: in racing to a conflagration, in getting a stream on the blaze, in amount of water pumped, in appearance of the fire apparatus. At first this was a friendly rivalry, such as stimulates endeavor in any field. But in time the competition became fierce. The men spent more energy in battling with other companies, in many instances, than they did in fighting fires."[v]

This earliest form of competition between fire department members was not just limited to inter-company rivalries; there were also gains to be had by arriving at a particular fire station faster than your fellow company members. While in reality firefighting is truly a team effort, it's no secret that more attention and praise gets lavished on those members who are actually handling the nozzle (or "pipe") at the end of the hose, as it could be construed that their efforts are more directly responsible for the extinguishment of the fire. By arriving at the firehouse before everyone else – and this still holds true in many volunteer firehouses across the country today – that person entitles himself to man the nozzle at the fire. This practice goes at least as far back as November 9th of 1808, where records from a New York City fire company state, "on alarm of fire the first person repairing to the Engine House was entitled to have the pipe."[vi]

To complicate matters, in the mid 1700's, money got involved. As if human vanity wasn't enough, there was now human greed to go along with it.

This came courtesy of local property insurance companies, who during that time began to stick their fingers in the fire suppression pie – perhaps innocently enough – by trying to protect their own financial interests. Weekly newspapers of that time were filled with accounts of fires; not so much the events and evolutions surrounding them, but more about the specific financial losses associated with them, in actual dollar amounts. At some point, it occurred to some underwriter that if the members of the fire department could be motivated to get to a fire involving one of their buildings more quickly – vanity notwithstanding – they could perhaps cut some of their financial losses. To do this they began providing monetary rewards to the fire companies, a practice that was common in London prior to the advent of the fire service in the United States. (After exhaustive research this author was unable to find any direct evidence – insurance documents, fire department orders, manuals, etc. – that this practice actually took place, but there are several direct references to it in historic literature.) And, as could be expected, it worked: firefighters strived even harder to be first to the fire, and competition between them intensified. Holzman writes:

"Oddly enough, it was the fire insurance companies and the most respectable citizens in town who innocently caused much of the rivalry. Such parties frequently offered cash prizes for the first company to reach the scene of the conflagration, or the first crew that actually got water on the flames. As early as 1740, Boston offered a premium of five pounds to the company whose engine 'shall first be brought to work' at each fire. A resolution of the Mutual Fire Insurance Company of Germantown, Pennsylvania, provided 'that $3 be paid to the Engine Company that shall be first in operation in any Fire in which the Company is interested, and that the Secretary be instructed to furnish a copy to each of the Engine Companies in the Township of Germantown.' This meant that 'heroic' means were used to keep one's adversaries from getting there first. It might mean street fighting. It might mean racing one's engines on the more smoothly surfaced sidewalks, at great risk to pedestrians. It might mean sounding false alarms, to wear out one's rivals so they could not run so swiftly to a genuine fire."[vii] Kenneth Holcomb Dunshee, in his book, As You Pass By, refers to this as well, in speaking about the New York Fire Department Protection Engine Co. 5, which was operating as early as 1762: "When the Thirty Dollar Prize for the first company at a fire was in vogue, the 'Honey Bees' won twenty of the prizes in five years, a record second to no other company in the city."[viii]

In retrospect, that was quite an accomplishment. In those early days, moving the apparatus from the firehouse to the fire was accomplished through manpower only, by using ropes that were attached to the front of the rigs and hauling them as fast as you could. While the very earliest firefighting machines were fairly small and manageable, the concept of "bigger is better" didn't take long to make its way into the fire service, and by the mid-1800's engines weighing at least a ton were not uncommon. It's not hard to imagine, then, how much work – and sheer physical effort – it took to get one of these machines to a fire, and for the same reasons how proud its company members might be if they were able to get theirs to the scene ahead of anyone else.

And, as far as competitiveness was concerned, it was not enough to just get to the scene of the alarm first. For every ounce of glory gained by a particular fire company for doing so, there was an equal amount of humiliation cast upon those companies that failed to do so. In addition, there were other types of humiliation that could befall firefighters as well: not making it to the scene first was bad enough, but if you happened to take the same route to the fire as a rival company and they actually *passed* you along the way, it was considered an ultimate put-down. Holzman noted, "If one company overtook another en route to a fire, the latter crew was said to be 'passed,' and this was a fearful disgrace. Sometimes a company, to avoid being overtaken, would weave erratically on the street, so there was no room to get by. Often a collision would be engineered, or brawlers would attack the men of a fast-approaching company."[ix] A similar reference is made in another historical text: "Sometimes when a company saw itself on the eve of being outrun its foreman would shout, 'Round to, and go back for more men!' or, 'Stop, and fix that wheel!' and in these cases there was no race. Occasionally, in trying to pass each other, engine or hose companies would be driven into piles of bricks, or

A firemen's "Washing Day" on Grand Street in New York City, ca. 1810, as rendered in oil by W.P. Chappel. This contest between two engine companies determined which could force the other back by the strength of the stream from its engine

holes, and men and machines would be injured. Companies used to hide behind walls, or piles of bricks, until rival companies came along, and would then rush out. This was a direct violation of the law, and was a common occurrence."[x]

So, one can begin to see that at the time what began as a truly altruistic goal of striving to be first to save lives and property was gradually evolving into a series of competitions that threatened to miss that point entirely. The goal was always there, but at the turn of the century, more and more it seemed that goal was being chased for the wrong reasons. "It was noted," Holzman wrote, "that 'the present generation can hardly imagine the spirit and endurance exhibited in a race between fire companies in those days. Talk about horse racing and rowing matches – they vanish like bubbles into thin air, in comparison with the race of the firemen, each endeavoring to outdo his fellows in the strife to be first at the fire, first at work, and have first stream on.'[xi]

Writer George William Sheldon, in *Harpers New Monthly Magazine*, quotes Mr. Charles Forrester, an old "fire laddie" of the volunteer fire department:

"'In those earlier times – say previous to 1836 – the city was not districted, and whenever there was a fire anywhere all the engines were out in a jiffy. The excitement kept us up, I suppose. One night my company went up as far as Fifty-third Street and Fourth Avenue. Another night, I was sitting at home with a bad cold, and taking a vapor bath. The fire-bell rang; I threw off my blankets, and though in a dripping perspiration ran with the engine up to Forty-second Street and Tenth Avenue. The next morning I never felt better. I was fit to run for my life. What was the inducement? Well, the love of excitement, and of excelling the rival companies. We were fully repaid when we could brag about our exploits, and make our neighbors feel jealous.'"[xii]

This loss of focus was not limited to the firefighters of New York City: in 1819 Boston fire engines carried water to the fire with them for the first time, but instead of welcoming the new technology, their fire officials "...cried foul play. They said carrying water to fires gave the company an unfair advantage over others in competing for first-water bonuses."[xiii]

To keep this in context, one must keep in mind what was going on in the rest of the United States and the world at the time. The United States had only recently declared its independence, and in 1803 the Louisiana Purchase had just effectively doubled its size. The War of 1812 with England highlighted just how unstable things still were at the time, and despite the New York City Volunteer Fire Department becoming officially incorporated in 1798, there was still a long way to go in terms of structure and discipline within its ranks. The number of different companies within it was gaining dramatically, however, as the city grew in leaps and bounds, and with that came more and more rivalries between them.

The fire apparatus being used in those years were still quite simple affairs, consisting essentially of a wooden box on wheels, which could contain a certain amount of water. This box had within it a smaller box containing a piston-pump that was attached on the outside to a proprietary configuration of wooden rails, called "brakes", that firefighters – from two to many – would hold onto as they worked the pump by pushing up and down on them. In this manner the firefighters could pump – or "play" - the water that was within the "tank" of their "engine" through the piston-pump and out through a nozzle towards the fire. The first wave of these machines operated independently of one another, but as time went on and technology allowed, hoses were invented. With this development came the ability to pump water from one engine into another, allowing water to be effectively moved from point A to point B, from as great a distance as the number of engines, used in tandem, allowed. (Sheldon notes that "At a fire in a hair-cloth factory twenty-one engines were in line, October 26th, 1824, and at a fire in a feed store at Eldridge and Rivington Streets, twenty-three engines, with 6000 feet of hose, were in line – a fact entered in the minutes of Engine No.13."[xiv])

What this new ability to pump water in series also created, however, was another way to bring humiliation to a fire company: if, in the process of pumping water from one engine to the next, the company receiving the water could not pump out the same amount (or more) of water that was being supplied to them, their "tank" would overflow, and they were considered to have been "washed". This was often the result of brute strength and manpower – it implies that the second company didn't have what it takes to keep up – and was considered a badge of honor for the "washer" and a source of embarrassment for the "washee". Sheldon makes reference to this in his book: "The first time an engine was 'washed' she was 'said to be no longer a maiden.' 'It used to make her men sick at the heart,' says an old fireman; 'it was a terrible thing to be "washed."' Peterson Engine Co.15 for a long time was not washed and received the popular title 'Old Maid.'[xv] Dunshee refers to this legendary company as well: "In its prime, No.15 had as large an array of distinctions and nicknames as any fire company could boast. It had one of the largest volunteer roles in the city, and its runners, divided into 'The Fly-By-Nights' and 'The Old Maid's Boys,' were said to be able to 'lick' any other company in New York. The men were often called 'Dock Rats' because of their speed in getting to the river to get their water first when the alarm sounded. They claimed to have washed more engines than any other company and were known as 'Old Maid,' because of never having been washed themselves."[xvi]

Such was the shame of being "washed" that often companies that had been tried any means they could think of to downplay it, or make like it didn't really happen: "All kinds of excuses were offered when an engine had been 'washed' by a rival; 'she was graveled' or 'she had a brick in her' or 'there was a stick under the valve.' In 1892 at a fire in Water street and Maiden Lane, No.13 Engine, while trying to wash No.15 Engine, complained that the latter, when within three inches of going over, stopped playing and would not go to work again. They accused her men of having bored holes in the box of the engine, in order to prevent her from overflowing. Companies were frequently accused of cowardice by rival companies for 'refusing to take our water.'"[xvii]

"The Beginning of a Race", a watercolor painted by S.W. Hill in 1830. New York City Hall is in the background, with Park Row to the right. Engine Co. #44 can be seen in the right foreground (also known as "Live Oak" as many of its members were shipbuilders) getting ready to race with Protection Engine Co. #5 (the "Old Honey Bee"), at left. Note the pedestrian at far left who was not quick enough to get out of the way in time!

This phenomenon of "washing", as humiliating as it may have been, may also hold a significance that is extremely important insofar as this book is concerned: it may be the first recorded instance of a *formally arranged* competition between two fire companies that was done in a friendly spirit – albeit with the serious ramifications of the day: "When Engine Company No.13 in New York obtained a new pumper in 1835, the firemen promised that the foreman of the first company to 'wash' them would get a new suit of clothes. Engine Company No.11 won its foreman this finery in a furious contest, during which the victor's men pumped from 130 to 150 strokes a minute; the pace was so terrific that no man could stay at the pump handles for more than fifteen seconds at a time."[xviii]

Up until this point, gatherings of firefighters and their apparatus, for any purpose other than extinguishing a large fire, were uncommon, to the point where there is little or no mention of any parades or formal contests between firefighters in historical literature. One of the very earliest references to firefighters marching points to a parade that took place in New York City on the 4th of July in 1824, where the volunteer firemen paraded "…in honor of Lafayette…"[xix] Two years later, on Independence Day of 1826, Long Island saw its first parade of firefighters, "in honor of the semi-centennial of the Declaration of Independence."[xx] The following year, on October 20th, 1827, New York City firefighters took part in another celebration, which began with a parade, and ended with what may be one of the very first Firefighter Musters to take place in New York. A muster is literally an "assemblage, or collection" of firefighters, who borrow the term from its traditional military use of the term, as in "go muster the troops." From a fire service standpoint, a muster represents any group of firefighters or fire companies, who gather together at a predetermined time and place to meet and exchange ideas, test or show off new equipment...or compete against each other. This early gathering does not appear to have been for that purpose, but one can infer from its description that there was no doubt a considerable amount of showing-off and one-upsmanship present that day: "On Monday last the celebration of the fire companies took place...they entered the Battery at the river gate, and formed on the outward path. Here again, 'as time and tide wait for no man,' another delay was occasioned by the ebb tide, but about half past four the water rose sufficiently to admit the suction pipes; and then, at a given signal, they commenced playing, at the right of the line, 'by companies.'"[xxi]

While the New York City Fire Department was growing, so were the fire departments of all the other major cities in the northeastern United States at the time: Boston, Philadelphia, Washington, D.C., Providence, Hartford and others were all up and coming as well. With both unique fire apparatus design and firefighters' pride and competitive spirits being what they were, it was not long before they were taking excursions to these other cities to show off their engines and test their abilities against those of their brothers in other locales. The "Old Maid", Engine Co.15, was one of the first to do so, in a reference to a contest in 1836: "Their last machine, built by Harry Ludlam, was taken to Philadelphia for a playing match, and was the only machine to succeed throwing its stream over the cupola of the Exchange Building."[xxii]

However fierce the competition between companies in New York City may have been during that time, to beat one another to a fire or have the best and/or most powerful engine, it was all conducted on a fairly gentlemanly basis. Sure, boasts and insults flew, and some underhanded practices took place on the street to gain an advantage, but for the most part the firemen kept it on a somewhat-civil basis. In 1830, however, that began to change...for the worse. Whether it was in conjunction with – or as a result of – a financial recession that was taking place in the city at the time, we may never know; all we do know is that things began to get nasty, and in ways that would ultimately bring about the downfall of the New York City Volunteer Fire Department and the creation of the paid Fire Department of the City of New York – the FDNY – that we know today.

The Era of Fighting had begun.

Chapter 3: The Era of Fighting

The first unpleasantness of record was noted by Sheldon to have taken place on July 9th, 1830, and which appeared in the books of New York Fire Department Engine 21 of that date. As rivalries between the different fire companies grew, so did the notoriety of their respective firefighters, to the point where each fire company in the city had by this time acquired, in addition to its crew, a group of "hangers-on." These groups of civilians were comprised mostly at the time of young boys, who, enamored by the bravery and exploits of the firemen, would hang around the firehouse, listening to the stories and helping out in whatever way they could. At first their assistance may have been limited to making coffee, running errands or sweeping the floor, but in time the firefighters began allowing them to respond to alarms with them, and they became "runners" for the companies, whose job it was to run down the street ahead of the responding firemen, shouting a warning to any pedestrians to move aside and make way for the fast-approaching apparatus. In their devotion to their heroes, their passion towards "their" fire companies often exceeded the passion of its actual members, so it's not surprising to find that it was these "runners", and not the firemen themselves, that were responsible for upping the competitive ante between rival companies. Sheldon relates Engine 12's book entry: "'...a complaint against the riotous conduct of boys from the neighboring engine-houses after alarms of fire, and request that a stop be put to it,' the boys, of course, being not members of the companies, but only 'runners' with the engines."[xxiii]

In spite of the complaints, these "runners" were continued to be put to good use by the fire companies by employing another dirty trick in order to gain an edge: as fire hydrants were introduced to the city, they would run ahead, as fast as possible, to the hydrant closest to the fire and place a barrel over it, and then watch nonchalantly as a rival fire company passed them by in search of the closest hydrant. An account by A.E. Costello, a New York Fire Department historian in the late 1800's, describes what may have been one of the first instances of this (although in this case it involves a fireman himself):

"Plug Boys" guarding a hydrant in the early days in New York City

"A volunteer, belonging to No.14 Engine, conceived what he thought a magnificent idea to become solid with the company. A new district had been opened up, and the firemen were not very well acquainted with it. But the volunteer, in rambling through the locality, noted a fire plug, and impressed its position in his memory. One night a fire occurred in this section, and the volunteer made a mad dash for the plug. He saw an old barrel near, clapped it over the plug, and sat on it. He began to smoke and assumed as graceful and careless an attitude as though he were an accomplished actor. An opposition company, dashing up and finding no hydrant, caused the volunteer to smile and chuckle. He was thinking of how proud Fourteen would be of him when they came along. He would then off with the barrel and reveal his treasure. But he didn't get the chance. Another company coming along saw the man on the barrel, and something about him made them suspicious. Perhaps he seemed too lazy and indifferent on the barrel. They rolled him off his seat, lifted up the barrel, and discovered the hidden hydrant. Alas, for that zealous runner and genius! He did not consort with No.14 for a long time after. It was weeks before he was able to run, and more weeks before he could even sit on a chair. As for hydrants he never again sat on one."[xxiv]

This practice of hiding the hydrants by these "plug boys" ran unabated, and would cause conflicts between responding fire companies that would eventually lead to blows. Engines 12 and 13 were among those who traditionally didn't get along: "The two companies were necessarily rivals, No.13 keeping its Engine in Duane Street near William Street, and No.12 its engine in William Street near Duane, so that whenever a fire alarm sounded, the struggle for precedence on the way to the scene of the conflagration was hot and vigorous. On their way to a fire uptown the two engines usually met at the corner of Roosevelt and Pearl streets, and when, as was often the case, the arrival at the hydrant or cistern was almost simultaneous, a conflict was the natural result. These rivals thus came to hate each other most cordially, and a single blow was enough to produce the explosion of a good deal of combustible matter."[xxv]

One such fight had a bit of a lighter side to it: "On one occasion a very funny, but withal, fierce, fight took place for the possession of a hydrant. The survivors often laughed over it. Hose No.14 was a popular company, and Henry A. Burr was foreman of it, and W.W. Corlies assistant foreman. One dark night there was a fire up-town, where the streets were then poorly paved and worse lighted – in fact, not lighted at all in some places. In one of those dark spots, not far from the fire, Burr, as the company dashed along, saw the outline looming up of a big hydrant. He instantly made for the hydrant, but found another man clutching at it also, a member of a rival company that had come along. Burr and the other fellow contended for that hydrant in the darkness; then, finding the fight going against him, Burr called for his men, who came. Then the other chap called out for his men, who came, and a fierce struggle took place in the dark for the supremacy and the hydrant. The fight was terminated in a curious way. One of the men, who was smoking while the rest were squabbling, lit a match at the end of his cigar and looked at the object for whose possession they were fighting all around him. Then he burst into a loud laugh, and no wonder; for what they thought was a hydrant wasn't a hydrant at all. It was a buried cannon, with half of it sticking out of the ground. This discovery ended the fight."[xxvi]

Not helping matters at all was the fact that these conflicts between companies were starting to take on an air of Sport with the general public. Despite numerous Letters to the Editor each week in *America's Own and Firemen's Journal, The Brooklyn Eagle* and other newspapers of the time decrying violence and fighting between the firemen, whenever there was a fire and companies would race each other to the scene, crowds of bystanders would gather on street corners to watch them pass by. These gatherings would often occur at certain known intersections in the city where, based upon the locations of the firehouses, the public knew from experience that two or more companies were likely to meet up and begin racing. As well, certain members of the fire service had already by that time gained reputations of their own for being especially strong, tough, competitive, or just well-liked; an analogy between them and today's major league star athletes would not be unfairly drawn. As a result, the yelling and applauding of the crowds as the responding companies went by served only to incite the firemen and their desire to be ahead even more: "Engines 6 and 8 had many a lusty struggle. 'Tweed was Foreman of six engine then,' says an old fireman, 'and crowds and crowds of people would congregate at Chatham Square to see "Big Six" and Eight Engine coming down town. Eight lay in Ludlow Street, and came down Grand, through the Bowery; Six lay in Gouverneur Street, and came through East Broadway, and thence into Chatham Square; and when they met, the excitement was intense and the cheering furious as one or the other engine gained in speed.'"[xxvii]

In 1831, a new Chief Engineer, James Gulick, was elected to oversee the New York Fire Department. One of the first things he tried to do when he took office was to create reforms within the department aimed at promoting greater discipline amongst the members and more cooperation between the companies, in an effort to quell the fighting between its members that was becoming more and more prevalent. He was extremely popular and well-liked among his firemen – there is a famous story of a fire in 1836 where they all lay down their hoses and refused to fight the fire when they heard City Hall was going to remove him from office - but his efforts to curb the violence were ultimately unsuccessful. One possible reason for this was that he might not have often followed the "lead by example" method of leadership. In Our Firemen: The History of the New York Fire Departments, Costello writes: "In 1832 No.33 got into difficulty with Engine Company No.11. A fire occurred in Pearl Street, near Maiden Lane. No.11 was taking water from the dock and supplied No.33. Both companies were noted for their strength. No.11 tried to 'wash' No.33, and failed after working all night. When daylight broke, the boys, tired out, went to breakfast, and then the runners or hangers-on of the engines took their places to keep up the supply of water. These 'irregulars' were animated with the same sentiments that characterized the great men they were permitted to follow, and again it was sought to get No.33 over, but in vain. Then No.11 charged No.33 with 'niggering.'[1] This meant only working at intervals and compelling the supply engine to take out its butt. Hot words followed, and it was decided to settle the matter by a fight. As a rule champions were selected for the companies, and the rule was observed on this occasion. The engines were deserted, a ring was formed, and at it the gladiators went. In the midst of the fun the gigantic Gulick appeared upon the scene. He was the Chief Engineer. He settled the contest immediately. A blow from his fist was like a kick from a horse. He sailed right in and scattered the ring, the fighters in a minute."[xxviii]

Indeed, it seems that at the time the path to being a respected leader among your men had more to do with being a tough guy than it did being a good leader. Golway writes: "Costello observed with approval that the company's elected leaders, their foremen, 'did their very best to maintain their reputation for courage and personal prowess.' In other words, they were banging rivals' heads alongside their men instead of trying to restore order and some semblance of professional decorum."[xxix] Costello, it seems, didn't think the fighting between members was anything too serious to be worried about. "There were many fights, and hot ones, too, in the old department," he writes, "but they grew out of a natural emulation and were not lacking in a certain rugged element of chivalry which promoted manhood, though somewhat at the expense of public order. The murderous revolver and assassin-like disposition which now marks its use were unknown in those days. The combats were fair hand-to-hand fights between man and man, and he who resorted to any other weapon than those which nature supplied was accounted a ruffian or a coward."[xxx]

One of the more well-documented fights between companies took place in the summer of 1838, between the "Old Maid," Engine Company 15, and the "Lady Washington" Engine Company 40. It must have been a real Donnybrook, as accounts of this fight appear in several different historical texts.

"Between No.15 and 40 an ill-feeling had been a long time brewing. Both companies when running to downtown fires used to frequently meet in Chatham Street, and fright the town from its propriety with angry wranglings, or with excited shouts as they would 'buck' the pavements, one on either side of the street, and engage in a furious contest of speed in their rival efforts to be first at a fire. On this remarkable occasion, however – it was a bright summer Sunday – the two companies, after several hours of arduous duty at a fire in South Street, near Wall, turned homeward about the same time, and passed up Pearl Street together, in dangerous proximity to each other. The ropes of both sides were – as usual on Sundays – fully manned. Probably five hundred men confronted each other on either side. It had been rumored early in the day that No.40 intended to attack No.15, and destroy its boasted invincibility. The presence in the ranks of the former, of several noted fighters from 34 and 44, lent color to this report, and the hostile companies were followed by an army of expectant lookers-on and partisans.

"The two wheeled into Chatham Street together, No.15 leading. No.15 turned eastward on its usual way homeward, via the Bowery. No.40 should have taken the westward course down Chatham Street to its house on Mulberry Street, but instead of doing so, followed its rival until the broad space of Chatham Square was reached. Here the two companies came abreast, and the satirical chaffing which up to this had marked their antagonism swelled into a chorus of mutual taunts and menaces. So they continued, until the Bowery was reached. At the head of 15's rope was 'Country' McCluskey, confronted by the formidable Jim Jeroloman at the head of 40. At the rear of the line Hen Chanfrau was opposed by Mose Humphreys. Both sides were by this time ripe for conflict.

"The two foremen, Colladay, on the part of No.15, and Carlin, on the part of No.40, passed rapidly up and down their respective, ostensibly endeavoring to preserve the peace, but in reality stimulating the courage and exciting pride of their partisans. 'Now, boys, no fight,' said Colladay, audibly, adding in an undertone, 'but if they will have it, give 'em good!' 'Be quiet, men,' said Carlin, loudly; but then in a lower tone, "till they begin, then go in!' The vociferations of the companies grew louder and more threatening, until at last

[1] This was a highly derogatory term at the time, used to characterize someone as being especially unmotivated or lazy

words being exhausted, both sides began to 'peel' for the conflict which was felt to be inevitable. Jeroloman took out his earrings and put them in his pocket.

"Passing from Chatham Square into the Bowery, the opposing companies, which up to that point had moved side by side, surrounded by a dense crowd, were, by the narrowing limits of the latter street, forced into collision. That proved the signal for the fray. The lines of each extended nearly an entire block. The instant McCluskey and Jeroloman came within striking distance of each other, they dropped their respective ropes and became fiercely engaged. The shock of battle rolled down the line, and quicker than it takes to write the words, one thousand sturdy, stalwart fellows were fiercely grappling each other in a hand-to-hand fight. The din was frightful; louder than the furious exclamations that filled the air would be heard the resounding blows. The fighters were so thick that there was scarcely room for one to fall. Combatants who, so to say, felt themselves knocked down, were upheld by the pressure of the surrounding throng, and those who were unlucky enough to find mother earth were nearly trampled to death by the feet of friend and foe before they could escape. The leaders on both sides did their very best to maintain their reputation for courage and personal prowess.

McCluskey, after a hot and desperate struggle, thrashed Jeroloman, Freeland triumphed over a noted fighter of No.40, known as 'Orange County,' and other champions of No.15 were making it all right.

"The fight lasted about thirty minutes. It resulted in the total defeat of No.40, who abandoned their apparatus and fled precipitately."[xxxi]

These skirmishes were not limited to Manhattan. On May 22nd, 1840, while Brooklyn still had its own fire department, separate from that of New York, the following story appeared in the Huntington *Long Islander*: "The inhabitants of Brooklyn were in a state of considerable alarm on Monday night, by the riotous conduct of the firemen who paraded the streets in large bodies threatening violence to the city authorities. Two of the companies, it seems, had agreed to play against each other for wager that afternoon; but this practice being strictly prohibited by ordinance of the Common Council, as soon as it became known to the authorities, they interfered to prevent it. The parties, however, were unwilling to lose their sport and resisted the officers who finally succeeded in obtaining possession of the engines. The whole body [then] made common cause, and a riot ensued, which at the time threatened serious consequences. They were, however, fortunately prevented by the timely intervention of the Mayor and a strong body of police, and the riot was immediately quelled. The whole of the engines are now in the possession of the authorities, where they will remain until the matter is settled by the Common Council."[xxxii]

The above passage is noteworthy because it illustrates just how serious the competition between fire companies was becoming at the time: never mind that the contests were illegal; when their apparatus were confiscated they were driven to riot against their superiors to the point where the Mayor himself had to get involved!

Another indicator of the escalation in the intensity of these conflicts comes from a passage written by Dunshee about a fight that took place in New York on July 26th, 1846: "Benjamin J. Evans of Engine Company No.31 related the following particulars of one of their battles, the big fight of Sunday morning, July 26, 1846:

'We were called out from our quarters in West Broadway on a still alarm…our foreman singing out "Come pull away boys, for West Broadway's away!" As we were running home opposite the park, we met Engine No.6, and she commenced to bark. Then along came Equitable, and she thought she would help No.6, but found that she was in a mighty pretty fix herself.

'Number 5 just turned up near Ann Street, and the "Short Boys"' cry greeting their ears, they said, "Let us go and help old 31 and make the short boys feel sick."

'Chief Engineer Anderson was standing, at the time, on the Astor House steps. Number 6, which then lay in Reade Street, began to "bear" at us, and a fight resulted. In the midst of it, No.1 Engine, which lay at the foot of Duane Street, appeared and sided with No.6. Then Engine 23 of Leonard Street turned up and sailed in with us. No.36 quickly followed, taking 6's side.

'Pipes, axes, and any weapon on hand was used in the fight. It was a terrific fight and lasted a long while. At last Anderson succeeded in putting a stop to it and made us go down Canal Street instead of through Chambers, so as to avoid our foes.

'Before the fight No.5 sent their engine home in charge of a few men and this precaution saved them from being disbanded.'"[xxxiii]

That "pipes, axes, and any weapon" were being used now in the frays was a dark and frightening enough development; it appears that as a result the authorities had no choice but to increase the severity of the consequences in kind. As Dunshee refers to in the last sentence, in an effort to quell the fighting entire companies were being completely disbanded if they were thought to be especially provocative or troublesome…even if it meant leaving whole sections of the city unprotected. Costello, in his book, talks about that same fight: "On Sunday morning, July 26th, 1846, a big fight occurred in Broadway between the runners of several of the companies. The companies involved were Engine companies No.1, 6, 23, 31 and 36. The fight lasted a considerable time, and many of the combatants received more than they bargained for in their desire to have some fun...On August 5, all the companies concerned were disbanded by order of the Common Council. In view of the magnitude of the combat, the Alderman considered they could not overlook the affair. The battle had raged all the way to Canal and Hudson Streets, and attracted an immense crowd of citizens. After the disbandment, there was not a single fire company left in the Fifth Ward."[xxxiv]

The fighting, however, was beginning to take its toll; not only on the firefighters, but on public opinion regarding the fire service as well. Once seen as champions on a pedestal who could do no wrong, firemen by the mid-1800's were now being seen as lazy, brutish, unrespectable lower-class citizens who would as soon drink and fight as put out a fire. An indication of this sentiment appeared in the May 9th, 1849 edition of *America's Own and Firemen's Journal*, in a fictional (no less) series entitled "The Book Folder, or The Firemen's Bride: True Sketches of City Life":

"Now, reader, come with me to another scene. We are now hurrying up Broadway in the direction of the fire. Look! See that engine dashing down Grand street – now they have turned the corner, and are almost flying up Broadway. See! Another is just turning Broome street corner. Now they are after the one that a moment before came out of Grand street. Behold how that man in a light-sack coat, with a trumpet in his hand, is cheering on the men. Hear his voice as it rises above the thunder of the wheels.

'Now boys! Every one of you – work her up men – don't let her lag. Pull away my hearties – all together men!'

See, he has now taken hold of the rope himself: they gain on the first engine. Yes, it is a race – now they are side by side. Both companies are putting forth almost supernatural exertions for the mastery. How madly they dash on! Pride spurs them to the contest – money could not make men exert themselves in that manner. The wheels scarcely touch the pavement, so great is their speed. Now comes the tug-of-war – the engine which came out of Broome street begins to pass the other. Hear the wild shouts of the victors, cheering to themselves, and which calls forth all the faculties of their rival, to compete for the superiority. They now strike the wooden pavements, and the noise of the engines is drowned with the cheers of the men. Still on they go, a dark mass of beings, heedless of everything, so they may reach the scene of conflagration first. Fiercer grows the strife – the engine which came out of Broome street has passed the other. Hear their triumphant cry, as they strain every nerve to distance their rival. The man with the trumpet is again urging on his company. You noticed the boy with the green coat and Oregon cap, and his companion, whom he called Greaser, has hold of the rope of the engine that is fast distancing its rival. Did you not perceive them, as the light of their torches and yonder lamp cast their rays upon them as they passed the corner? But listen; hear what those two persons just ahead of us are saying.

'And do you really mean to tell me, Mortimer, that these men get no pay for dragging those engines through the streets? You must be joking.'

'No, Traxton, they do not receive a farthing for their labor, but they are low fellows and call it sport. They most always get into a fight after the fire is extinguished, and sometimes on the road to it.'

'Strange kind of sport, that; you would not find people in England exerting themselves in such a manner, without they were remunerated for their trouble.'

'No, English people are not quite such fools as the Yankees, but this is a little of their republicanism. Why, Traxton, there is not one of those fellows who have just passed us with their engines, but thinks himself as much of a gentleman as what we are. Fact, I assure you.'

'Why, Mortimer, you astonish me. I should kick one of those fellows out of doors if he should attempt to put himself upon a par with either of us.'

'That is easier said than done, Traxton.'"[xxxv]

America's Own and Firemen's Journal was a weekly paper that was first published in New York in 1849, whose journalistic goal, among others, was to support and promote the fire service, despite all that was going on at the time. The very first issue of the paper, published on April 14th, included this editorial: " It will ever be our aim to dispel the clouds of animosity now hovering about the different companies, and endeavor to awaken a feeling of love and kindness towards all, and we hope that the fire of harmony once kindled will never be quenched."

As one reads through subsequent issues, it becomes apparent that they were fighting an uphill battle from the start. Two issues later, on April 28th, this appeared: "Owing to the time occupied in collecting evidence, in relation to a charge brought against us for alleged libel, we have not been able to devote much attention to getting out our paper, but the case comes up for trial before the Court of Sessions in the May term, and if we have the least show of justice, we will be acquitted. We then hope to present the Firemen with a Journal worthy of their patronage; but for the present we ask a little indulgence." It seems that their support of the fire service was not well-liked, and apparently some of those who were in power took steps to try to put them out of business, as indicated in their July 7th issue: "We have had many difficulties to encounter since starting this Journal, and persons of wealth and influence are striving hard to crush us, owing to the stand we have taken on behalf of the working classes." Despite what looks like an honest attempt to report on the fire service fairly, each issue was filled with numerous Letters to the Editor complaining about the fire service and all the fighting within it, and it was only a few years later that the paper finally gave up.

So too did New York City, and on August 1st, 1865, for reasons that very likely included racing and fighting between companies, the New York Volunteer Fire Department was officially and entirely disbanded. As the Civil War ended, a new, fully-paid fire department took its place, "...as a sorrowful, war-weary city adjusted to peace and mourned the death of Abraham Lincoln."[xxxvi] It was not a joyous time, and the transition, however smooth, was not an easy one for the firefighters. The first thing to be outlawed – not surprisingly – was contests of any kind between firefighters. "The state commissioners who governed the new Department clearly were determined to make a permanent break with the excesses of the volunteers, a point that was driven home in the agency's rules and regulations. Not surprisingly, 'racing to and from fires' was prohibited, as were 'spirits and intoxicating liquors' and 'any game of chance.'"[xxxvii]

Costello, ever the champion of the fire service, acknowledged these new changes, but still held a torch for the earlier days, and the competitiveness therein:

"It is just to say here that, as a whole, the Paid Department of this city has not in heroism, intrepidity, or disinterestedness, excelled the active Volunteers, when all things are taken into consideration. The same company rivalry, which, under the old system led to racing and sometimes brawls, still exists, and despite iron-clad rules and public sentiments, crops out now and then, showing that, after all, all firemen are human. The very rivalry which existed between Volunteer companies was in the end an advantage. Had they been apathetic, few apparatuses would have reached a fire in time to prevent serious loss. The racing was based on a point of honor, a test of endurance, skill and strategy. It kept up the *esprit de corps*, and left laggards in the rear. The anxiety in running to a fire to prevent a rival company from passing was that of an English or American university crew of oarsmen to avoid getting 'bumped.' The system of competition taught firemen that the shortest distance between two points is a straight line, and that the alert and fleet-footed get to the goal first. So that in these days the company which has the 'hitching-up' drill well mastered, and leaves the quarters the moment a home signal is ticked off, may reach the station indicated before a less drilled and not so wide-awake company much nearer to the rendezvous, and so shame the laggards into better performance of duty another time."

To Costello, the new commissioners, by banning all forms of competition between its members, were being hypocritical:

"If the naked truth were told, few outside the Department were gravely anxious about them at all times. To many it appeared that until New York paid for fire service it would be impossible to have such service at all efficient without at least winking at the competition which pitted one company against the other, and resulted generally in a gain in point of speed in reaching a fire. Men who did their utmost to create a Paid Department were members of companies whose reputation for zeal in distancing or vanquishing rivals by feet or hands was the best or worst, as the partisan or critic may determine, and were as eager as their comrades when the apparatus was rolled out and the race began, or when the goal was reached and others were found to have the ground of vantage. How many staid citizens of the Volunteer time who yet live will affirm that, even on Sunday, when the races either on genuine or false alarms, were along Broadway or the Bowery, and the ragamuffins ran to see them, and the often consequent unpleasantness between companies, they did not gird up their loins and become interested spectators? And how many, under the pressure of conscience, will fail to admit that at a pinch they vented their preferences, apathies or interest in a shout or an exclamation, if they did not take a hand in the results of a collision after a war of words?"[xxxviii]

Although the demise of the volunteer fire service in New York City – and the fist fighting that went along with it – marked the closing of one chapter in the history of firemen's competitions in New York State, another chapter – that of more formal, and significantly more friendly, contests – was just beginning.

Chapter 4: The Dawn of the Muster

While the New York City fire service was going through its growing pains in the late eighteenth and early nineteenth century, other areas of New York State were slowly starting to become colonized as well, and as these new settlements were formed, new fire departments were created to protect them. Like New York City, they first manifested themselves in the form of Fire Brigades – such as the one created in Hempstead in 1726 - but as time went on and populations grew, these organizations became more formalized, and true fire departments were chartered into being. The village of Hudson, located on the Hudson River roughly 125 miles north of New York City, claims to be the oldest volunteer fire department in New York State, having been chartered on March 19th, 1794,[xxxix] with the small whaling village of Sag Harbor, out on the eastern end of Long Island, following suit not long after. December 15th, 1831 saw Hempstead's firemen formally organized as well, as from one end of the state to the other the fire service grew from being haphazard groups of untrained individuals into solid, if still unsophisticated, groups of citizens who were dedicated to protecting their communities from fire.

As could be expected, with this onset of organization – and for the same reasons as in New York City – came the urge for the firefighters to compete with one another, both on and off the fireground. Happily, however, it seems that for whatever reasons – small-town country manners vs. big-city discourtesy – competition between members of the fire service outside of New York City never reached the cut-throat proportions as it did in the metropolis. That is not to say that these competitions were without controversy - on the contrary! One of the very first recorded rural contests, in Utica, resulted in the *death* of a firefighter:

"Fulton No.3 was proud of its own machine and a strong rivalry grew between the two companies [Osceola No.9] as to which was the better machine. On June 12th, 1843, the two companies met on the canal near Broadway bridge to test the relative merits and prowess of the firemen, concerning which a bet of $50.00 was pending. A large concourse had assembled, judges were appointed, and the contest began. An unexpected catastrophe ended the test. Daniel Nestle, a volunteer who asked to work on No.3, stimulated by the excitement to exert himself beyond his physical endurance, collapsed and died.

"As a result, the Common Council, on June 26th, 1843 passed an ordinance to the effect 'That any Foreman, or Assistant Foreman, or Member of any Fire Company, who shall bet, or offer to bet on the performance of any fire engine, or shall cause strife to either company, shall be discharged from the Fire Department, and any Foreman, Assistant Foreman, or Member of any Fire Company who shall take out, or cause to be taken out, any engine, except at fires, or at the regular monthly meetings, without the consent of the Chief Engineer, or of the officer for the time being in command of the fire department, shall for every such offense forfeit and pay the sum of twenty-five dollars"[xl]

What's important to note here is that despite the "strong rivalry" that grew between the two companies, they decided to settle the score not by fighting, but by holding a fair contest between themselves to see who was the best. Contests such as these would often be in the form of a competition to see which company could pump the most water through their engine, by either "washing" their opponent or throwing a stream of water the farthest. In Troy, an 1844 entry makes reference to the latter: "Henry Clay having been nominated by the Whig party for the Presidency of the United States, the followers of the Sage of Ashland, in Troy, erected a white-ash pole of three lengths, about one hundred and sixty feet high, near the north-west corner of River and Albany streets. For several years the different fire companies were accustomed to test the operative action of their engines by throwing streams of water against the pole to various heights."[xli] According to Stan Dixon, in his book The Firemen's Muster – Americas Sport, "This was a simple winner/loser type of competition, with those judging such events being in no position to judge distances accurately. A sumptuous meal at the loser's quarters, or a round of liquid refreshments at a local inn were usually the only tangible rewards for the winners."

Another entry from Troy records, five years later, describes a specific contest that took place there:
"After Eagle Engine Company No.10 received its new engine, made in 1845 by John Rogers of Albany, Niagara Engine Company No.7 became ambitious to be regarded as possessing the better engine. In order to obtain an opportunity to test the merits of the two engines, the last-named company challenged Eagle Company No.10 to make a public exhibition of the water-throwing power of each company's apparatus. On Saturday afternoon, April 28th, 1849, the two companies, in the presence of a large number of firemen and citizens, displayed the action of their engines at the foot of Broadway. The decision of the judges was the following:

'The undersigned judges, appointed by engine companies No.7 and 10, do hereby assert that No.10 washed No.7 in just four minutes and thirty seconds on the first trial; No.10 playing through fifty feet of hose, and No.7 playing out of one hundred feet. At the second trial, No.7 played through fifty feet of hose from the suction, No.10 taking the water and playing through one hundred feet of hose, leaving eleven inches of water in No.10's box, after playing the specified time, as by agreement, five minutes. The trial was made April 28th, 1849, and decided in favor of No.10.'"[xlii]

Apparently, however, the members of the Niagara Company No.7 felt afterwards like they didn't get a fair shake during the contest. But again, rather than getting worked up and angry about it, they simply issued a new challenge...and even raised the stakes:
"The members of Niagara Company having intimated that they had been unfairly treated, those of Eagle Company made them the following proposal on May 7, that year: 'We would most respectfully tender you an invitation to give us another chance to try the respective qualities of engines No.7 and No.10, as the one, had on the afternoon of April 28, did not, according to your opinion, give you satisfaction; and to make it more interesting, we will deposit $1000 (one thousand dollars) in the hands of some one we may mutually agree upon, you placing the same amount in said person's hands, and then work the engines under the same, or similar regulations, as they were worked at the former trial, - and as

The Kennebec team from Bath, Maine, winners of the first organized group muster in 1849

Governor Marcy said, "to the victors belong the spoils" (alias $2000). The proposed trial to come off in this city within three months from this date.'

"No answer was returned, and the Eagle Engine Company made no further effort to bring about a second trial."[xliii]

That year appears to have been a watershed year for the development of fire department competitions. According to Dixon, a historian of early musters, it was in 1849 that the very first recorded organized *group* muster, consisting of five different engines, took place in Bath, Maine, with other musters following that year in Ashland, Massachusetts, Utica and Troy. A June 19th, 1849 *Utica Daily Observer* blurb stated, "Independent Fulton Fire Co. #3: That this company attend the Annual State Fair, to be held in Syracuse in September next, with their new engine, armed and equipped as the laws direct, ready to meet all, and either give or 'take de butt'" and on August 29th, the *Troy Daily Post* ran this story: "If the weather favors, the Firemen with their Engines and other apparatus will make a splendid and highly attractive display on Washington Square this afternoon. The exercises are always (to the credit of the gentlemen composing our unsurpassed Fire Department) conducted with good humored rivalry, [and] will be of an exciting and highly interesting character. The several companies will present their apparatus in the best possible order. The Washington volunteers have had their Engine rebuilt in the very best style, and expect to astonish all by its beauty and admirable working qualities. Take it all in all the picked young men of the city, all in costume, astir with good humor and commendable conviction working their rattling Engines, with all the loveliest girls of the city present among the spectators, and the public may feel assured that it will be by far the most splendid gala-day of the season."

From that point forward, the "muster phenomenon" (as Dixon put it) "blossomed" throughout New England. It was not limited to New England, however; it also began spreading throughout the rest of the country, with tournaments being held in California, Illinois, Indiana, Louisiana, Michigan, New Jersey, Ohio, Pennsylvania, South Carolina, Tennessee, Virginia, Wisconsin, the Dakotas and even Canada.

It's unclear just what exactly promoted this sudden explosion of interest in these competitions. One theory, and from what is evident from accounts of that time, is that despite the serious nature of the competition between fire companies, these musters were more and more becoming *social* affairs, to which the public began to attend in growing numbers (note the last sentence of the *Troy Daily Post* story, above.) In light of the history of inter-company rivalries in New York and other large cities, and the violence and resulting public outcry and derision therein, it's somewhat ironic that the competitions outside the cities should suddenly start becoming such positive and welcomed things. Keep in mind that by this time – the mid-1800's – the fighting in the cities between firefighters was reaching its worst point, and it was not just confined to New York City: a correspondent from the *Utica Daily Observer* forwarded a notice that appeared in a Philadelphia newspaper, actually advertising an armed brawl that was *planned* – no less – to take place that night:

"NOTICE – The Millerites of Moyamensing, from ten years old upwards, are requested to meet this evening on business – the Western Division will meet in the market house on Eleventh Street; and the Eastern at Eighth and Fitzwater Streets. Those having guns or pistols, will bring them along. Those not having these useful weapons are requested to bring as many brickbats and stones as they can carry. The police and watchmen will be on the ground to see fair play. Huzza! Franklin, go it!"

A subsequent account in the *Observer* related the outcome:

"During an alarm of fire the Franklin and Moyamensing hose companies came in collision, when a fight ensued. The parties, after separating, again commenced at 10:00 o'clock in Fitzwater Street, with more deadly violence – Here firearms of every deadly character, including rifles, long guns and pistols, were brought into requisition, and horrible to relate, one young man, a member of Franklin Hose Company, named Alexander Gillies, aged 21 years, was almost instantly killed by a ball in the heart.

"Water Street, or at least the lower sidewalk, is stained with blood, and the street itself is lined with brickbats and other missiles. It is impossible to ascertain the number of persons injured, but it must be great."[xliv]

America's Own and Firemen's Journal included stories of riots taking place in Boston as well, and their September 1st, 1849 issue included references to several different conflicts that had taken place throughout New York City between different companies, including a large brawl between Engine Co.'s 14 and 20 "…which proved to be the most serious riot that any Fire Companies have been engaged in for some years past." When the city's Chief Engineer gave his report two days later, it outlined no less than five different companies who had brought assault charges against one another!

To be fair, not every rivalry between New York companies was settled with violence; there are also records of engines competing against each other at the time in less hurtful ways…although they would still sometimes use some underhanded tactics:

"When [Boss] Tweed was foreman of Americus Engine Company No.6 in New York, his men participated in a spirited contest to see whether any pumper could project a stream of water above a tall new Liberty pole in front of Riley's saloon, at Franklin Street and West Broadway. Tweed's boys may have been politically inspired, but they were just about as good a company of firemen as the city then had. Their stream reached higher than those of competitors, but it was still three feet from the top of the pole. Amidst heavy wagers, Tweed promised to have his men shoot a stream over the pole on the following day. That night, the boss hired a sailor to climb Riley's pole and to saw off the top six feet; this was the day of the clipper ship, when seamen were familiar with lofty riggings. Tweed promised to pay ten dollars for the job, five dollars being paid on the spot. That night, the sailor surreptitiously climbed the pole. And on the next day, Americus Six again was able to wet the pole three feet from the top. The consistency was remarkable but Tweed's men lost

A pumping contest at Market and Fifth Streets in Philadelphia in 1850

their bets. In a towering rage, he sought out the sailor, who calmly explained that inasmuch as only half the fee had been paid, only half the pole had been severed."[xlv]

All the while, the popularity of firefighting musters and tournaments outside the cities began to grow, almost exponentially. Dixon relates that by the 1850's, an average of nineteen musters were being held in New England each year, with an average of ten engines competing at each. From 1850 to 1855 *America's Own and Firemen's Journal* describes tournaments in Rennselaer County, Newburgh, Geneva and Schenectady in New York, as well as Newark, New Jersey; Providence, Rhode Island; Worcester, Lowell and Springfield, Massachusetts, and Hartford and New Haven, Connecticut, where in 1853 "Engine Co.38 of this city challenged the Gasspee of Providence as follows: - 'To play them with the engines then and there used, at New Haven, through four hundred and fifty feet of hose, for the sum of two thousand five hundred dollars…'"[xlvi] There are stories about a "Great Southern Match" between the Washington Fire Co. No.8 of Mobile, Alabama vs. the Neptune Fire Co. No.4 of New Orleans, as well as a "Trial of Engines" at the Firemen's Festival held in Chicago, that drew contestants from several different states all around, (including an engine from Buffalo, New York) which was eventually won by a team from Racine, Wisconsin.

As indicated from the last sentence, by 1855 musters were no longer just local events; teams were starting to travel some great distances to attend some of these competitions. One explanation for this could be that by this point in history many new cities and communities across the land – and especially in New England – were starting to develop and come into their own, and so, by extension, were their fire departments, who were

THE LIFE OF A FIREMAN.
The race.-"Jump her boys, jump her!"

"The Race" by Currier & Ives, 1854

taking deliveries on new fire apparatus on an increasing basis. Naturally, they would be quite keen to test the merits of their new engine against those of other fire departments – after all, their new machine, being the newest, must be the best! - so – why not issue a challenge to those firemen over there who you've heard about and have been bragging about how great their engine is, and take a fun trip while you're at it?

These "excursions," (as the newspapers referred to them then) appear to have been somewhat formal affairs – at least on the outside. Visits such as these were sometimes planned months in advance, and based upon newspaper descriptions, were filled with much pomp and ceremony. This was probably as much due to the mores and mannerisms of the day as it was to wanting to be polite hosts and make a good impression on their guests, but nonetheless there was often a great deal of welcoming, touring, feeding, speeches and toasting that went along with simply testing the engines against one another. No doubt the firemen, being as firemen are, relaxed a bit and had a couple of brews between them after the formalities ended, and for sure once the contests were on some niceties were laid aside in lieu of competitive spirit, but for the most part it seems these visits were as much about goodwill between the two cities as they were about the testing of fire engines against one another.

An example of this can be found in the June 9th, 1858 issue of the *Niagara Falls Gazette*, a weekly newspaper of the time:

"The regular monthly parade of our Fire Companies took place on Monday evening. Our fire department was out in larger force than usual, and made a fine display. They were visited by No. 1 of Niagara City, with their new engine. It is from the manufactury of Hunneman & Co., Boston, and differs but little from

No. 2 of this place. It is a very fine machine, and manned by a finer appearing company than we often see. Most of them are unused to running 'wid der machine,' and of course can not be expected to show the full capacity of their engine as readily as when they get used to the business. Some fair hands had prepared a beautiful wreath of flowers, which crowned the machine. They were met halfway by No. 2, and escorted through the village, to the race above the Cataract House, where a trial of engines took place. The principle strife seemed to be between No. 1 of this place and No. 1 of Niagara City, and in justice we should remark that neither of them were in perfect working order, and that, owing to a community of feeling between the Niagara City Company and No. 2 of this place, the former was much more fully manned than our No. 1 – the two companies assisted each other. Very little difference was, however, to be seen in the height of the two streams – that difference being in favor of the Niagara City Company. No. 2 threw the most beautiful stream, and as near as we could guess (for she played alone) considerably higher, but she was in perfect working order and fully manned. The result as usual in such cases, only convinced each company that their machine was a little better than the others."[xlvii]

It is only after reading several written accounts of these musters that one can begin to get a true sense of just how wonderful – indeed, *grand* – these occasions were in those days! Unfortunately – or fortunately, as the case may be – there was no television or radio back then, so it was the responsibility of the journalist to "paint a picture" with words, as vividly as possible, of the events before them, so that readers who could not attend could get a true sense of what it was like to be there. Some descriptions were somewhat dry, listing only the facts and the events in the order that they occurred, but others, written by talented correspondents who clearly had a flair for bringing a printed page to life, were quite incredible in how they brought the reader back to that day, and made them feel like they were actually there, experiencing all the sights and sounds of the occasion. One of the first examples of this comes in 1858, when in September of that year Albany held a large muster that was attended by teams from New York and five other states. It was covered in depth by two different local area newspapers of the time, the *Atlas & Argus* and the *Albany Evening Journal*, as well as being mentioned in the Troy Fire Department's history:

"An interesting event of the year was the firemen's parade and trial of engines at Albany, on Wednesday and Thursday, September 29 and 30. As Premier, Empire State, Niagara, and Eagle Engine companies of Troy intended to participate in the parade, and to contend for the prizes of $500, $400, $300, $250, $200, and $150, that were offered for throwing streams of water to the greatest distance, they made considerable preparation, and exercised their members in working the engines before going there."[xlviii]

<u>The First Day</u>

"To-day the town wears the aspect of a gala day. The streets are thronged with strangers and citizens, the hotels and public buildings draped with flags, the sidewalks and windows on the main streets crowded with spectators of the show. Red shirts and glazed caps are omnipresent, and give an exceedingly gay appearance to the scene. The uniforms of the different Companies vary greatly in details of caps, ornaments, badges, and etc., but the appropriate red shirt is universal, except with two or three, who have a neat and serviceable dress of blue. The bands alone are numerous enough to form a Regiment, Dodworth's alone, with its fifty-two members and its military-looking dress, being larger than most uniform companies. The Firemen muster between 2500 and 3000 strong. We have heard more than one wish for a conflagration, of a wild type that they might all be seen at work at once – capable as they are of throwing more water than comes down Cohoes falls. Since this cannot be had, however, why might we not have 'Hail Columbus' or some other air performed by all the bands in unison? So grand a performance would eclipse the Monster Concert at the Crystal Palace, or any other ever heard in this country.

"The engines both of the visitors and of the city companies are dressed in their best. They are of all shapes, patterns, sizes and varieties, upright and horizontal. The brass and steel upon them is polished to the brightness of silver and gold, and wreaths of flowers tastefully decorate them. As they traverse the streets one does not wonder at the zealous care and just pride with which each Company guards its 'machine.'

"Banners, signals and devices suitable to the occasion were borne by several of the companies.

"Six different states are represented – Connecticut, Rhode Island, Massachusetts, New Jersey, Michigan and New York. Providence on the East, Detroit in the West and Newark on the South, being the most distant points from which our visitors have come."[xlix]

The weather that day, apparently, was beautiful:

"The weather must have been satisfactory to the most querulous on that point. Not a cloud was visible, the sun shone out brightly and with just sufficient warmth to call out the expression from everybody – 'what a delightful day it is.' It was not too warm to parade over our hills in red shirts, not too cool for the ladies, who thronged the walks, to appear in light dresses."[l]

To say that it was a large affair, especially for its time, would be a complete understatement. There were 24 local and 40 visiting companies in line, as well as 24 marching bands:

"Averaging the Companies at sixty men (not a high estimate), there were in line 3840 firemen. Add to these the Bands, containing from six to fifty-two members, respectively, and it will not be extravagant to say that there were over 4000 in the procession."[li]

The Second Day

"The firemen have had another beautiful day for their exercises. Notwithstanding their fatigue of yesterday, and the late hour into which many of them pushed their festivities, they appeared fresh and muscular this morning.

"Soon after 8 o'clock they began to move, in isolated detachments, to the Fair Grounds, where there was to be the 'tug of war.' Before 10 o'clock, all the contestants were on the ground, with many thousands besides. The scene was highly animated and picturesque. The immense Flag Staff, 200 feet in height, upon which the painted figures loomed out distinctly, was the center of attraction. Some 200 feet to the west was the Judges' stand, 60 feet high; and near where the engines were stationed to play, was the Chief's stand, at whose signal the contestants stripped for the work."[lii]

The contests that day – as they were at the time - were divided up into three categories, based – we believe - upon the size of the pump being used. So-called First Class engines had a pistons sized greater than 7 inches; Second Class were those with pistons between 4.5 and 7 inches, and Third Class with pistons smaller than 4.5 inches. (Modern-day musters that still use the original apparatus divide their classes into the same three classes, which they refer to as Class A, B, and C, but the modern-day Class C category wasn't even created until the 1970's, so it is unclear if the Third Class contests were the same thing, as at the time the three classes were also sometimes defined by the number of men working at each engine.)

Competing that day in the First Class group included New York teams from New York City, Brooklyn, Buffalo, Lansingburgh, Little Falls, Utica and Troy, as well as visiting teams from Stamford, Connecticut, Springfield and Lee, Massachusetts, and even a team from as far away as Detroit. Teams in the Second Class division, in addition to those above, included West Troy, Whitehall, Williamsburgh, Yonkers, Hartford, Waterford, Norwich and Newark. Third Class teams included a team from Providence as well.

"During the entire day the grounds was thronged with visitors, not a few of whom were of the gentler sex, and we heard the number of visitors estimated as high as ten thousand. There could not have been less than three thousand firemen on the field, and it is hardly necessary to say that they, of all others, enjoyed the occasion. From the Judges' stand the scene presented to the beholder was decidedly picturesque. A field of ten acres, perfectly level, thronged with thousands of spectators, thickly sprinkled among whom were three thousand men, in red shirts and black shining hats, and all moving through and fro, constituted a picture most pleasing to the eye."[liii]

From all accounts, the competition, though fair, was quite fierce:

"The strife was exceedingly animated, and the excitement ran much higher than during the contest between the First Class machines. The thousands who were watching the playing cheered vociferously whenever the streams reached an unusual height, and these manifestations, with the almost frantic calls and gestures from the commandants of engines mounted upon their machines, induced such Herculean efforts on the part of those at the brakes, that our only surprise is that scores did not drop exhausted or do themselves some fatal injury."[liv]

In the end, it was Franklin Company No.3 of Brooklyn that came out on top in the First Class division with a stream of 152 feet, followed by Stamford with a 149-foot effort. Whitehall took First Place honors in the Second Class division with a 165-foot throw, and New York's Company #8 took the Third Class win with 156 feet.

"Those of our citizens who failed to visit Washington Parade Grounds yesterday lost much that was both instructive and amusing. All who desired to see the fireman in his *element* – praising, and loving, and working his pet machine – should have been there. All who have ever heard or seen the 'Scenes of New York life' as mimicked by Chanfrau and Miss Albertine, and who would like to have seen these scenes represented to the very life, should have been present at the Muster yesterday. They would have seen there, in all his glory, the big-fisted, big-hearted 'Mose,' ready to wager his existence on the success of his favorite machine, and ever on the alert to use the most convincing arguments upon any who might venture to express a doubt as to her ability to cope with the best. They would have seen there, the genuine 'Syksey' ready to back up, in his own peculiar way, whatever his tried friends might advance, no matter how extravagant. They would have, too, the veritable 'Lize' – indeed a score or two of Lizes – dressed in all the amplitude of the fashion of our day – with plaid silk dresses of the broadest and brightest styles, low in the neck, and short at the bottom, as if the extremes were endeavoring to meet – wearing shed bonnets and sunny countenances, fringed with beau-catchers. The would have seen 'Mose' and 'Lizey' and 'Syksey with his lady,' the boys with the bottoms of their pants rolled inside their boot legs, promenading arm-in-arm, and occasionally halting at a booth and demanding 'lager, bologna and Switzer cheese, for two boys, do yer hear me.' They would have learned the way to address the boys to induce them to greater exertion when the pet machine was in play; such – 'Now, start easy boys, every man stick to his work and don't look at the stream. Steady now – steady, and let the heavy men stand by to "spell" off. A little faster – faster – faster – *now*, then sail in my hearties, sail in; shake her up lively – lively I say; give way my boys. Every man o' you work now or never' – etc., etc, to the end of the chapter, spiced throughout with somewhat stronger language.

"All this they might have seen and learned, and much else that cannot be well set down on paper."[lv]

If nothing else, the event was timed well: later that evening a severe thunderstorm passed through the muster grounds, knocking over the top part of the flagstaff used for measuring the height of the streams.

Back in Brooklyn, Company No.3's win in Albany was treated with somewhat reserved praise:

"Engine Co. No.3 of this city, the winner of the First Prize at the muster of the firemen in Albany, did not arrive this morning as expected. We learn from a gentleman from Albany this morning, that "Franklin" accompanied Engine Co. No.11 of Troy, as the escort to that city. No.3 will return to-morrow, and the Department of this city should give them a hearty reception upon their return."[lvi]

That the Big City should still be so lukewarm about the muster was typical; it seems that the metropolitan population still had a ways to go to catch up to the amount of enthusiasm shown towards firefighter tournaments compared to that of their rural brethren. It wasn't as if they weren't popular there, judging by an account of a January 24th, 1855 contest: "A huge crowd turned out on January 24th, 1855, to watch the water-playing trials at Tom Riley's Hotel at Franklin Street and West Broadway. Companies No.53 and 54 supplied the hose to Engine No.42, the 'Mankiller,' most powerful engine in the city. The famous liberty pole, which had been marked off to indicate the height of the competing streams, showed that the Mankiller, winner of the muster, played to a height of 137 feet."[lvii] But understandably, based upon all the problems they'd had in the past, the public was still skeptical: "We hear that a race is to come off on the Union Course, some day not yet named, for a purse of $500, between Hook and Ladder Companies, Nos. 1, 3 and 4. However novel this species of amusement may be, we think that the excitement and contention consequent therein will be productive of more harm than good. The consequences will not be contained to the race course, but will thereafter be made evident in the streets of the city, in a manner that will not likely add much to the present fair reputation of the Fire Department. We are rather inclined to look upon the whole affair as a 'sell,' got up to see what effect it would produce. The Department would certainly not allow so flagrant a violation of the laws by which they are governed – at least we think not." [lviii]

Their skepticism notwithstanding, fire department musters were here to stay, and would only increase more and more in both size and popularity. Dixon reports that by the mid 1850's there were an average of 19 musters per year throughout New England alone, with an average of 10 "handtubs" (as they were called) at each. One noteworthy event took place in Worcester, Massachusetts on September 4th, 1857, where 60 different engines gathered together "...to flex their muscles and prove to the world that they were the best." Dixon also recounts a story of a Mr. Fred Burrell of the Weymouth, Massachusetts Volunteer Fire Department, who "...would let homing pigeons go from various muster locations, announcing to the people back home how their handtub had done at the muster."[lix] He did this from as far away as Lisbon Falls, Maine – a distance of over 150 miles!

As the popularity of fire department musters grew and spread out across the land, it seemed that this form of competition was finally starting to gain a bit of respectability amongst the members of the public, especially across the New England states of the northeast. But with growth often comes conflict, and there is evidence that these new, exciting competitions were not without problems of their own. Even as the firemen who were on "excursions" to other cities were on their best behavior as they showed off their apparatus and took part in "friendly and spirited" contests with their hosts, the scourge of fighting between firefighters of different companies still existed – no matter what city they were in. And while these fights for the most part didn't occur to the severity and extent that they did in New York City, it seems that despite their being in different locales they occurred for pretty much the same reasons: pride, greed, and an out-of-control competitive spirit.

One of the first references to this appears in Dixon's text, regarding a handtub muster that took place in Lockport, N.Y. on September 1st, 1855: "...all the engines played at the same time which resulted in much fighting, ending up in a full scale riot. Adding fuel to the fire was the fact that the judges had made arrangements to have the Wright Bros. Engine win the contest."[lx] Troy seemed to have a particularly difficult year in 1859, starting in February, with "...a number of vicious, quarrelsome men, whose connection with the companies was considered by the other members highly derogatory to the service of the department":

"As remarked by one of the daily newspapers regarding the ill-will causing such encounters [fights between companies], 'a large majority of the firemen detest all feeling of the kind, and it is too bad that they should be disgraced by the continued connection of loafers with the department, or made even indirectly responsible for the acts they abominate. Action should be taken in this matter at once, and before the prevalent feeling reaches a culminating point of brutishness similar to that which injured the department several years ago.'"[lxi]

Apparently, no action – or at least not enough action – was taken, since this entry appears not long after:

"A disagreement having arisen between the members of Eagle Engine Company No.10 and those of Neptune Engine Company No.2 respecting a prior right to use a certain hydrant at the fire which burned about six o'clock on Sunday morning, March 20…a fight followed for the possession of the hydrant, in which a number of the participants received facial disfigurements and bodily injuries from the violent use of trumpets and wrenches in the hands of excited officers and firemen. Two firemen were also hurt by bricks thrown from the top of a house near the fire. 'Several aldermen and police officers, who were on the ground,' as related by one of the daily newspapers, 'used all their power in quelling the disturbance, and some of the offenders were taken to jail.'"[lxii]

By July 10th of that year, things had still not quieted down:

"The pipeman of Hibernia Engine Company No.12 having turned the stream of the engine upon the members if Niagara Engine Company No.7, at work with their engine, orders were given to the pipeman of the last-named company to turn its stream upon those of Hibernia Engine Company. In the fight that followed, wrenches, axes, and clubs were violently used. The members of Hibernia Engine Company also fiercely attacked the policemen who were trying to quell the fight, and several were badly hurt."[lxiii]

A mere six days later, over in Massachusetts, there was "…a timed event where each handtub was required to fill a 1,500-gallon tank. The winner was the 'Eagle #5' of Danvers, MA who accomplished the feat in 6 minutes and 43 seconds. Unfortunately, this muster was marred by much drinking, gambling and fighting."[lxiv]

Oddly enough, firefighters then - as now – would as soon stick together as they would fight amongst themselves if they ever felt threatened as a whole. In Springfield, Massachusetts on October 12th, 1860, "…Sheriff Frederick Bush arrested one of the bands at the muster who were disturbing Judge Lord of Salem, Mass., who was presiding over a trial nearby. After the muster, the judge was nearly mobbed by a group of infuriated firemen at the hotel in which he was staying."[lxv] (Much later, in 1947, this same mindset would contribute to events that would cause significant changes in the way firefighters would compete in New York State from that time forward.)

It is apparent that firefighters then, again as now, could be somewhat of a rowdy bunch, in spite of their honor and dedication to the duty of saving life and property. But as the 1850's came to an end, there are fewer and fewer historical references to firemen fighting amongst themselves; from reading them it almost seems as if the men were getting tired of not getting along, and the bad rap that came with that. Although it would end up being too little too late, even New York City showed some evidence of trying to clean up its act. A log entry from Chatham Engine Co. 27 is notable in that it describes what seems to be a contest that *didn't* include any fighting: "Sunday, December 25th, 1859, 9-1/2 p.m., 7th District: Fire Duane & Wm., W. Walsh in Command. Number of men out – 38. Both officers out. On the next day we took our large engine to Reillys pole and Beat every apparatus playing that was there."[lxvi] As well, an 1863 Letter to the Editor in the *Brooklyn Eagle*, from (presumably) a fireman named only "Franklin", tries to put a positive spin on a obviously horrific event: "It is unnecessary to call attention to the fact that your reporter frequently of late has been compelled to chronicle in large letters, 'Another Row Among the Firemen' or else record in a more conspicuous place in your widely-circulated journal, 'that as two rival engine companies were racing each other they knocked down and crushed to death some unfortunate individual who was at that time crossing the street'. All this, Mr. Editor, tends to detract from our fair name, and we cannot in justice reproach the outside public for setting down our organization as a 'loafing institution'."[lxvii]

This sentiment was not limited to New York City, as it appears in newspaper accounts in western New York State as well. After a "grand parade" in Lockport, in 1867, a "trial of engines" took place, during which "...A splendid show was made. The skill and celerity with which the various engines were handled showed that no tyros were handling the brakes. Everything worked in fine order, the only mishap being the bursting of hose once or twice, which was soon replaced. Quite a spirit of friendly strife as to which should throw the best and highest stream was manifested, and no little excitement was consequent. As there was no committee to decide, and as the friends of each company contest the honor, it would be uncalled for, for us to give even an opinion. It is sufficient to say that they all did themselves great credit."[lxviii]

Perhaps in keeping with, or as a result of, this desire of the firemen to put forth a better image of themselves, there is also an increase in written mentions of firefighter parades by the early 1860's. One might infer that current events of the day may have played into that as well, as 1861 saw the start of the Civil War, and a mentality that engendered large groups of men marching before the public (after all, the fire service is often considered to be somewhat paramilitary) would have been prevalent. On September 14th, 1860, in Huntington, Long Island, "Our village was visited yesterday by an excursion party of firemen, with their families, from the city of Bridgeport, Conn. The whole party, numbering some 200, two companies, (an Engine and a Hose Company,) paraded our streets for a short time with their Machines, accompanied by a Band of Music, and made a fine display"[lxix] and the following month, in New York City, during a visit to the city by the Prince of Wales, "...the firemen gave a tremendous torchlight procession in his honor on the night of October 13. There were six thousand smoke-eaters in line, and Prince Edward reviewed the marchers from the balcony of the Fifth Avenue Hotel. This was considered to have been one of the city's finest displays."[lxx]

The Civil War took a toll on the fire service, and the practice of musters and competitions during those years, as scores of firemen were recruited into military ranks to fight. Many died, and for those who didn't their absence affected tournaments back home. Regarding a muster in 1861 Dixon writes: "At the Annual Tournament of the active engines of the Newburyport, Mass. Fire Dept., only the Protector #3 entered. All the other machines failed to appear due to many members being in the Army and away at war."[lxxi]

Another significant change that came about during this time was the invention of steam engines. Although it would eventually be seen and remembered as a boon to the fire service, its introduction to the fire service was anything but auspicious; in fact at first firefighters felt they could do a better job of firefighting without them. Interestingly, though, it was a contest – a form of tournament, if you will – that helped facilitate the introduction of the steam engine as a viable piece of firefighting apparatus. Holzman writes that in 1855, "...New York saw its first working steamers. A manufacturer, Lee & Larned, furnished two steamers to the city, but the firemen at first would have nothing to do with this equipment. A delegation of volunteers petitioned the City Council to reject the machines. On February 9, a contest was held in City Hall Park between one of the newfangled steamers and a tremendous hand pumper of the type known as a 'man-killer.' The latter stream was slightly the longer, and most of the firemen shouted happily: human muscles were better than a machine. But two fire department officials noted that whereas the exhausted firemen dropped at the brakes, the steamer steadily pumped its water."[lxxii]

Ever the champion of the olden-days of firefighting, Costello didn't seem too pleased about the advent of the horse-drawn steamers either, as they all but eliminated any kind of competition between two companies as they raced to a fire:

"The horses attached to the engine did away with the rope, and with it the yearning of the rabble to take a hand and 'jump her,' and the horses did not yell and urge each other to exertion on the way to a fire, but attended strictly to business and left the camp followers who were not the fleetest of foot in the lurch, so that the pranks and depredation of the 'sidewalk committee' were done away with. The horses, too, had no fancy for hurrying to rush past a rival company, or halt and fight out a rancorous sentiment, partly jealousy, partly rivalry, while the fire which had started was getting ahead."[lxxiii]

The "Steam Against Muscle" contest in New York City in 1855

A postcard depicting a Firemen's Muster held in Manchester, N.H. on September 16th, 1859

The mid-1860's also saw another significant development in the practice of firefighting musters: as cities expanded and more fire departments moved out into the countryside, the format of the competitions changed from that where hose streams were shot straight up into the air, with the distance of the stream (in this case, the height) being measured against a flagpole or other tall object of a known height, to horizontal streams, where the length of the stream was measured by where the farthest drops from that stream fell onto the ground. This change is generally considered to be as a result of the teams simply having more room to work with during the competitions, as they were no longer "hemmed in" by buildings or other restrictions such as you'd find in a larger metropolitan area. There is some evidence that this changeover occurred somewhat gradually, based upon a reference in Dixon's book from the muster that took place in Springfield in 1860: "Five handtubs played against each other, each pumping one vertical and one horizontal stream. Both streams were added together with the winner being the 'Niagara #1' of Springfield with a total of 261 feet."[lxxiv]

Chapter 5: The Birth of FASNY and the Dawn of Competition on Long Island

In August of 1872, a group of local firefighters in Auburn, N.Y. were watching a parade of a state benevolent association move along their Main Street, when someone brought up the idea, "why not have a firemen's convention in New York?" At their September 6th meeting of the entire department, they resolved to form a national firemen's association, with the object being "...cooperative in its nature and with the view to improvement in the general government of fire departments, the discussion and adoption of modern and improved fire apparatus, also a general interchange of ideas and to discuss important questions pertaining to the various duties of firemen."[lxxv] It was in this manner that the Firemen's Association of the State of New York – or FASNY, for short – was born. A convention was scheduled for October 1st and 2nd of that year, for which there appeared to be a great deal of excitement: "As the day for the opening of the convention drew near, interest and enthusiasm grew by leaps and bounds."[lxxvi] Eleven different states sent letters stating they would be interested in attending, but when the actual day arrived, only two other states – New Jersey and Iowa – were represented.

The primary directives of the FASNY conventions were ostensibly geared towards fire service administration and the exchange of information, but records show that demonstrations of and competitions between engines were not far from the organizer's minds as well: the Schedule of Events indicated that "...the second day would be devoted to a parade and amusements. Visiting steam fire engine, hose and other companies would compete in tests, and a trial of the city water works would be given. As a special feature, arrangements were made for a half-mile race between Hose Co. No.7 of Ithica and Niagara Hose Co. No.3 of Auburn."[lxxvii] Syracuse and Oswego were also invited to participate, but it is unclear if they actually attended or not. It is also unclear if these contests, despite being scheduled, even took place per se; there is no mention of them in any descriptions written after the convention, other than to say that after dinner everyone gathered to "...witness the trial of the capacities of the steam fire engines and the city water works. Water was taken from several hydrants and as many as ten streams being thrown at the same time. In the meantime the trial of the

Ithaca steamer was in progress at another point drafting its own water. The distance thrown with a single stream by the steamer was 225 feet and that thrown by the water works was 180 feet."[lxxviii]

In 1873, plans were made to hold a second gathering, but they were cancelled due to the sudden death of Thomas Towne, the inaugural President of the association. As a result, the 2nd Annual FASNY Convention was held in 1874 in Oswego. According to the Secretary's notes, plans to hold a firefighter's tournament at that convention "...were abandoned, the Committee on Subscriptions reporting 'poor encouragement in getting money for the purpose.'"[lxxix] Despite this setback, it appears that a contest of sorts took place that year at the convention anyway, with a local Oswego newspaper reporting that "...the coming Firemen's Convention is the principle topic of conversation among the firemen. The challenge of Cataract Hose Co. No.4 of Oswego should be accepted, the race to take place the day of the annual review, August 19th."

If the prospect of holding tournaments at the state conventions was off to a rough start, the conventions themselves were not. Additional Secretary's notes from the second convention reported that one of the organizers, a Mr. A.F. Smith, "...reported receiving 10 or 15 letters per day from companies; he thought there would be 300 delegates present at the Convention, and probably 500 other firemen before the end of the parade and review."[lxxx] He was not far off: in addition to Oswego, participants from Batavia, Weedsport, Auburn, Peekskill, Baldwinsville, Frankfort, Hudson, Utica, Ilion and Mexico, N.Y. paraded and took part in the festivities.

Much of what there is to be learned about firefighting tournaments that took place at the FASNY conventions in the years that followed comes from Secretary's notes, but after thoroughly reviewing them they are maddeningly sparse when it comes to references to these contests. We know – or presume, anyway – that tournaments were held in association with the conventions from pretty much that point forward, but for many years they are referred to only in passing, and then usually only if there any items of administrative significance that pertained to them.

Fortunately, however, the references that *do* appear are very often quite significant within the context of the history of firefighting tournaments in New York. For example, in the minutes of the 3rd Annual Convention that was held in 1875 in Poughkeepsie, the only relevant entry was, "Mr. Johnson, of Hamilton, thought and suggested that the association should establish some rule for tournaments, and on motion it was voted that the whole matter be referred to the Executive Committee."[lxxxi] It may have been only one sentence, but it tells us that the interest and participation in the musters had grown by that time to such a point that more organization was needed for the contests.

(An interesting side note regarding the convention that year: the minutes from the morning session of the first day note that "...Mr. Ware, of Elmira, offered a resolution that the Executive Committee be authorized to petition the next Legislature to regulate the number of threads to the inch and the size of couplings for fire hose, throughout the state, so that when one city or village is called upon to aid another in case of fire, there will be no difficulty in connecting together the hose of the different places. President Poucher stated it would be an unconstitutional act and would amount to nothing. This resolution was referred to the Executive Committee."[lxxxii] Their refusal to adopt this resolution still haunts the fire service to this day, where there are still at least three different thread standards being manufactured and sold, resulting in neighboring fire districts that own two different standards having to spend thousands of taxpayer dollars for thread-adaptors so they can work together at a large fire.)

It appears that at the 4th FASNY Convention, held the following year in Elmira, the subject of creating rules at tournaments that had been brought up the previous year was addressed by the membership. A committee had been formed to study the problem, and during the proceedings they presented their findings on the topic, "Are tournaments beneficial and should the association establish rules?" In spite of what seemed to be an increasing popularity in the competitions at the conventions, the committee gave the following report: "Your committee to report on tournaments beg leave to report that, in their judgment, tournaments are not beneficial to the Volunteer Fire Departments, as it tends to engender ill feelings between companies and departments."[lxxxiii]

Once again, the Secretary's notes from the 1876 convention include no mention of the tournament held in association with the convention that year, but thankfully an account does appear in the August 16th issue of *The Elmira Daily Gazette and Advertiser*, and it describes what may be one of the very first Cart Hose contests ever recorded: "The contest for the Drake Silver Cup took place Wednesday morning, August 16th, on William Street, being a run of one fourth of a mile and return, each cart carrying three hundred feet of hose, at the end of the run to un reel, uncouple, couple up and reel. There was a large crowd present, both sides of the street being greatly crowded, so much so that the exercises were much impeded."

There was talk of another potential first for the sport - teams wearing actual racing uniforms:

"Lots were drawn for the order of running, which resulted in No.1 taking the first hack at it, followed by Nos. 5, 2 and 6. Nos. 1 and 2 surprised their friends by the odd appearance they made in their speedy uniforms.

"The following is the time of the respective companies, No.1 beating the next best by 4-1/4 seconds:

No. One – 4:40; the run was made in 3:35; fifteen men.
No. Two – 4:44; running time 3:53; fifteen men.
No. Five – 5:34; running time 4:02; thirteen men.
No. Six – 6:15-1/2; running time 4:08; twelve men.

"The result when announced was greeted with much applause, and beaten companies joined heartily in congratulating the winners."

And, no less, the article includes one of the very first references to a "juvenile" contest, involving young adults under eighteen:

"The hose race was followed by Hook and Ladder race participated in by two juvenile companies, and which resulted in the Young Actives winning the prize silver cup presented by Col. E.R. Abbott. The cup was borne on their carriage in the parade."

As a finale, the writer makes a mockery of the FASNY Committee's conclusion that "...tournaments are not beneficial to the Volunteer Fire Departments, as it tends to engender ill feelings between companies and departments":

"The best thing of the whole contest was the friendly spirit in which the result was received. With rousing cheers for each other, for the impartial judges who so satisfactorily conducted the contest, and for Mr. Drake, the generous donor, the two companies and friends separated. It was an event never to be forgotten in the history of the department."

In 1880, however, at the 7th FASNY Convention in Auburn, it seems that the association's skepticism towards firefighter tournaments still hadn't changed. However, parades were another thing: at the convention, Topic No.15 was, "Annual Parades and Reviews – Are They Beneficial?"

"Your Committee would respectfully report that fire parades and reviews are an excellent thing for the purpose of furthering better discipline and improving the appearance of Companies and Departments, and that they create a friendly spirit of rivalry in drill, and fire matters generally, between fire companies, and in our judgment they are very important and exceedingly beneficial. We are also of the opinion that drills should be encouraged, as they intend to improve the efficiency and discipline of fire organizations."[lxxxiv]

So, why the popularity of parades and not tournaments? There is no one clear answer, but one thing that should be kept in mind is that all this was taking place during what historians refer to as the Gilded Age, from roughly 1865 to 1901, during which America saw unprecedented economic, industrial and population growth and expansion. One description put it this way: "During the 'Gilded Age,' every man was a potential Andrew Carnegie, and Americans who achieved wealth celebrated it as never before." In spite the fact that the division between the rich and the poor was wider than it ever had been - by 1890 11 million of the nation's 12 million families earned less than $1200 per year - "...the opera, the theatre, and lavish parties consumed the ruling class' leisure hours." It's not hard then to figure that firefighter tournaments, considered as large groups of people gathered together to watch a spectacle of pomp, ceremony, parades and competition, would be a perfect opportunity to throw a "party", and romp in the excesses of a festive, extravagant atmosphere. So, while much of the rest of the world felt we were on a fast track to hell (when French prime minister Georges Clemenceau visited, he said the nation had gone from a stage of barbarism to one of decadence - without achieving any civilization between the two), the advent of the Gilded Age brought with it a period of unprecedented growth in the size and popularity of firefighter parades – and tournaments - as well. It's a bit sad, and perhaps ironic, that this sudden growth had little or nothing to do with the firemen themselves – the fire departments were almost certainly made up entirely of middle- or lower-class citizens, who fell into the "11 million" bracket – but for the events to become as large and as popular as they did then, based ostensibly (at least) on the bravery and prowess of the firemen is at the very least a nod and a compliment from the "elite" in their direction.

But grow they did, and in ways and places that they never had before. Additionally, and significantly, a divergence began to form during these years in the practice of firefighting musters and tournaments in the northeast: in the New England states – Connecticut, Massachusetts, Rhode Island, Vermont, New Hampshire and Maine – "musters" were still taking place between fire departments, using traditional apparatus such as handtubs, etc. to compete for pumping power/length of streams with, while in New York State the competitions during the 1880's – now described more and more as "tournaments" - started to include other types of firematic competitions as well, such as the Cart Hose race described earlier. Why this divergence took place isn't clear, except to say that the concept of Parade and Drill may have had something to do with it: whereas the musters that were taking place outside of New York State appear to have been for the most part gatherings of firemen to "test their engines" and compete with little or no ceremony involved, the New York State events were more and more becoming huge affairs where, in addition to just parading, the concept of "drill" was increasingly popular. During the parades, the firefighters would march or perform maneuvers together in unison, and be scored by a panel of judges for such things as their uniforms, the number of men in line, the precision of their movements, and so forth. It was another form of "competition" between firemen that was proving to be very popular amongst both the public and the firefighters themselves, as indicated from the minutes of the 10th Annual FASNY Convention, held August 15th – 18th in Rochester: "…a Competitive Drill, described as 'one of the most pleasant and interesting features of the Rochester Convention' was held on the last day with seven companies displaying their marching and drilling skill."[lxxxv] The Rules and Regulations that day were as follows:

> "I. Companies shall be called for drill by one tap of the bell in the order named, commencing at 9 o'clock a.m., sharp.
> II. Each company must be on hand to the minute of each bell, and if more than five minutes elapse thereafter such company will be considered to have withdrawn, and the next company in order will be called.
> III. Twenty minutes will be allowed each company to drill, and the second tap of the bell will announce the expiration of the time, when the company will be immediately marched from the ground.
> IV. Companies may drill with or without music, as they may elect.
>
> Finally, appreciating the fact that no standard tactics have been adopted by the fire companies of the state, we have decided to allow the greatest latitude to all companies participating in this drill, reserving the right, however, to criticize in detail the various movements presented, the same to be an element in making the award of prizes. General perfection in appearance, bearing and movements, will be considered as a basis of award, and the earnest desire of the judges will be to make such award with

perfect fairness and equity to all. All may rest assured of our best consideration, and no element of favoritism will enter our councils."[lxxxvi]

The Knickerbocker Steamer Company of Waterford took First Place honors with a score of 95-1/2 out of 100, followed by Palmyra, Auburn, Penn Yan and Geneva. The FASNY Secretary refers to a Competitive Drill in his report again two years later, but states that only two departments entered – Waterford and Satterlee – with Waterford again taking the prize.

As the popularity of Drill grew, a logical next step might have been someone's idea to include other things within the fire service that would require precision to do well – such as unrolling a hose and hooking it up to a hydrant at a fire – in the tournament as a form of Drill competition. Whatever – and whenever – that first took place we will never know for sure, but it can be stated that without a doubt it changed the way tournaments in New York were run from that point forward, and would eventually make New York State unique from any other state – or indeed, anywhere in the entire world – in how firefighters would compete with each other.

That fact seems to have played a part in the regional popularity of musters and tournaments in the northeast as a whole during the 1880's, for during that time as the vogue of the competitions grew in New York State, their celebrity actually began to lessen in the New England states. Dixon writes, "This decade saw a steady decline in muster activity. Averages fell to 10 musters per year, attracting 5 handtubs each. Small, local contests seemed to be the trend, with regional musters almost nonexistent…An omen of how poorly the sport would fare in the next few years emerged in Fitchburg, Mass., when the judges arbitrarily gave the handtub Rescue of Grafton, Mass. a score of 170 feet even though it had broken down and never played a stream."[lxxxvii] In contrast, New York tournaments were breaking ground not only in the types of contests that were being held, but also *where* they were being held. In Niagara Falls, it appears a rivalry had developed between the firemen there and the firemen across the river in St. Catherines, Canada, that if the *Suspension Bridge Journal* is any indication, both sides took pretty seriously:

"For many years Niagara Falls Hose Co. of Niagara Falls had in its possession the figure of a fireman known as the 'iron man.' Some years ago it was stolen from them by members of the St. Catherines fire department and was placed on a tall hose tower in that city. There has been much badgering about this trophy between the firemen of the two places and it was well understood that the lost property would be replevined at the first opportunity. That opportunity came to the Niagara Falls Hose boys on the recent tournament in St. Catherines and they brought home with them the 'iron man' and it again adorns their headquarters. The Niagara boys felicitate themselves on the fact that they captured the trophy while the festivities were at their height and while two watchers set to guard it were on duty. Joy reigneth in the camp of the Niagara and the St. Catherines mourners go about the streets and refuse to be comforted."[lxxxviii]

Almost as if to add insult to injury, and in an unusual departure from what would normally be seen in newspapers at the time (high praise for the hosts of a tournament), that same issue of the paper also carried a letter from a member of the Suspension Bridge Fire Department complaining about how badly they had been treated at the tournament:

"No space that I am warranted in asking would be sufficient to fully express the disgust felt by Suspension Bridge firemen on account of our treatment at the recent tournament in St. Catherines. I shall only give voice to the unanimous sentiment of our firemen when I declare the management of the tournament to have been simply disgraceful. Without commenting at length on the failure of the committee in charge to receive the visiting firemen, or their neglect to assign us to a position in the line of march, we want in justice to ourselves and all concerned to denounce the managers of the tournament in failing to extend the common courtesies due to men who had come from a distance and at large expense had contributed toward making the tournament a financial success. We were invited to go to St. Catherines by the firemen of that place, and were urged to compete for certain prizes offered – prizes which, by the tone of their advertisements, were to be as much for improved apparatus as for number or equipment or speed of men. The liberality of our people at home has given us a Hook and Ladder Truck so much the superior of anything in St. Catherines or vicinity, that the managers of this tournament found it necessary to break away from their published program in order to retain in Canada the prizes they had offered. And the judges? Will some firemen outside of St. Catherines pretend to endorse the selection of men so wholly unfit for the position as were the men who passed judgment, by favor of the managers, on the different contests. Not one of them has now, nor do we believe ever had, any connection with a fire department; and while we are ready to concede to the wise men, who acted as judges that day, ability to 'take time' at country horse races accurately and satisfactorily, we submit that such qualification does not make them competent to decide as to equipment or drill of firemen. We are, however, under obligations to several citizens of St. Catherines for courtesies extended during the day and evening. This was all the more appreciated and pleasing to acknowledge because of seemingly studied neglect on the part of the St. Catherines firemen."[lxxxix]

It is doubtful that the Suspension Bridge firemen were really treated that horribly; more likely, it was just an example of just how formally developed tournaments had become in New York, and how spoiled (in a sense) her firemen had become as a result; to expect the same treatment and conditions that they were now used to at New York tournaments in other places – like St. Catherines, Canada – was probably an unfair demand.

But St. Catherines notwithstanding, and despite the growing "maturity" of tournaments, there was still one area of New York State that up until this point was still way behind the times as far as firematic competitions were concerned.

Long Island – aptly named, for it runs over 120 miles long, from New York City to Montauk – is an area of the state that remained, up until the late 1800's, strangely devoid of any real tournament activity, despite its relative proximity to New York City. The area boasts some of the very earliest settlements in the New World, dating back to the 1600's, and the fire services there got an early start as well – from Sag Harbor (the second oldest in New York State) to Hempstead (1831), Riverhead (1836), Greenport (1845) and Rockville Center (1875), but oddly enough the fire departments there seemed to be not only completely disassociated from the firematic activities that were going on everywhere else in the state prior to the early 1880's – which can be understood, considering the distances that needed to be traveled and the methods of transportation available to them at the time – but there are also no records indicating that any contests of their own took place on the island at all up until that point, which is rather puzzling. The first recorded firematic gathering of any significance, in fact, wasn't a tournament at all, but simply a parade of firemen. The event retains a significant amount of importance, however, as it marks the birth of firematic drills and tournaments[2] on Long Island, which today stands uniquely as the center for the sport, and where more firefighter competitions are held than anyplace else in the entire world.

The auspicious occasion during which this took place was Decoration Day (soon to be re-named Memorial Day) of 1882. The Patchogue Fire Department, which had been founded just two years earlier, was holding a firefighter's parade to celebrate the day, and their Chief, Edwin Bailey, took it upon himself to invite the members of other fire departments to march along in their parade with them. Strangely, the local newspaper, the *Patchogue Advance*, made no mention of this parade, but the events of the day were described in some detail in other publications. A somewhat dry blurb in *The New York Times* stated:

"Yesterday's observance of Decoration Day in Queens and Suffolk Counties was more general than that of any previous anniversary...At Patchogue there was a parade of the members of the Grand Army and the Fire Departments of the villages of Babylon and Islip. All business was suspended during the afternoon. At Sag

---

[2] Although the term "drill" originally referred only to regimented movements and formations associated with parading, the term has since grown to include all other firematic contests and the events at which they are held. Unless otherwise specified, the terms "drill" and "tournament" will henceforth be used synonymously.

Harbor the Fire Department paraded and the church bells were tolled for an hour. There was also a parade at Riverhead and an address by the Hon. Thomas Young."[xc]

The Times' story indicates that the Islip Fire Department took part in the parade, but this may have been a mistake on their part, as none of the other references found made any mention of them being there. One even states that Patchogue was "the only one to respond":

"The Grand Army of the Republic, as represented in Wm. J. Clark Post, No.221, together with the Fire Department, had arranged for a parade, and invitations were sent for different Fire Departments to join in the procession. Babylon department was the only one to respond, and with a special train arrived in Patchogue at about 11 o'clock, where they were received in good style and escorted from the depot by the Patchogue Fire Department."[xci]

A more descriptive account appeared in Babylon's *South Side Signal*:

"The Babylon Fire Department took an active part in the celebration at Patchogue on Decoration day, notwithstanding the disappointment they had suffered in the failure of the proposed steamboat excursion, which was brought about by the boat not being ready as promised. As soon as the fact was ascertained arrangements were made to go by rail. Accordingly it was arranged to be taken down by the mail train and return at 7 p.m. by special train in charge of Conductor Ross Miller. All three of our companies were well represented, numbering 65 men in line. They made a fine appearance in their bright new uniforms, and in marching and in general discipline did credit to the village they represented. Our 'boys' were received on their arrival by the entire Patchogue Department, headed by the Cornet Band and Drum Corps of that place. Chief Edwin Bailey and Foreman John Roe Smith left nothing undone. To say that they were well taken care of would state but half the case; they were treated and otherwise entertained in a royal manner, their stay forming a real 'red-letter day' in the history of the department. The friendships formed and the generous emulation incited cannot fail of having good and lasting results. Preliminary arrangements are already being made for having the Patchogue Department return the visit of the Babylon boys, when an old fashioned 'firemen's day' is promised. It is through the interchange of courtesies like these as much as through any other means that the enthusiasm and efficiency of the volunteer force is to be maintained."[xcii]

The Budget echoed the same sentiments. After describing numerous speeches by Edwin Bailey and others, they went on to say:

"The Fire Department entertained its visitors by a trial of the engine at the mill pond, and after regaling them with a further feast of good things, escorted them to the depot, where the platform was made the rostrum for several complimentary speeches from representatives of both departments. The visit has cemented a firm friendship between the two departments, and will incite both to a pleasant rivalry to excel in their work for the people."[xciii]

Apparently the gathering was such a success that, as implied in the articles, plans were immediately made for Babylon to host a parade, to which Patchogue would be invited, as well as other fire departments from Suffolk County (the county in which they were located) and Queens County (which had not yet by that point become Nassau County as it exists today). That parade was held on August 30th of the same year, and was attended by not only Patchogue, but by firemen from Rockville Center, Freeport, Islip, Hempstead, Huntington, Sayville, Pearsalls (now known as Lynbrook) and Breslau (now Lindenhurst). "The Department feels pleased with the success of their first annual parade," wrote *The Budget,* "and from expressions of pleasure and approbation, the citizens were pleased with the Department."[xciv] (This despite "certain mistakes in the line of march.")

This author's research has indicated that the above-mentioned teams did in fact take part in the Babylon parade, but there are other historical documents in existence that confuse this finding. The New York State Volunteer Fireman, the official publication of FASNY, implies otherwise in its June, 1937 issue. The FASNY author contradicts himself on at least two occasions in his recount of the history that year. First, he relates how Babylon invited Patchogue to march with them in their August 30th parade of that year,

"...extending the guest list to include several of the then Queens County fire companies." Then, in the same article, while relating a story about a subsequent parade that took place on a rainy day in Hempstead on October 19th, 1882, he contradicts himself by saying, "Hempstead, now the largest village in Nassau County, which was then part of Queens County, invited the Patchogue and Babylon fire laddies to help celebrate the Hempstead department's fiftieth anniversary, which is the first time the firemen of the two counties had joined in a parade." Had "joined in a parade" in *Queens*, perhaps? There is no doubt that the Hempstead parade took place on that date, but this author maintains that it was Babylon, and not Hempstead, that takes honors as being the site of the first inter-county gathering of firemen on Long Island.

Back upstate, at about the same time, FASNY was still trying to figure out what to do about firefighter tournaments. Although they were still not an official part of the convention program, it's likely that they were going on anyway, despite the continuing lack of any description regarding them in the Secretary's minutes. (For example, an 1885 issue of the *Lockport Journal* lists the results of a Hose Contest between Lockport and Albion.) What *is* indicated, though, is that there was as much hemming and hawing with bureaucracy then as there is now. Even after deciding in 1880 that parades were a good idea, it took two more years for FASNY to reverse its earlier opinion about tournaments, when at the Rochester convention in 1882 a committee, reporting on the topic "Are tournaments beneficial and should Associations adopt Rules and Regulations?" concluded that "Your Committee on Tournaments recommend the holding of such, and for the following reasons: First, it stimulates the boys in the absence of fires; Second, it influences outsiders to take an interest in fire matters in general."[xcv] True to form, however, they abstained regarding the "Rules and Regulations" portion: "On the second part of Topic no report." The convention itself was an obvious success: the minutes also state, "Friday, August 18th, which was known as 'Parade Day,' dawned auspiciously, and from an early hour crowds poured into the city by every avenue and approach. At the hour fixed for the procession to move it was estimated that 80,000 people were on the streets."

And, Long Island had finally made the connection to Up North: at the previous year's convention in Saratoga, the minutes reflected attendance at the convention by the very first delegate from Long Island, a member of the Empire Hose Company of Flushing.

The Island, it seems, was on a roll. Another milestone for firefighter competitions occurred in Babylon on September 21st, 1883, when Babylon Chief Joseph H. Cummin, as much of a firefighting "buff" as his friend Edwin Bailey of Patchogue, introduced the Tournament feature to firefighters on Long Island at a gathering there that day. The *South Side Signal* issue of September 29th recorded the festivities, reporting that twelve different teams – Greenport, Persalls, Riverhead, Freeport, Rockville Center, Jamaica, Sayville, Hempstead, Patchogue, Islip, Huntington and Babylon – all competed in contests such as Individual Ladder, Hook & Ladder, a Hose Race and a Steamer Competition. "The hook and ladder contest opened the hall in the exhibition of speed and skill in handling fire apparatus," they wrote, and "...the hose race formed one of the most exciting contests of the day." Later on, "...the contest at running 50 feet and placing a hand on top round of a 30-foot ladder was well contested." Babylon won a silver trumpet by beating the other teams in the 300-yard Hook and Ladder contest and the Relief Hose Company of Greenport won the Hose Race in 57 seconds. The Steamer competition was also won by Greenport, with a stream thrown of 150 feet, and Henry Baylis of the Babylon truck company won the Individual Ladder contest with a time of 7-3/4 seconds. "Awarding the prizes was a pleasant

The parade in Babylon in 1884

feature of the day," the paper wrote, "and took place at the department building. Owing to some misunderstanding the orator selected for the occasion failed of appearance, and this duty devolved upon Chief Cummin, assisted by L.H. Fischel, Esq. All hands were really too tired to listen to long speeches, so the honors were distributed with a few well-chosen words in each case and as heartily received. In the distribution of prizes there was very general satisfaction, and none of the growling and protesting that too frequently characterize similar contests."[xcvi]

Well...almost. Apparently Greenport won the Hose Contest only after a re-run, since (again) "owing to some misunderstanding" they were never timed during their first run down the course...and in another section of the very same issue, the following also appeared: "It is to be regretted that any dissatisfaction should arise as to the manner of awarding the prizes contested for at the firemen's parade and tournament, as the following paragraph, clipped from the Jamaica *Democrat,* indicates:

> 'The opinion of those who witnessed the contest was that, if the prizes had been justly awarded, Degrauw Hose Company, of Jamaica, would have taken the first prize.'

This reflection upon the fair-mindedness of the judge in the hose contest is ungentlemanly, ungenerous and uncalled for. Just why Degrauw Hose should have been awarded the first prize is beyond our comprehension. Their time was 62 seconds – four seconds behind the record of Phoenix Hose, of Babylon, which captured the second prize. If our young friends from Jamaica fondly entertained the opinion that victory perched with certainty upon their banner, they were disappointed – an experience with which several other companies met. We can hardly believe that any one of the young gentlemen connected with Degrauw Hose allowed himself to become so disaffected as to mislead Editor Brennan. Just here we take occasion to remind our friends in Jamaica that the Babylon boys 'shook the hay-seed from their rustic shirts' without Queens county help. 'Let it be recorded.'"[xcvii]

It is unfortunate and ironic that there should have been a protest at the very first recognized firefighter tournament to take place on Long Island, but in light of the history of the sport, it is almost fitting; controversy has traveled hand-in-hand with the sport since its inception, and continues even until this day.

Whether it was by word-of-mouth of the goings-on in Long Island, or after a decision (finally!) that firematic tournaments were "o.k.", FASNY appears to have formally introduced firefighter competitions to the "official program" of their conventions at their August 21st, 1884 gathering in Utica. Nine teams – one each from Newark, Cortland, Clinton, St. Johnsville, Waterville, Homer and Long Island City, and two from Amsterdam – took part in hose races at Riverside Park on that Thursday, the 29th. "The contest was over 900 feet of ground," the notes state, "running 600 feet, laying 300 feet of hose, attaching to hydrant, breaking connection at the finish, and attaching pipes."[xcviii] The Emerald Hose Company of Cortland won the day with a time of 49-1/2 seconds, and took home a $150 cash prize; St. Johnsville and Amsterdam took second and third, respectively. Ironically, in what appears to be the first reference to a team from Long Island ever attending a tournament upstate, the Tigers of Long Island City came in a dismal fifth place with a time of 75 seconds – a not very promising first-representation from the one area of the state that would soon become the hub of the firematic racing world!

The FASNY notes from that year are interesting as well in that they relate a "parade of the old firemen" that took place the afternoon of the same day as the hose races, after which a pumping contest was held, much in the same manner as the musters of New England. Interest in this type of competition was obviously still high, as the notes state: "Both sides of the street were packed with people and it was exceedingly difficult to obtain a view of the contests."[xcix] What's interesting is that the participants in the contest were the "old firemen", implying that even back in 1884 the method of extinguishing fires using hand-pumped "enjines" was starting to be considered "old fashioned"!

Back on Long Island, the success of the Babylon tournament had no doubt inspired the little village of Greenport, located out on the east end of the north fork of the island, for on September 18th, 1884 – just one month after the FASNY convention in Utica - they decided to hold a tournament for both Suffolk and Queens County firefighters. "Perhaps never before in its history were as many persons present there at one time," wrote *The Sag Harbor Corrector*. "It is estimated that from 4,000 to 5,000 people took part in the Firemen's Tournament either as spectators or as interested parties. Seven hundred firemen were in line."[c] That figure may even have been a conservative estimate, as *The Long Island Traveler* of neighboring Southold, in its front-page story, portrayed the event as being much larger: "Two special trains, one from Jamaica with seven cars, one from Babylon with ten cars, brought the firemen, apparatus and citizens from those two places as well as all the intervening stations on the main line and branch. In addition to these two train loads, people flocked in from all directions from far and near, so that about eight thousand people thronged the streets by afternoon." Moreover, they described a "...grand parade, consisting of nearly one thousand men...the whole line extended nearly a mile." And then: "At the conclusion of the parade, the crowds that had been scattered throughout the village began to congregate on the lower part of Main Street, where the contests for prizes took place. From the post office to the wharf was one solid mass of human beings, with every available window, piazza and housetop thronged."[ci]

There were five contests that day: a Hook and Ladder contest, wherein contestants would run 300 yards with a ladder, place it against a building (in this case "Webb and Corey's Building"), climb to the top and place a hand on the top rung; a Hose Carriage contest (not described), two "contests between engines" to see who could throw the longest stream and the best two streams, and an Individual Ladder contest wherein contestants would be timed running 50 feet, climbing a 30 foot ladder and placing a hand on the top rung. Of the twelve teams that attended the tournament – Huntington, Hempstead, Jamaica, Freeport, Flushing, Rockville Center and Long Island City of Queens, and Babylon, Patchogue, Breslau (Lindenhurst), Riverhead and Greenport of Suffolk – only Greenport and Babylon took part on the Hook and Ladder contest, which was won by Babylon with a time of a fraction over 53 seconds. "The Babylon boys did the best running," stated the *Traveler*, "but the work done by the Greenporters on the ladder could not be beaten. The Babylon truck was much the lighter of the two."[cii]

The Hose Carriage contest was equally exciting, but was apparently once again filled with mishaps and controversy. Teams from Babylon, Riverhead and Greenport competed: "The Riverhead boys came first, but had a bad balk by dropping their nozzle. Next followed Eagle No.1 of Greenport. They ran over the whole ground, but owing to a mishap the judges failed to mark their time, so they had to run again, as did the Riverhead boys. The Babylon boys then ran and Relief No.2 of Greenport followed, making the best time in running of anything, but as they were breaking couplings and putting on the nozzle someone _____ [*illegible*] them and bothered them. Riverhead was awarded 1st Prize and Eagle No.1 of Greenport 2nd Prize, time 51 sec. and 54 sec."[ciii] Jamaica won the Engine Contest, and Rockville Center won the Individual Ladder climbing contest in 8-1/2 seconds.

While researching this event, it seemed odd that firemen from Sag Harbor – who had only two short ferry rides to get there – had not attended the tournament. There is no indication why they didn't, but such being the increasing fervor over firematic tournaments during that time, it was amusing to see their local paper call them out onto the carpet for not being there:

"Why was Sag Harbor not represented by some organized company at the Firemens tournament at Greenport on Thursday?

"This question was asked a hundred times by persons from all over the County. Sag Harbor is the largest village in the County. It has the oldest organized Fire Department, yet no one of its companies were at Greenport. This we think is very much to the discredit of the department as well as the village.

"Babylon, and Patchogue, and Riverhead were all on hand with apparatus and uniformed companies, presenting an excellent appearance, marching well, and contesting nobly for rewards of merit. But Sag Harbor was not there. Why not? It is a question we ought to ask, and it cannot be answered favorably. Either the department is inferior and shuns competition with other places, or there is a lack of interest in our Fire Department which does not speak well for the place. There were a great many Sag Harbor men at Greenport, and they all regretted that their village was not represented in its department.

"As an indication of how much interest other villages take in this matter we may mention Riverhead, which came down with one hundred and twenty five uniformed men. Among them some of its best and most substantial citizens. We saw in the ranks such well know residents of that village as Treasurer Perkins, Architect Skidmore, Lawyer Stackpole, Justice Millard and others.

"The interest such men take in their department does them credit, and Sag Harbor may well emulate their example. We hope at the next tournament, which will probably be held at Jamaica, that Sag Harbor will not be missing."[civ]

It's unknown for sure if Sag Harbor attended the next tournament that was held in Jamaica, on September 9th of the following year, but if *The Brooklyn Eagle* was any indication, it doesn't seem as if they did: "…Long Island City, Flushing, Hempstead, Greenport, Riverhead, Patchogue, Woodside, Breslau and the old Continental Company of Brooklyn were present," they wrote.[cv] If they weren't there, it was a shame, as the event turned out to be the largest of its kind to date: the *Brooklyn Eagle* described this third two-county tournament as having "fifteen hundred uniformed men in line", and the *Hempstead Inquirer* wrote that "unusual interest had been manifested in the tournament and a much larger number of fire companies and members were in attendance than on any occasion of a similar character in the history of the island."[cvi] Even the

1776--INDEPENDENCE DAY!--1885

MONSTER CELEBRATION

—AND—

TOURNAMENT

At Lockport, N. Y.

SATURDAY, JULY 4th, 1885

Under the Auspices of ACTIVE HOSE CO. No. 5.

GRAND PARADE at 10 o'clock A. M., SHARP

FIRE ORGANIZATIONS!

FANTASTIC BRIGADE!

TRADE REPRESENTATIONS.

Grand Tournament at the Fair Grounds

COMMENCING AT 2:30 P. M.

MAGNIFICENT MUSIC AND A GLORIOUS TIME

Grand Display of Fireworks in the Evening.

Admission to Fair Grounds, 25 Cents. Full Particulars in Official Programme.

R. & G. & P. D

New York Times, which traditionally has been known to be somewhat "highbrow" and as a result usually disinterested in the goings-on of ordinary people, saw fit to write a small blurb about the event:

"In a hose-laying contest each company was allowed 18 men, and to use either a two or four wheeled cart, capable of carrying 500 feet of hose, and was required to run 200 yards, lay 300 feet of hose, break coupling, and put on a pipe. The first prize, a racing cart, was won by the Tiger Hose Company of Long Island City, and the second, a pair of silver bells, by the Relief Company of Greenport. Ladder companies were required to run 300 yards and raise a 25-foot ladder against a building. There were two entries only, and the Union Truck, of Breslau, was awarded the championship belt, and the Woodside truck a pair of truck lamps. Protection Hand Engine, of Jamaica, won the first prize, a pair of silver pipes, and Neptune No.2, of the same village, the second prize, a silver lantern, in the contest for first-class hand engines. The Patchogue engine won first prize, a silver trumpet, in the second-class contest."[cvii] Their interest was probably helped by the fact that the parade was unique as well in that Harry Howard, a famous Chief of the early New York Fire Department, was at the head of the marchers, who were led by a group of 100 veterans of the old New York volunteer department.

But despite its being a milepost in the size of firefighter parades and tournaments, it was not the most successful event ever hosted. Fred Flugrath, in The Suffolk County Volunteer Firemen's Blue Book, notes that the day was "...marred by a heavy rain storm," and that "muddy streets irked the paraders and upset the firemen in several contests."[cviii] Moreover, the by-now-traditional inconsistent judging of the contests, and its resulting confusion and controversy, seemed to once again rule the day:

"In the contest for prizes the judges and time-keepers got all mixed up and no attention was paid to the rules and regulations as printed in the official programme. The programme stated that the Foreman of companies entered in contests would select judges and time-keepers. The judges were, however, selected by some one else. The programme also stated that 'no allowance shall be made for mistakes or accidents,' yet Tiger Company, of Long Island City, in the Hose contest, was allowed to start three times. Protection Company, of Hempstead, was unquestionably entitled to second prize in the Hose race and every fair-minded person acknowledged it. The judges' time was – Greenport 30 seconds, Protection 33 seconds, and Tiger Hose of L.I. City – after the third trial – 25 seconds. No one could tell how they were timed as it was not taken at the start – it could not have been considering the time given."[cix]

As large and popular as the tournaments were becoming, as indicated above, it was becoming just as obvious that something needed to be done to help with the *organization* of the tournaments. As quoted from *The Argus* of Patchogue, "Between 1882 and 1886 the good feeling engendered between the firemen of the counties by these annual gatherings had developed a progressive spirit among the several villages, and where there had been but few organized fire companies in 1882, there had sprung into existence by 1886, so many fire companies that a one day's tournament engaged in by the departments of the two counties was necessarily unsatisfactory and unwieldy."[cx] A *Brooklyn Eagle* article from 1902, while echoing the same basic facts, even goes so far as to suggest that the sudden growth of fire companies in the Babylon area had more to do with the interest in firematic tournaments than it did with a need to provide adequate fire protection for a growing

The tournament at Jamaica in 1885

population: "...within a radius of twenty miles, Babylon had the only fire department, by 1885 there were six, the increase being due to the stimulus set forth by the gatherings of each succeeding year."[cxi] (It's a startling claim, but being that it was a newspaper, and thus ostensibly charged with reporting information in an accurate and un-biased way, it's tempting to ask, "so new fire companies were actually formed just so they could compete with each other, fire safety notwithstanding?" From a fire historian's point of view this would be outrageous fun if it were true, but common sense dictates that it probably wasn't the case.) Whatever the reasons, it was becoming more and more apparent that something needed to be done. In response to this need, it was our old friend, Patchogue Chief Edwin Bailey, who once again came up with an idea that would forever shape the course of Long Island firematic contests to come: that Suffolk County firemen should form their own, separate, organization, with the view of having a strictly Suffolk-teams-only tournament. It was in this manner that the Suffolk County Volunteer Firemen's Association was born.

Chapter 6: The SCVFA is Born

In truth, by 1886 the idea of forming a separate association of Suffolk firemen was not a new one; in fact it was already a few years old. Patchogue's Chief Bailey had conceived the idea after the success of having Babylon march with them in their parade in 1882, and had in fact called a meeting of local chiefs on June 24th, 1882, in Patchogue, to present his ideas. It stands to reason that the subsequent parades and tournaments only acted to strengthen his case, until finally on June 24th, 1886 – four years to the day – another meeting of the chiefs was held in Patchogue to formally discuss the formation of the Suffolk County Volunteer Firemen's Association. The Suffolk County Volunteer Firemen's Blue Book lists nine fire departments that attended the meeting that day: Patchogue, Breslau (Lindenhurst), Islip, Babylon, Bay Shore, Riverhead, Greenport, Blue Point and Amityville. It was not until a third meeting however, held two months later on August 25th, 1886, that elections took place, offices were filled and the association was "formally perfected". Stephen A. Titus, a prominent Brooklyn newspaper man, was elected as its first President, "...and there is not a member of the association who cannot recall the name of 'Steve' Titus without many pleasant recollections. Mr. Titus was a hard worker, and from his record as a member of the Babylon Fire Department his associates found the best material to lead the new organization."[cxii] (Author's note: a "Brooklyn newspaper man"? Perhaps involved with *The Brooklyn Eagle*'s claim that "new fire departments were formed due to the popularity of tournaments"?) A Mr. O.A. Terry of Riverhead was voted to be 1st Vice President; Walter L. Suydam of Blue Point was 1st Secretary, and John N. Frazier of East Islip was awarded the post of Custodian of Funds. Curiously, in a show of modesty, Chief Bailey refused to accept any office.

As far as membership in the association went, there were only six charter members of the SCVFA that day...but unfortunately who they specifically *were* isn't very clear. The Suffolk County Volunteer Firemen's Blue Book is the only reference that could be found that goes into any detail about the formation of the organization, and its descriptions concerning the fire departments that were involved at its inception are somewhat confusing: at the June 24th meeting, from the nine fire departments it lists as having attended the meeting, only four different fire departments seemed to have joined the association: Bay Shore's Hook and

Chief Edwin Bailey

Ladder Company, Sayville's Hook and Ladder Company, Islip's Hook and Ladder and Engine and Hose Companies, and the Minnehaha Hose and Gazelle Hose companies of Sag Harbor. However, "...the six companies reported on the membership roll" at the August 25th meeting don't mention Islip's Engine and Hose or Sag Harbor's Minnehaha Hose companies, but instead include Amityville's Hook and Ladder Company and the Protection Engine Company of Southold (whose fire department had only just been formed that year). It's entirely conceivable that during the interim between the two meetings the first two changed their minds and dropped out and the latter two decided to join, but this is only speculation; what actually happened will always remain somewhat of a mystery.

What is abundantly clear, however, is that the SCVFA's First Annual Tournament, held in Patchogue on September 8th of that year, was an event of spectacular proportions. The *Patchogue Advance* ran a special issue touting the event, and provided detailed coverage of the entire day's proceedings, which included morning preparations: "In hundreds of dwellings the ovens were heated and great piles of breads and biscuits and meats were forthcoming. Tables were made ready at the Lyceum and dishes and provisions were gathered, and as much was done as possible, anticipating the activities of the day.

"The morning dawned with an overcast sky. A gray murky veil rested over the village, but it soon began to lift, and promised a pleasant though warm day. The village was astir. In some parts of the line of march, busy hands were putting up bunting, and suspending banners, yacht colors and other adornments, and a few of our places of business were finely decorated. Written on the countenances of the people and pervading the very atmosphere, was expectation ominous of scenes about to transpire."[cxiii]

...and the reception of the visiting firemen:

"At the appointed or advertised hour the special trains from the west arrived, in one of which, by actual count, there were 1,500 persons. One of the specials from the east was delayed about one hour, which caused a little delay in carrying out the program. All the trains, both special and regular, brought great crowds of firemen and citizens, to participate in and witness the sights and enjoy the competitive scenes of the tournament. There were in all between 8,000 and 10,000 persons in our streets."[cxiv]

The *New York Sun* made the tournament the lead story in the first column of the front page:

"Extra trains with flat cars loaded with fire engines, trucks and hose carts on which red white and blue coated firemen clung like flies on sugar barrels, rolled into Patchogue station from east and west all forenoon. Brass bands kept the air in the quiet town tremulous all day. Brilliant uniforms dazzled the eye in all directions. As a display it beat general trainin' day all hollow."[cxv]

They went on to write, "...The sidewalks were thronged with the fairest daughters of the County, who came by the special trains, and the streets were jammed with farmers' vehicles, many of them driven from twenty-five miles away. Ginger ale and ginger cookie stands sprang up like mushrooms on every street. The

BUY YOUR DRY GOODS FROM S. Wechsler & Bro., THE GRAND CENTRAL DRY GOODS HOUSE! 293 to 299 Fulton Street, 266 to 268 Washington st., (REAR ENTRANCE,) BROOKLYN.

PATCHOGUE ADVANCE

THE LEADING JOURNAL OF THE LARGEST VILLAGE ON THE SOUTH SIDE OF LONG ISLAND.

INDEPENDENCE, NOT NEUTRALITY.

VOL. XV.} SUBSCRIPTION PRICE, $1.50 Per Annum. PATCHOGUE, L. I., SATURDAY, SEPTEMBER 4, 1886. SINGLE COPIES, 5c. {NO. 51

FIREMEN! Who attend the Tournament at Patchogue, and have friends with them, can find good accommodations at the Lake View House, Where Dinner can be had at all hours. Also, Clam Chowder, Tea and Coffee. E. H. TERRY, Prop.

PIANOS AND ORGANS TUNED AND REPAIRED BY L. MULLER, PATCHOGUE, N. Y.

The Day We Celebrate!

First Annual Tournament of Suffolk's Firemen.

OUR FIRE LADDIES.

The First Annual Tournament of the Suffolk County Firemen's Association takes place in our village September 8, and the event will mark an interesting epoch in the history of Patchogue. For the entertainment and comfort of the [illegible] "The Day We Celebrate," the Patchogue Fire Department have made every arrangement possible to make all that is pleasing to the sight and a comfort to the inner man a success. And it simply remains for our citizens, hotel keepers, and all concerned to make this gala day one to be remembered by affording every convenience to the visiting companies and their friends. That there is no lack of interest manifested by our citizens is shown by the handsome and useful array of prizes offered to the contestants. Messrs. Ioller & Price have contributed their share by giving a Silver Trumpet, which will be contested for as a first prize in the Engine Test—First Class. The second prize, a magnificent Tilting Silver Ice-Pitcher and Goblet is offered by the citizens of Patchogue, also in this class. And the last prize, in this class, is that of a beautiful engraved Silver Pitcher, offered by the citizens of Patchogue. In the contest of engines—second class, a Silver Trumpet is the prize presented by our citizens.

CHIEF ENGINEER EDWIN [illegible]

M. SCHULZ & BRO., 504 FULTON STREET, (NEAR BOND,) BROOKLYN. The most reliable house in the trade to buy FURNITURE, CARPETS, &c. ALL THE LATEST STYLES OF PARLOR and CHAMBER SUITS CONSTANTLY ON HAND. Also a large assortment of the Latest Designs and Petterns of Brussels, Ingrain, and other Carpets TO SELECT FROM. Special Terms of Credit if Desired. M. SCHULZ & BRO., 504 Fulton St., Near Bond.

THE Most Complete and Choice Stock of Groceries and Provisions IN PATCHOGUE, Is now offered for sale by GEO. D. GERARD, OCEAN AVE., NEAR MAIN STREET. Perfect Satisfaction Guaranteed. Every Purchase, or Money Refunded.

McNEIL'S Meat and Vegetable Market IS THE PLACE TO FIND YOUR Choice Cuts and Vegetables FRESH EVERY DAY. MAIN STREET, Patchogue.

A. FISHEL, DEALER IN [illegible]Y GOODS [illegible] Furniture, etc. [illegible] and Pine Sts., Patchogue.

[illegible]EO. J. KIRK. ([illegible]essor to T. Dar[illegible],) DEALER IN [illegible]toves, Ranges, Heaters and [illegible]OUSE FURNISHING GOODS. Plumbing, Roofing, Well Driving and Pump Repairing promptly attended to. Main St., Patchogue, L. I.

ROSE'S Baggage Express. Packages, Baggage and Merchandise, of every discription, delivered in any part of the Village with care and dispatch. General Baggage Delivery to and from Railroad. Orders left at Conklin's News Depot.

BARGAINS IN WATCHES DIAMONDS AND JEWELRY AT ROBINSON'S NEXT DOOR TO BANK.

[illegible] "RED CROSS" Cotton Fire Hose, FOR ALL PURPOSES. LINEN HOSE, PLAIN AND RUBBER LINED. ALSO GARDEN HOSE. 13 Barclay St., New York.

FARMAR, McQUILLAN & CO., Ladie's Tailors. RIDING HABITS, COATS AND SUITS. Mourning Outfits a Specialty. 150 Atlantic Avenue, BROOKLYN, N. Y.

[illegible] Stoves, Ranges and Heaters. Pumps, Pump Repairing, and Well Driving. Tinsmith, etc. Main St., Patchogue.

GEORGE WILSON, Furniture, Carpet, Oil Cloth, AND BEDDING WAREHOUSE. [illegible] and 42 Myrtle Ave., Cor. [illegible] St. BROOKLYN, L. I. Goods Warranted as Represented.

G. C. GRIOT, DEALER IN Household Goods, Notions, Woodenware, &c. Fancy and Staple Groceries. Large Assortment of 10c. Goods. Jenning's Block, Patchogue.

WILSON G. TERRY, DEALER IN GENTS' FURNISHING GOODS, HATS! CAPS! JENNINGS' BLOCK, MAIN ST., PATCHOGUE, N. Y.

GLEASON & BAILEY,

EQUIPPERS OF FIRE DEPARTMENTS. New York Show-Rooms cor. Mercer & Houston Sts. | Chicago Show-Rooms, 98 Franklin St. Modern Fire Apparatus A Specialty. Send for Estimates.

GRAND PARADE and TOURNAMENT.

THE FIRST ANNUAL TOURNAMENT —OF THE— --SUFFOLK COUNTY FIREMEN'S ASSOCIATION,-- —TO BE HELD AT— PATCHOGUE, —ON— WEDNESDAY, SEPTEMBER 8TH, 1886.

Procession will Form on Main Street, Right resting on Pine St., at 11 A. M.

LINE OF MARCH.

Main Street to West Street, to Cross Street, to Railroad Avenue, to Church Street, to Ocean Avenue, to the Bay; Counter-March on Ocean Avenue to Baker Avenue, to Ryder Avenue, to Main Street, to Medford Avenue; Counter-March on Main Street to Pine Street, leave Apparatus and march on Ocean Avenue to the Lyceum.

DINNER AT 12 O'CLOCK.

SEE FOURTH PAGE FOR DAY'S PROCEEDINGS.

FABRIC FIRE HOSE COMPANY, Nos. 1 and 1½ BARCLAY ST., N. Y. MANUFACTURERS OF UNIQUE. KEYSTONE. PATROL. ARROW. BRANDS OF Mildew-Proof Rot-Proof Balanced Woven RUBBER LINED COTTON FIRE HOSE. J. E. Gillespie, President, H. F. Wheeler, Gen'l Agent.

T. H. HORTON & CO., Fulton Avenue and Smith Street, BROOKLYN. One of the Largest, Cheapest and Best Assorted Stocks of FANCY AND STAPLE DRY GOODS! In the City. N. B.—No Tin Pans or Wash Tubs! You can Save Money by Shopping at our Establishment. [illegible] T. H. HORTON & CO.

We Recommend our Friends to Visit COWPERTHWAIT CO., THE "OLD RELIABLE" House Furnishers, Where you can select your Furniture, Carpets, Bedding, Etc., From the Largest Assortment at the Lowest Prices. Cash or Credit. 408 to 416 Fulton St., Brooklyn.

When You Want New CARPETS, OILCLOTHS, MATTINGS, RUGS OR MATS, TRY Geo. H. Titus, [illegible] Fulton Street, BROOKLYN. Reliable Goods AT LOWEST PRICES.

In the issue, date of December 19 last, the ADVANCE published an article headed "Our Fire Department," the writer of which gave a full and well written description of its history. As no better description of our Fire Company can be afforded the readers of this journal, we reprint the same:

On February 20, 1880, active measures were taken for the formation of the Patchogue Fire Company; and from its organization to the present time, the company has fully demonstrated its worth to and received the highest commendation from the citizens of our community.

In the first part of January, 1880, a fire occurred in the bakery of Nicholas Van Aradale, this village. The flames, which were controlled with difficulty, imperilled the entire business section of Patchogue. But the event served to crystalize the demand long entertained for a local fire department, and was immediately followed by a call of a meeting to discuss the importance and feasibility of such provision. This meeting was progressive and resolute. Mr. Jesse C. Mills was appointed to correspond with Mr. E. C. Lefforts, the inventor and proprietor of the chemical engine known as Storm King, and arrange with the latter to bring his engine to Patchogue and exhibit its working power. Mr. Mills, accompanied by Mr. George Smith, visited Huntington, the place of Mr. Leffort's residence. The latter responded by bringing his engine over, and a pracical trial of its merits was given by firing a mass of very inflammable material, boxes, &c., which had been heaped up for that purpose. As a result, the chemical spray instantly extinguished all parts of the fire on which it could be thrown, but as other parts proved to be beyond its reach, it was judged to be inadequate for the requirements of the village.

Public interest was by this time more thoroughly aroused, and at an adjourned meeting Messrs. G. D. Gerard and H. C. Losee, were appointed to

ASSISTANT CHIEF E. H. TERRY.

procure funds for the purchase of an engine, and Messrs. Jesse C. Mills, G. D. Gerard, A. Fishel and Edwin Bailey were appointed in conjunction with Mr. G. H. Ketcham, (expert) of Riverhead, a committee with power to purchase an engine such as might seem to them to be suitable. The committee's attention [illegible] inspected this engine which they found capable of sending a copious stream to the sixth story of a building and to be in thorough repair. It was put in operation for them and having satisfied themselves of its adaptation to their want they purchased it. This machine the company still has and it has in every instance given entire satisfaction. The receipt for payment for it bears date of April 5, 1880. It has a history, and has unquestionably more than once saved the business and other closely built portions of the village from being devastated by fire.

Since the organization of the Patchogue Fire Department, it is estimated that no less than $500,000 have been saved to property owners by this promptness. For the value of the Department and in appreciation of its service the Company was the recipient of a substantial reward from the insurance companies in recognition of their services at the burning of the Oaksmith property for saving the cotton factory which occurred in the early part of the summer of 1881.

(CONTINUED ON FOURTH PAGE.)

The Gutta Percha and Rubber Manufacturing Company. MALTESE CROSS BRAND Carbolized Rubber FIRE HOSE. BAKER FABRIC RUBBER-LINED Solid Woven Cotton FIRE HOSE 2, 3 and 4-PLY. THE BEST IN THE WORLD. COR. WARREN AND CHURCH STREETS, NEW YORK. BRANCH STORES:—Chicago, San Francisco, Boston, Toronto, Portland, Oregon.

M. KARSHICK, Late with Devlin & Co., Broad'y, N. Y. Merchant Tailor, OCEAN AVE., PATCHOGUE, N. Y.

CHICHESTER'S RESTAURANT Meals at All Hours. Steaks, Chops, Sandwhiches, Clam Chowder, Pies, Cakes, Oysters, every style. Cold Cuts. Finest Ice Cream Rooms in Town. Arcade Building, Ocean Avenue. GEO. L. CHICHESTER, Prop.

WYCKOFF BROS., 1[illegible] Flatbush Ave., opp. L. I. R. R. Depot. REAL ESTATE. City and Country. North and South Side Farms and Village Property For Sale, Exchange or Rent. Co-operating with: J. FRED. HEGEMAN, Hempstead, opp. the Depot.

MILLS & HAMMOND, —DEALERS IN— Dry Goods, Fancy Goods, Furniture, Carpets, Shoes, Groceries. See Our Complete and Elegant New Stock—Cash Bought It—Low Prices Will Sell It. OUR SPECIALTY—To Please Our Customers. OUR AIM—To Save Money For Our Patrons. OUR INTENTION—To Do Better by You than Anyone Else.

flower of Suffolk County politicians mingled in the procession with even more distinguished Queens County politicians."[cxvi]

Once all the firemen – and spectators – had arrived, it was time for the parade. The *Advance* lists twelve different fire departments as having marched, in this order: Patchogue, Bay Shore, Breslau, Babylon, Southold, Hempstead, Sag Harbor, Greenport, Riverhead, Islip, Amityville and Sayville. "The march was a long one," they reported, "the heat intense and the highways dusty, but the men bore it bravely and maintained the most perfect order."[cxvii] As was fashionable at the time, the newspapers described the apparatus and uniforms of each marching company using quite a bit of detail, so that the reader – or anyone else who could not be there that day – would have a perfect mind's-eye view of what those who did attend were able to see and experience. For example, Patchogue's "Honey Bee" engine was "covered with wreaths and surmounted with a massive pyramid of flowers, artistically arranged"; Babylon's "Storm" engine was "drawn by four beautiful white horses", and Hempstead was represented by no less than three companies, a 19-piece band, and "…a company of five axe-men, with axes on their shoulders and fans in their hands, which were a necessity due to the midday heat and the immense bearskin headgear worn by each man. The leader of the axe men carried in one hand a Roman sword and a fan in the other, and wore a large corsage bouquet."[cxviii]

The same attention to detail is true for the descriptions of the luncheon put on at the conclusion of the parade by the host firemen for the visiting teams:

"A long table ran through the center of the Lyceum, from near the door to the stage. Shorter tables were ranged along each side of the spacious hall, and in the most economical manner to secure room for the largest number. The tables were beautifully set and on them were large and tastefully arranged bouquets.

"Nearly 500 firemen sat down to the bountiful repast, and after fully satisfying their hunger the tables were occupied by others. It required the third setting of the tables before the wants of the great multitude to be fed were satisfied. When dinner was over, which was witnessed by large numbers in the gallery, there was abundance of bread, chicken, ham, tongue, etc. left to feed a regiment, so bountifully had the people responded to the call from the fire department for assistance in feeding the guests."[cxix]

After everyone had eaten, it was time for the contests to begin. As could be expected, they were a huge draw:

"During the contest a great crowd had gathered. Windows that commanded a view were occupied. Balconies and the roofs of projecting porches were filled."[cxx]

There were four different contests that day: The Hook and Ladder Contest, wherein contestants would run with their truck 300 yards, raise and climb a 25-foot ladder against a local merchant's store building and grasp the top rung; a Hose Contest, wherein contestants would run 200 yards, lay 300 feet of hose, break the coupling and attach a nozzle; a "Test of Engines", which was essentially the same type of competition as which took place at the New England musters, to see which company could throw water the farthest, and an Individual Ladder Contest. These were all run in the same fashion as they had been during previous tournaments…including the protests and controversy. The Hook and Ladder, Engine and Individual Ladder contests went off smoothly, but, according to the Patchogue *Advance*, "The only trouble in these trials occurred at this time. Sag harbor claims that when they had laid the hose and adjusted the pipe one of the Greenport men removed it and, consequently, it was put on the second time. This occasioned a delay, while they claimed to have done it all in 48-1/2 seconds. We give the judges report as given us.

"'Sag Harbor protests against Greenport hose company taking the second prize for hose contest.' This was overruled under rules and regulations as four of the judges decided. If the protest was entered on account of mistakes or accidents it could not be allowed under the rules."[cxxi]

The Riverhead and Islip teams carried the day, with Riverhead winning the Hose event with a time of 48 seconds, taking home with them an engraved silver trumpet with cord, while Greenport, after winning the protest, kept second place. Islip won a Tilting silver ice set for clinching the Hook and Ladder Contest;

Riverhead's Engine #3 took First in the Engine Contest with a throw of 175'6" and got a silver trumpet with cord of their own, and Islip then finished first in the Individual Ladder Contest with a time of 6-3/4 seconds.

There was also to have been a Steamer contest that day, but "...There was no steamer on the ground. Hempstead expected to bring one, but the trustees of the village refused to have it taken away. But as the company was present, the prize was given to her, a handsome silver trumpet with cord."[cxxii]

The *New York Sun* had this description of the contests:

"Chief Edwin Bailey was Master of Ceremonies, and to his energy is due the finish of the contests before dark. It was as exciting as a sprint race to watch the uniformed men urging their truck piled with ladders down the dusty street through a lane of spectators, until it dashed in front of a dry goods store which was supposed to be on fire. Then the spectators held their breath as the twenty-five foot ladder was yanked from the truck, its base planted, and while it trembled in the air an agile fireman scrambled like a squirrel to the topmost rung ere the top had struck the wall. Another event was a contest between hose companies. This contest did not go off as smoothly as the others did. There was a dispute, a gathering of 500 firemen and outsiders, a little scrapping, in which the fists of a dozen firemen could be seen gyrating like windmills above the heads of the multitude. Next came a contest of the throwing power of hand engines. A special prize for the best drilled and equipped company in the procession was voted to stay in Patchogue. Finally members of the different companies contested for prizes offered for the men who could best run 100 feet and climb to the top of a 25-foot ladder. The prizes were delivered amid cheers in the evening in Tower Hall."[cxxiii]

One little-known fact of that day is that prior to the running of that first tournament, a rule was adopted by the new Suffolk association that the contests during the competition be confined to Suffolk County companies only. According to J. Fred Flugrath, author of <u>The Suffolk County Volunteer Firemen's Red Book</u>, which was published in 1907, 26 years prior to his "Blue Book", the adoption of this rule "...brought forth a storm of protest from Queens County, to which [Babylon] Chief Titus, then editor of the Babylon Budget, replied as follows:

"'The Queens County firemen feel hurt that they are not included in the tournament in Patchogue, and some of the papers intimate that the action of the Suffolk firemen is ungenerous and unfair. This action is the outgrowth of the experience gained at every tournament. Inaugurated in Babylon, with the newness and, of course, consequently smaller attendance, and continued in Greenport and Jamaica with larger attendance, it was found impossible to crowd all the exercises into one day. Inability to carry out the programme at Jamaica, and consequently hard feeling, decided the Suffolk firemen to take this step, and form a separate association. Queens will doubtless do the same thing, and Suffolk departments will be happy to visit and take part in the parade and be interested spectators and not participants in the contests. There is no feelings against Queens County departments, or disposition to slight them, but they will readily conceive the position and necessity of curtailing the number of entries in the various races. Since the organization of these annual tournaments, departments have been formed in many places, and with only Suffolk County companies admitted to the contests, it is anticipated there will be as many entries, if not more, than were at Jamaica or Greenport. The tournament is now confined to one day and, of course, the contests to a few hours.'"[cxxiv]

Roughly two weeks later, basking in the success of their first Annual Tournament, the SCVFA held their first Annual Meeting on September 29th. Two more members were added to their roster that day: Islip's Alert Hose & Protection engine, and Sag Harbor's Minnehaha engine, and by August of the following year their membership would grow to include 20 of the 35 fire companies in Suffolk County, paving the way for an association that would go on to help define the sport of drill team competition as it exists today.

Regarding the tournament, there are several aspects of that first SCVFA gathering in Patchogue that bear further discussion, as they were iconic of what went on at many of the musters and tournaments of that period, and were significant in and of themselves.

The first involves the preferred method of transportation that both the firefighters and the public used to travel to wherever the tournament was being held: the railroad. As the popularity of the competitions grew,

A Long Island Railroad train carries fire apparatus to a tournament during the early days

and attendance at them began to swell, a reliable method of getting everyone quickly and safely there and back became paramount. And, as there were sometimes long distances that needed to be traveled, transportation by rail was an obvious choice. Understandably, seeing the potential for profit, the railroad companies were all too willing to pitch in, and in turn the relative ease by which one could now get to a tournament made the decision to attend that much easier, so it was a win-win situation for both parties: "The fire departments would hire special trains, under a guarantee to the railroad company, and thus help induce a large attendance from the villages along the line. The firemen also used to load their apparatus on flat cars, furnished many times without cost by the railroad company, some of which were attached to the special passenger trains, while other special trains of these loaded flatcars preceded the fire boys to the scene of the tournament."[cxxv] The same was true for the old-fashioned musters of New England: "Until the advent of the motor truck, trains were the basic means of transportation for handtubs and their crews. For many years, New England League musters were held on Thursdays, as Saturday schedules were not as accommodating. It was not unusual for a muster train to start in Portland, ME, boarding the groups from Portland, Brunswick, Bath and Topsham, ME; stop in Portsmouth, NH to pick up their association along with those from Berwick, ME and Somesworth, NH; go on to Newburyport, MA to load up the locals as well as those from Amesbury, MA; then on to Salem, MA to pick up the Marblehead, MA and Salem contingents, then on to the muster in the Boston area. One can only imagine what must have transpired on this train on the return trip, loaded up with highly competitive and not-too-sober firemen."[cxxvi]

On Long Island, the fare for traveling to a tournament was inexpensive, at least by today's standards: the *Brooklyn Daily Eagle* announced in 1890 that "...The Greenport Fire Department will go to the [Islip] tournament by special train, having chartered six coaches and two flat cars for $375", and excerpts from minutes of the Babylon Fire Department in 1889 reflect that they "...chartered a train of six coaches and two flat cars to run from Amityville to the tournament at Sayville on September 4th. Fare from Babylon – 50 cents." But the railroad's "potential for profit" led to some skulduggery in September of '86:

"Daniel S. Hermance, the agent of the Long Island Railroad Company, at Hempstead, is to be prosecuted for Grand Larceny if he can be caught. Detective Rogers has been on Hermance's track, and has strong hopes of apprehending him. An examination of the agent's books shows him to have defaulted with almost $1,000, instead of $300, including the amount paid by the Hempstead Fire Department for a special train to the Patchogue tournament."[cxxvii]

By and large, however, the railroads played a very large and valuable part in the ability for tournaments to grow and prosper during the latter half of the eighteenth century, and would continue to be the primary means of transportation to the gatherings until the invention and advent of the automobile in the early 1900's.

As successful as its first tournament was, Suffolk County association's tournament was not, however, the "only game in town" on Long Island at the time – in fact, it was far from it. In spite of the fact that the New York City Fire Department had become a paid department, and in doing so had banned firematic competitions, that didn't stop neighboring areas from starting their own associations and respective tournaments, and as the excitement over these kinds of competitions spread, so did the number of different tournaments each year. It was only one week after Suffolk held their first tournament – on September 16th, 1886 - that the Eastern District Exempt Firemen's Association of Brooklyn held a tournament in Ridgewood Park. While little was said about the competitions, this humorous description appeared in *The Brooklyn Daily Eagle*:

"The second and last day of the tournament…was begun yesterday morning at Ridgewood Park and successfully brought to a close at midnight amid a shower of red fire, a booming cannon and the din of a number of brass bands….'Chippie' catchers abounded in numbers, while many well known faces, who in days gone by assisted with holding up the bight could be seen among the crowd relating stories of their experiences when they were ordered to 'pick up' and run to a fire…A number of individuals who objected positively to sitting down for a moment ran up and down the main path of the park, dragging a long rope and shouting 'pick her up, boys'. There seemed to be nothing to pick up until one of the runners espied, when it was too late, a little stone about a foot high in his path. To the unsophisticated it seemed that this is what the men were waiting for, for when they had picked up the unlucky individual the rope was put away and the sport was dropped, to the extreme comfort of a large number of spectators."[cxxviii]

A year later, in upstate New York, FASNY held its 15th Annual Convention in Troy. There was nothing particularly interesting about the convention and tournament itself - except to note that that in itself was the case: if there was a point in time that could be said to be when the focus, or "hub" of firematic competition shifted away from upstate to Long Island, it may very well have been at this point. Whereas on Long Island, where tournaments were being attended by thousands of spectators and several types of firematic competition were being developed, including ladder events such as the Hook & Ladder and Individual Ladder contests, the tournaments that were associated with FASNY by this point in time seemed to be (almost logically) more of an addendum to the conventions than they were events unto themselves: Troy's tournament consisted of nothing more than Hand Engine Trials and Hose Races, that were comprised of no more than six and thirteen entrants, respectively. It is interesting to note, however, that of the latter, two of the teams competing were from Adams, Massachusetts and Fairhaven, Vermont…and who took 5th and 6th

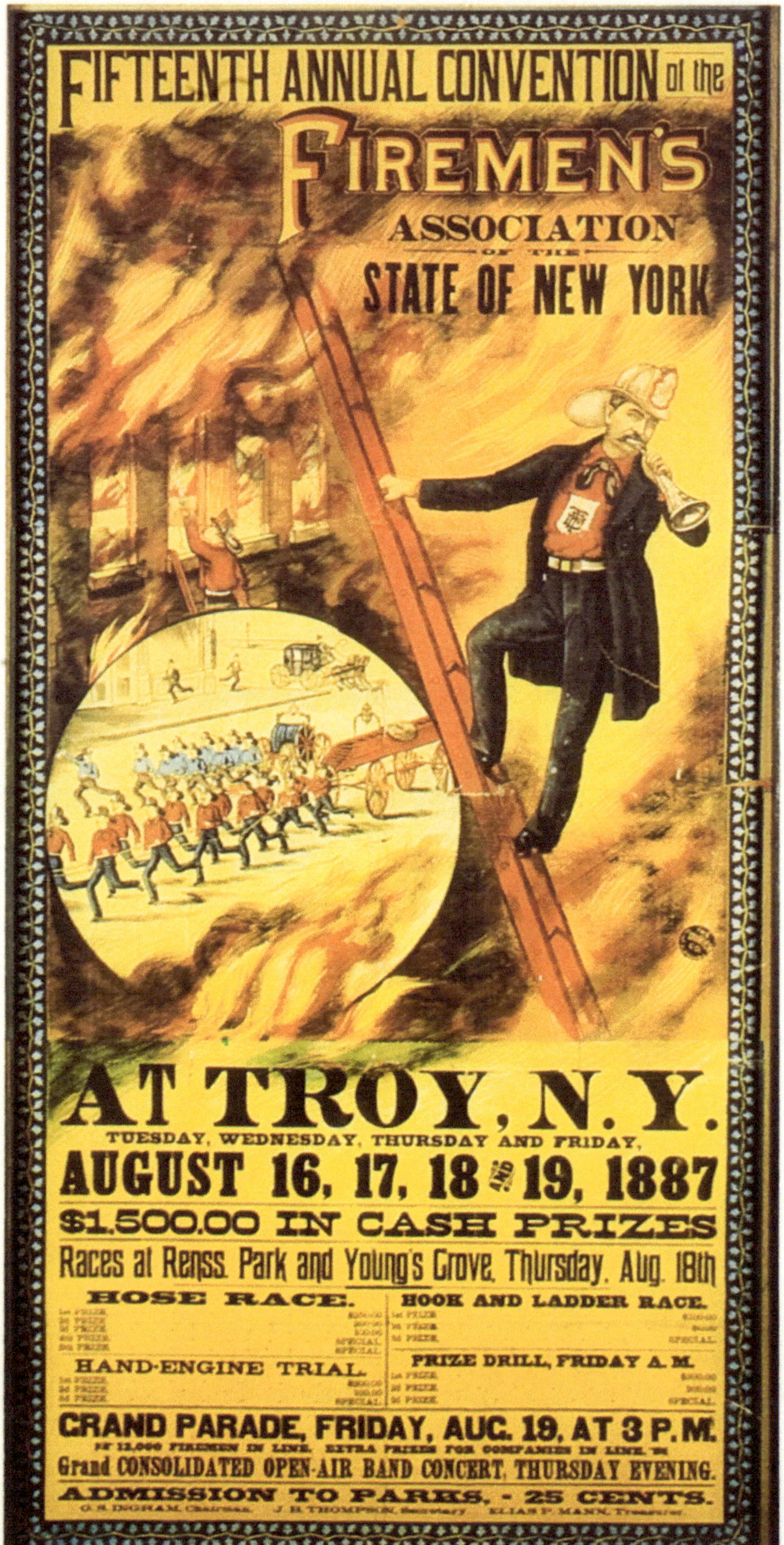
FIFTEENTH ANNUAL CONVENTION of the
FIREMEN'S
ASSOCIATION
OF THE
STATE OF NEW YORK
AT TROY, N.Y.
TUESDAY, WEDNESDAY, THURSDAY AND FRIDAY,
AUGUST 16, 17, 18 AND 19, 1887
$1.500.00 IN CASH PRIZES
Races at Renss. Park and Young's Grove, Thursday, Aug. 18th
HOSE RACE.
HOOK AND LADDER RACE.
HAND-ENGINE TRIAL.
PRIZE DRILL, FRIDAY A.M.
GRAND PARADE, FRIDAY, AUG. 19, AT 3 P.M.
Grand CONSOLIDATED OPEN-AIR BAND CONCERT, THURSDAY EVENING.
ADMISSION TO PARKS, - 25 CENTS.

Place, respectively!

FASNY's fascination with Parade and Drill was still in evidence that year as well. Perhaps as a result of the emerging popularity of fire service events on Long Island, 1887 marked the first year that a fire department from Long Island marched in a FASNY parade, when Long Island City sent a contingent to attend.

Back on Long Island, on September 15th of that year, the SCVFA held its 2nd Annual Tournament in Riverhead, and once again it was a large, splendorous affair, covered extensively by the local press. Apparently the day got off to a late start:

"The trains came in late, that from the south side consisting of over twenty cars not arriving until noon."[cxxix]

...and the parade was gigantic:

"The line was nearly a mile long, took twenty minutes to pass a given point, and an hour and a quarter to finish."[cxxx]

...after which "...the finish was at No.1 Engine house on First St. and then a bee line was made for Riverhead Hall, formerly the Rink, where loaded tables and a seating capacity for over 600 men met them. They were filled in a jiffy and chowder, roast beef, boiled ham, and all their accompaniments were at their disposal. No one called for anything the second time, the system of waiting upon them being so good. All were filled, the tables were reset and over 300 more sat down to the second table. They were filled and many citizens generally strolled in and no one went away hungry. Between six and eight p.m. fully one-half of the firemen returned for supper, got all they wanted, and still there was room for more."[cxxxi]

Apparently the Riverhead Fire Department had come up with a unique idea for preparing and serving coffee that day: "A large coffee pot in the back of the Hall was the object of much curiosity. It was an original idea of Chief Terry and Asst. Chief Hallett. The coffee, etc., was cooked by means of steam pipes from the engine connected with the hall. Other pipes and faucets made it convenient for ready use on the floor."[cxxxii]

The Breslau firemen were the biggest victors after the competitions had ended, winning three of the six contests held that day: Hook & Ladder (55-1/2 seconds), Special Ladder (which appears to be the same as Individual Ladder)(8 seconds) and a new contest called the Special Engine Contest, which they won in 42 seconds. Patchogue won the Engine trials with a throw of 170-1/2 feet, Babylon won the Efficiency Contest in one minute three seconds, and Sag Harbor's crew took honors on the Hose Contest with a time of 51-1/4 seconds.

This new Special Engine Contest really *was* special. Participants ran their engine 300 feet, placed a suction hose into a well, laid 100 feet of hose, broke a coupling, attached a nozzle pipe and then sprayed water onto the side of a building. Breslau won this contest (in 42 seconds) by default, as they were the only team entered, but that is overshadowed by the fact that this may very well have been the very first "wet hose" contest to have ever been held at a tournament. During all hose contests held up to this point – including those held by FASNY – although hose was laid out, water was never actually run through it, and "time" was taken when the coupling at the end of the hose was broken (from what is now called the "dummy length") and the nozzle was attached. With the introduction of this contest, water – along with its physical nature, and the myriad set of peculiarities and problems that it brings with it – would become an integral part of firematic competitions from this date forward.

By all accounts, though, Riverhead outdid itself in hosting the tournament that year. In spite of the fact that the last contest of the day, the Special Ladder Contest, wasn't finished until it was almost completely dark, and that some of the teams had a rough time getting home afterwards ("That night the firemen's railroad special ran out of water and coal oil, so the 55-mile trip home was dark and dry"[cxxxiii]) there was nothing but praise for how well the event was organized. From *The Sag Harbor Express*:

"Everyone that was met was profuse in his praise for the marked hospitality received and the good order and success of all the exercises.

"On this occasion, Riverhead did her best. She had invited the whole of Suffolk County's 'brave fire laddies' to be her guests; they had assembled about one thousand strong, with twice as many lookers on, and from the whole concourse of people no complaint was made or no cause of complaint was offered; and it appears to be the universal exclamation, from all present, a perfectly enjoyable occasion."[cxxxiv] The *Express* was apparently sufficiently impressed with the affair as to even suggest that Sag Harbor host the Suffolk County tournament the following year: "One thing the Sag Harbor boys now want, and should have, and that's the next tournament here. Patchogue and Riverhead have borne the honors; why not Sag Harbor? We presume that the fire-laddies from the other parts of the County would be pleased to come here, and we are well assured that our people would like to have them come. Why not, therefore, have them? It requires work; it requires funds; but from the work and the funds one of the greatest gala days which our village ever saw will be thrust upon it. If not, why not?"[cxxxv]

Well, that editorial apparently put a bug in someone's ear, as the following year the 3rd Annual SCVFA Tournament was indeed held in Sag Harbor, in September of 1888. That event was also a resounding success, with one notable, albeit somewhat humorous, difference: the amount of alcohol. The Riverhead tournament, it seems, may have been unique in the history of firematic tournaments in the amount of alcohol that firefighters *didn't* consume, as indicated in a clipping found in the *Sag Harbor Express*:

"Some of Riverhead's caterers to those that indulged their cups, made great preparations to fill all that should attend with beer and sundry other drinkables. They were a sadly disappointed lot. To their credit be it said that scarcely a fireman or any other citizen was to be seen 'out of the way,' and the shipment of beer back to New York this week is something that has never occurred before."[cxxxvi]

In contrast, the consumption of alcohol was far from absent at the Sag Harbor tournament, as indicated in a description of the day's proceedings published in a New York City newspaper called simply *The Star*. Written by a prominent writer of the time named Prentice Mulford, it has to be one of the most beautifully descriptive accounts of a Tournament Day ever written:

"At 6 o'clock the town was all astir. Flags were going up, the canvas coverings for the firemen's wharf dining hall and the band stand were being erected, tournament truck red stakes were being driven in the main street; some colored fishermen had already commenced firing up; barkeepers were busy getting their ammunition in order for the day's engagement, and sundry town fossils were crawling down to get a good seat on which to grumble and prophesy disaster. At 7 o'clock vacant stores began to crop out as temporary restaurants, and banners were hung out indicating that pies, cakes, sandwiches and lemonade could be had for a trifling pecuniary consideration. At the same hour temporary works went up hastily for the sale of fermented and other stimulants in public, and in retired village localities where such appearances were never seen before and never will be again. In one case one of this skirmish line fortified himself in a rear street with one keg of beer in a disused pigpen. It was his intention to open on the procession on their return march, which later intent was frustrated through their taking another route.

"At 8:45 a.m. the first fight of the day occurred in a saloon, where it had been brewing all the previous night. The crowd now coming in continually from the Hamptons, the Springs and Acabonac poured into the place with a mighty rush of numbers, choked it up and stopped the feeble outburst of alcoholic belligerency by not granting it air or room enough to go on…Soon after came the steamer Shelter Island, full of Orientals from Orient and natives of Greenport and Southold. Besides, the thrifty and far-seeing Greenporters chartered a tug, which towed a barge over to New London, with whose natives they filled up at a price per head, and returning to their native town they took on board their fire department, consisting of forty hoosiers, G.F. Tuthill, foreman; the Greenport Band, John Geeling, leader, accompanying.

"Thus far I was able to keep the run of the inpouring crowd. But as the morning sun ascended and its beams came down hotter and hotter, the masses from all points north, south east and west got away from all orderly calculation; and even had they not done so, the statistical array of companies and military bands and delegations would have formed a mass of names and figures fit only for a government report. Such a tramping and braying and banging of brass bands has never before been heard on our streets. There were the Hempstead band, the Breslau band, the Greenport band, the Patchogue band, the Sag Harbor band – all making music during the day; sometimes all at once and sometimes in patches variously located about the town."

Then came the feast, where Sag Harbor seemed to borrow a page from Riverhead's playbook:

"What I dwell upon with the most pride regarding the many features of this notable day is the arrangements, the system and the executive ability shown in feeding the 1,000 visiting firemen. With the quick perception of genius, some one of the Executive Committee fixed upon the new and still unfinished wharf as the proper site for the dining hall. Over this was quickly raised a light timber framework, 180 feet long, and the top was covered with cloth. Under this were the tables, sun-sheltered, but free to the breeze from the harbor's broad expanse. At the pier head, and twelve feet above it, stood on supports, looming in bas relief against the sky, an immense copper retort in which 500 gallons of coffee were made, and in which at the appointed time men clambered on ladders and tossed in bag after bag of ground Mocha. Under the great coffee urn were great iron pots, in which during the morning, and preparatory to the feast, barrels of clam chowder fumed and bubbled and sent but far and wide upon the air a vigorous chowder aroma so solid and substantial that it might have satisfied any delicate and refined stomach alone, without tasting its more material part. Close by was a tin cauldron of potatoes.

"Steam generated from the contiguous steam pile-driver engine and conveyed in flexible hose cooked all these various fluid and solid compounds. All the morning young ladies were bearing from the adjoining flour mill huge dishes of baked beans, platters heaped with slices of ham and corned and roast beef, and basket on basket of bread for the tables, while the coming visitors by the hundreds, who were not firemen, and who therefore had no part as guests in this feast, wandered up and down and around the tables and sniffed in, snuffed up both with eye and nostril, the appetizing sight and tantalizing aroma of the viands prepared and in the course of preparation, with appetites of their own already in an imperious state of demand after their morning trip hither by land or water.

"Then they went away sadly, to wander with thousands of others up and down the main street, wondering when the parade would begin, and wishing something extra would turn up, or that the band would play, or that a horse would run away and break his own or a man's neck, and then they rushed to the railroad station to see the next relay of incoming visiting firemen, or down the old wharf, a few yards distant, to watch the unloading of another army of visitors from yacht, barge or steamer, who immediately joined this grand army of wanderers, and went with them again to sniff the vast feed in preparation. So, with the crowd passed the morning. The excitement and the multitude swelled. It culminated as two imposing trains rolled in at 11:00 and 11:30 a.m., bearing the variously uniformed firemen of Islip, Hempstead, Patchogue, Moriches, Center Moriches, Amityville, Setauket, Babylon, Breslau, Farmingdale and Southampton, one thousand in all, with bands and banners and brooms to sweep everything clean in the coming contest."[cxxxvii]

If this account alone is not enough to give the reader a true sense of just how large the meal was, there is also this description that appeared in the *Brooklyn Times*:

"...the hungry firemen went for it in a way that predicted a famine. The largeness of the supply may be guessed from an enumeration of a few of the edibles such as a barrel of cut-up potatoes, 2 bushels of onions, 20 bushels of clams, 550 pounds of ham, 75 pounds of pork, 8 bushels of beans, 275 pounds of corn beef, 500 pounds of roast beef, 3,000 biscuits, and 1,000 loaves of bread. Forty dishwashers were kept busy after the 'laddies' got to work eating, immediately following the parade."

The *East Hampton Star*, in an earlier edition, also lists Riverhead, Huntington, Greenport, Bay Shore and Sayville as having also participated that day, but oddly enough, neither provides any description of the contests themselves. The *Sag Harbor Corrector*, however, in its September 8th issue, wrote that during the Special Engine Contest "...the Breslau laddies did excellent and quick work" and that the Special Ladder Contest "...was one of the most exciting and best contested contests of the day. Hawkins and Brady [both of

Islip] tied each other three times, making each time 6-1/2 seconds. After the third trial the judges asked them if they were willing to divide the prizes. They said 'yes', and it was so awarded."

Then, the next day, when it was finally over:

"To-day how different is the scene! How profound the quiet! The flags are disappearing. The temporary restaurants are closed; the streets are quiet, the last howling inordinate imbiber of exciting fluids has been laid at rest; tired landladies are gathering up fragments and setting tables in their accustomed order; and the dining hall is a wreck, for a September half gale from the north last night tore the cloth from its roof and tore down most of the balloon framework.

"What with the crowd, the music, the parade, the dinner, the contests of skill between the firemen, the winning of prizes, the presentation of prizes and the varied sights and sounds attendant on the coming of 2,000 strangers into a small town, there was even too much of delight and festivity for human nature to endure comfortably. It lasted even until dusk, and until dusk the tired thousands – men, women and children, whelps and brats – stood on their aching limbs on the street or sidewalk wanting more, and yet wishing the fun would end; anxious to go home, yet afraid to go for fear something might happen never to be seen again if they did. There was enough of it, all crammed into one day, to last three."[cxxxviii]

So successful was the Sag Harbor tournament that less than a month later the SCVFA began making plans for their next year's tournament, at their annual meeting that October 3rd. Huntington was the prime candidate, however "It was thought...that as Huntington had suffered a severe loss in the burning of its churches, it would be too much of a financial burden to finance the tournament, including the dinner, and a new church in one year"[cxxxix] so Sayville got the nod instead.

It should be mentioned that by this point "tournament fever" was not just limited to New York State; records have been found praising the skill and prowess of fire department teams from many different cities and states across the union, including Belleview, Ohio; El Paso, Texas; Gillette, Wisconsin; Spanish Fork, Utah; Grand Junction, Iowa; Plymouth, Indiana; Aberdeen, South Dakota; Peabody, Kansas and Polo, Illinois, where in 1890 "Polo won first in the free for all hook and ladder race of 200 yards in 32 1/2 seconds. Polo boys added to their laurels by winning the State Championship Hook and Ladder Race. "[cxl] Interestingly, although on a more poignant and somber note, records show that on June 18th and 19th of that same year 500 American Indians from the Sioux Nation Humps and Bigfoot camps competed in a fire department tournament in Pierre, South Dakota. "The Indians will enter the tournament with one hose team of the swiftest runners in the Sioux Nation, who will have a hose cart loaned to them by the Pierre Fire Department. They practiced yesterday and one heat they made astonished everyone, being five seconds faster than the best time on record."[cxli] It was a wonderful example of two different cultures putting aside their differences so that they might compete with each other in the spirit of friendly competition; sadly, it was not six months later – on December 29th of that year – that the Wounded Knee massacre took place, forever changing relations between them.

A Hub & Hub Race in Deadwood, South Dakota on July 4th, 1888, between the only two Chinese hose teams in the United States

Chapter 7: Growing Pains

While the team from Polo was being crowned as Illinois State Champions, the concept of a "State Championship" had also been introduced back in New York at just about the same time. At their 17th Annual Convention in Plattsburgh in 1889, the FASNY secretary's minutes state that "The Hose Race for the championship of New York State was held on closing day." From this account, the members of the Forbes Hose Company of Canastota, N.Y. were ostensibly the very first New York "State Champions" of a fire department tournament. But while this may have been technically true, it must be noted that up until this point there had been virtually no participation in any of the FASNY tournaments by teams from Long Island, at least based upon the information provided by those same secretary's minutes. For whatever reasons – a lack of communication, or perhaps the distances needed to be traveled – it seemed as if the development of the sport of firematic drill team competition was developing, independently, along two geographically different lines. In 1888 the Brooklyn Exempt and Long Island City teams swept the old fashioned Hand Engine Contest at FASNY's 16th Convention in Cortland, but it would not be until 1891 that a team from eastern Long Island – Sag Harbor's Gazelle Hose Company – competed in a FASNY Ladder or Hose Contest, at their 19th Annual Convention in Herkimer. One might assume that based upon the profuse development of the sport on Long Island that the Sag Harbor team would have had little trouble trouncing the competition "up north", but such was not the case: they only managed a 5th Place in the Hook & Ladder contest out of eight teams entered, and did even worse in the Hose Contest, placing next-to-last out of fourteen teams!

Parades also continued to be a form of competition at the FASNY events, and were apparently extremely popular: at the same 1891 convention that Sag Harbor attended, 2,765 men were listed as having participated in a parade that took over three hours to complete. In addition to the standard marching awards, however, some rather oddball "Special Prizes" started being offered. In 1887, in addition to the usual "Largest Number of Men in Line" and "Company Coming the Longest Distance", etc. kinds of awards, there were prizes for "Heaviest Fireman in Line", "Youngest Married Fireman" and "Fireman Having the Longest Beard". The 1888 convention even took it up a notch, by offering additional prizes for "Handsomest Fireman in

Line", "Sweetest Looking Fireman in Line", "Hungriest Looking Firemen in Line", "Fireman Having the Most Mouth", "Fireman Having the Largest Foot" and even "Homeliest Fireman in Line" (what a distinction!)[3]

This point in time in the history of firematic competitions - the year 1890, and the years immediately following it – should be marked as being especially significant in the development of the sport, which was now growing in leaps and bounds. Indeed, the period between 1890 and the end of the second decade of the 20th century could be called "The Golden Age of Tournaments" without stretching the truth. America – especially the northeast - was entering into an historic time of its own, marked by never-before-seen growth, prosperity and opportunity; a time often referred to as The Gay Nineties. "The decade was a period of extremely exceptional economic expansion, and, in particular, of rapid wealth gains in New York City and Boston. The American empire of trade was at one of its zeniths, and cities were growing rapidly. While the same decade saw an explosion of immigration to the United States from less economically prosperous lands, and consequently no gaiety for the working classes, it was a period of vast wealth for a newly emergent 'society set'. The railroads, the agricultural depression of the Southern United States, and the dominance of the United States in South American markets and the Caribbean meant that industrialists of New England were doing very well."[cxlii]

From a tournament perspective, this burgeoning happiness and prosperity translated into a rapid growth of the sport, as indicated by an effusive description of the 4th Annual SCVFA gathering at Sayville in 1889, where even Labor Day took a back seat to the tournament:

"In Babylon, Bay Shore and other places Labor Day was not observed in order that to-day might be a holiday. In these places to-day the stores are closed, and everyone who could get out of town has come to the tournament. Hundreds of city people have driven into the village and the array of handsome turnouts to be seen is one of the attractions. Special trains from Amityville, Babylon, Sag Harbor and Riverhead have landed several thousand people...The feats of strength and agility performed in the bayside village were enough not only to invoke enthusiastic plaudits from multitudinous spectators, but to attract the smiling approval of hundreds of lovely maidens. The red shirted heroes measured up perfectly to the standard of exalted anticipation. Never before were machines whisked about with greater rapidity or manned with greater skill or precision...the event was healthful and helpful. Its influence will be manifested for months to come through many wholesome channels."[cxliii]

These types of descriptions appeared in newspapers of this time on a regular basis, each expounding the events of each day with more and more superlatives as time went on. The reason for that was not just editorial one-upmanship; the truth is that as the decade – and the popularity of the sport – progressed, there was more and more pressure for each succeeding tournament host to "out-do" their predecessor in terms of Grand Scale when hosting the event, resulting in a vicious cycle that very nearly ran out of control as the size and scale of the events became more and more unwieldy.

One problem was the sheer number of people attending. At the 5th SCVFA Tournament held in Islip on September 3rd, 1890:

"The way people kept arriving on trains and in carriages at Islip yesterday morning filled the natives with wonder. Each long special which pulled into the Islip station had a bulging freight of human beings: firemen in red shirts and characteristic hats; girls in summer dresses and smiles; children and babies all inextricably mixed up together. The train loads scattered at the

[3] A complete list of parade prizes representative of that era can be found in Appendix XI on page 307

station, only to come together again in a solid, impenetrable jam along the main street when the contests had begun."[cxliv]

This in turn caused problems with getting the track ready for the contests to begin:

"It took some time to clear a course along the crowded street. This was done finally by stretching two lines of heavy wire behind which the crowd was fairly well restrained."[cxlv]

...Which in turn caused the day to run long, prompting the writer to make the conclusion that these types of events had finally reached a point in size and scope that they could not go beyond. Regarding the Islip tournament, he wrote:

"It is likely that it is the largest that will ever be held, for at the present rate the tournaments are being so much enlarged and the expenses so much increased that they are bringing a limitation on themselves. The present festival has required an immense amount of labor on the part of a hard-worked committee and the expenditure of nearly $2,000. The programme is so long that, with the best speed, it can hardly be completed before dark."[cxlvi]

The writer, apparently, was not the only one who felt that way that day. If the following quote is to be believed, rumors abounded that since the manifestation of fire department tournaments had reached its pinnacle that day at Islip, and since there was no way it could be any better, the SCVFA should just quit holding tournaments altogether (although these rumors were no doubt fabricated by the hosts, who were trying to go out "on top"!)

"Very general was the remark yesterday that the festival might wind up the history of the Suffolk county association's celebration, and it was proper, were that the fact, that the festivals should wind up in a blaze of the reddest kind of splendor. Whether yesterday's tournament will turn out to be the last one or not does not really concern the present occasion. Anyway, the association has not officially announced the end of its existence, and until then it must be considered still living. The rumor concerning the possible falling through of all future tournaments may have arisen from the mild satisfaction or pardonable village pride of the Islipians, who undoubtedly felt, and with justice, too, that they had surpassed all previous records and established a standard for firemen's tournaments which no other village in the country would be willing to match."[cxlvii]

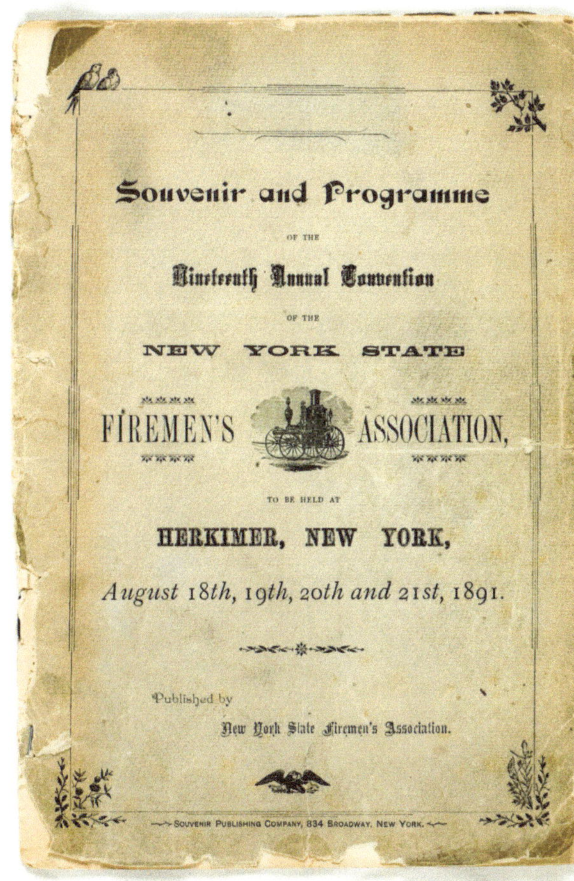

Elsewhere, in the greater New England states, the dawn of the 90's saw the old fashioned handtub musters evolving and going through their own resurgence as well. Dixon wrote, "This decade showed a remarkable resurgence of the sport with an average of 26 musters per year attended by an average of 7 handtubs each. The 'Gay Nineties, let's have fun' mentality of the nation at this time surely had some bearing on this increase of activity."[cxlviii] On September 5th, 1889, the last recorded Perpendicular Muster was held in Providence, R.I., as new rules went into effect in 1890. On November 20th of that year the New England States Veteran Firemen's League (NESVFL) was formed at the Boston Veteran Firemen's Association, "...the object of which was to encourage and perpetuate the oldest sport in the Country, assist outside interests in the promotion of Musters and serve as arbitrator in settling the questions arising among the several Associations affiliated with the League."[cxlix] There, the firemen "...without any hesitation voted to organize for the betterment of the sport." The NESVFL still exists today, and continues to act as an overseeing body for all handtub musters held in the northeast.

The issue of how to effectively handle the demands of (and demand for) huge firematic tournaments on Long Island was not going to go away anytime soon. The SCVFA, however, took a step forward in that direction when "By a close vote, at the meeting of June 9th, it was decided 'that at future tournaments each fireman provide his own refreshment.' A county paper, in commenting on this action, said: 'The dinner has always been a big item of expense, and the cause of no end of annoyance, and the lack of facilities for feeding 1,000 firemen in an hour has deterred many villages from inviting the firemen. This resolution relieves the smaller villages from their greatest drawback and will allow for the offering of more and better prizes."[cl]

As the size of the tournaments grew, so too did the pressure on the teams to perform well in them, to the point of recruiting non-fire department members to compete on teams. In reference to the SCVFA tournament in Greenport in 1892, the *Brooklyn Daily Eagle* ran an article commenting on the practice of western Long Island companies bringing in outside help to win contests: "That there should be great rivalry is to be expected. Some villages have been so anxious for local success that frequently outside talent has been secured to help win prizes. It is a fact however, and it is to the credit of the country boys that in every case where companies have resorted to this practice they have met with defeat."[cli]

The *Eagle* was by no means itself innocent when it came to promoting competition within the sport. At the Sayville tournament of 1889, they created the Brooklyn Eagle Trumpet Prize, which was given to the winners of the Hook & Ladder Contest at each SCVFA tournament. The prize – first won by Sag Harbor that year – became a highly coveted award, and in the following years teams competed fiercely to win it. Sag Harbor took the trophy again the following year, only to lose it to Patchogue in '91 and then regain it again at the Greenport tournament in 1892. Apparently, at the Patchogue tournament in '91 there must have been some taunting of the Sag Harbor team by the Patchogue team, for "...When the Sag Harbor hookies won the contest, at Greenport, the following year, they filled their extra long truck with hay, in exulted defiance to the fellows who had dubbed their truck the year before as 'the old hay rack.' From thenceforth, the Phoenix truck was known as 'The Hay Rack."[clii]

Another indication of what the competition was like in Greenport comes from the minutes of the Bayport Fire Department, which attended the tournament that year:

"At first there was no contest in which we cared to enter. For the hose companies there was only a long run of 200 yards and screw on a nozzle with no water. We did not have enough fast young runners to compete with the big villages and, furthermore, it was little more than a foot race practice which could not be considered fire drill in any sense, except for the man who screwed on the nozzle."[cliii] The passage suggests a bit of "sour grapes" attitude going on ("Why should we enter? It's a stupid contest anyway"), but at the same time it does give an indication of just how intimidating the competition could have been.

The 1892 tournament was notable in other ways as well. The village of Breslau had recently changed its name to Lindenhurst, with their fire department team now appearing on the roster as such, and a new, exciting contest was introduced to the tournaments that day: the Hook & Ladder Contest #2, wherein a barrel on top of a structure was filled by buckets carried up a ladder. This contest, later called the Bucket Brigade (or "Buckets"), would go on to be one of the most popular and enduring contests in the sport, and which still exists today as part of "The Big 8".

But other writings show that it continued to be a difficult task to competently host a tournament. The *Brooklyn Daily Eagle* reported that the day's events were delayed in getting started – the contests didn't even begin until 4:00 p.m. – and then ran long. An estimated 10,000 – 15,000 people "thronged Greenport's streets", and "The parade was not up to the usual standard. There was mismanagement somewhere, and many of the companies dropped out of the line before the march was half over."

TOURNAMENT NOTES.

SEPTEMBER, 1892.

The firemen are going to Greenport with the old Hay-Rack,
With the intention of bringing the EAGLE TRUMPET back.
The old HOUNDS will have their Engine in the ring,
And when they come home first prize they will bring.

The GAZELLES, as champions, will not contest,
But if their record is beaten they will have no rest.
The MINNIES this time no machine will take,
But will work with a zeal for the others' sake.

The JUVENILES go under the name of KING—
When they get their prize they will make things ring.
There will be with us the WHITE HAT BRIGADE,
Who will take part in the grand parade.

The FAHYS and BABYLONS play ball at ½ past 3,
And to go and see them will be a small fee;
The Fahys will beat the Babylons, it 's said,
And the BALL TEAM and FIREMEN will paint Sag-Harbor red.

In 1886 the first County Tournament they did hold,
At Patchogue, where the Gazelle boys got sold,
But in 1887 at Riverhead the first prize they won,
And the boys came home happy and full of fun.

So in 1888 they held it in Sag-Harbor,
And our Citizens and Firemen never worked harder;
Surely it was a grand success,
But none of our teams did contest.

At a meeting called in Sayville a delegate went,
Where an offer of a Trumpet the *Brooklyn Eagle* sent.
By decision of the meeting an H. & L. contest,
Three successive times would win—it's hard we confess.

But in 1889 we turned out in full force,
Each man when he went felt as strong as a horse.
Our entries we made thinking some prizes to bring back
With the Gazelle, and our runners on the old Hay-Rack.

Down the track the boys with Hay Rack came raving,
Beating the best record, and brought it below 47;
The Gazelles came down the road very slick,
Some said they had a snap coupling, it was made so quick.

Our boys arrived home that night about eleven,
And brought with them prizes amounting to seven,
The Phœnix and Gazelle, first and special prizes,
Sexton and Sylvia, according to sizes.

In the year '90 they thought to beat us in I-slip,
But found we were there and not very flip,
So as usual we lead them, and sung tra la-la,
And took second prize with Minne-ha ha.

But in 1891 we were called to Patc-Hog,
And surely they deserve the name,
We beat them like falling off of a log,
Although they acted rather insane.

Old Phœnix went up in full bloom,
Hoping to win the Trumpet once more,
But the scheme had been made, I guess in the Bee Hive room,
To skin us, and make us feel sore.

At that time in our Department there was a split,
But the men on each side had plenty of grit,
They were formed in line, each with their band,
And when at the Tournament went hand in hand.

The Phœnix they ran with plenty of opposition,
But of course they had to abide by the Judges' decision.
The Trumpet we won as fair as could be,
And Patchogue thought the scheme we could not see.

But our Lawyer, E. A., had a longer head than Bailey,
And could see they had been planning daily.
Although Patchogue got it, they did not win it fair
Which all other companies were ready to swear.

The Minnies and Gazelles had to run twice,
And the Gazelles came in for their usual slice.
To beat Sag Harbor they must get up in the morning,
Past record shows, and that is fair warning.

Last came the Engines for their prizes,
Ed. Bailey's Bee Hive goes in two sizes;
But when it came time for old Hound to play,
They took first prize and walked away.

But this year we go to Greenport with intentions of winning,
Aad know with the Judges there will be no skinning;
So boys let us rally around Phœnix and Hound,
And all things will come out safe and sound.

Our Juveniles last year won first prize,
And this year can beat any boys of their size,
So look out for them, when they come down the track,
For they will run a light Truck, not the old Hay-Rack.

Upstate, at this time, there were continued indications that the sport of firematic competition was developing along different – and much slower – lines. That same year, FASNY held its 20th Annual Convention in Niagara Falls, and in direct contrast to the huge crowds attending the tournaments on Long Island, the crowds at the FASNY tournament were anything but spectacular. A Ladder Race and Prize Drill were held in the morning, with the Hose Races taking place in the afternoon. The *Niagara Falls Gazette* reports that out of 13 teams entered in the latter only nine contested, and it is especially interesting to note that the winner of the Hook and Ladder Contest was actually a team from Canada: the Alert Hook and Ladder Company of Niagara Falls, Ontario, with a time of 62-2/5 seconds. Nevertheless, it appears that in spite of the fact that "...The day was a fine one for racing" the event itself didn't draw all that much interest: "The attendance at the track was not what it should have been, very few citizens being present."[cliv] In addition, "...Another bad feature was the time lost between each race."[clv] It is noteworthy that it was a problem even back then, as delays between contests continue to haunt firematic tournaments even to this day.

In spite of this, the sport – and the science of fire fighting - continued to grow ever more popular, and not just in the United States. On June 12th through 17th, 1893, an International Congress of the fire service was held at the Royal Agricultural Hall in London, England, which was attended by many different countries from around the world. Representatives from FASNY, including Statistician W.E. Churchill, attended the Congress, where interestingly enough firematic demonstrations, very similar to the contests in the United States, were held as part of the conference program. Exhibitions by Continental teams included:

- Lightning Hub Horse Hitching
- Ground Hub Horse Hitching with Harness laying flat on the ground
- Dry Line Hose Laying and Fast Coupling
- Fast Hose Line Laying, Coupling, and Throwing Stream
- From Bed into Bunkers down sliding pole, and Fast Hose Line Laying, Coupling and Throwing Stream
- Rapid Ladder Climbing, with Line of Hose
- Rapid Ladder Climbing, one line ascending and the other descending
- Rapid Ladder Work, Casting String and Drawing Up Life Line
- Bringing Down Rescued on Life Line
- Short Ladder Work
- Inclined Aerial Truck Work, and
- Rescue by Jumping Into Life Saving Net

The British firefighters had their own set of drill exhibitions. Among them were:

- One man to get the engine into complete working order with one length delivery and one of suction, with dam
- Two men to get the engine into compete working order, etc.
- Four men to get the engine into complete working order, etc.
- Fire Escape and Life Saving Drill

There was also an exhibition by the Cork Corporation Fire Brigade.

Although he was obviously influenced by national pride, Churchill entered the following into FASNY's minutes, regarding the proceedings:

"This program, with slight changes, was repeated every afternoon and evening, except Wednesday, which was entirely given up to exhibitions by the London Fire Department. Some features of the program were very commendable, but on the whole I feel justified in saying that the English people are fully fifty years behind America in the science of fighting fire, and this applies with equal force to all kinds of apparatus, and I am confident that Young America would not tolerate one-half of the foreign apparatus twenty-four hours."[clvi]

Whether or not the United States was years ahead of the rest of the world in terms of fire science at the time is debatable, but it cannot be denied that the Americans – and New Yorkers in particular – had everyone, anywhere beat hands-down when it came to throwing a firefighter's convention and tournament. The 21st Annual FASNY Convention and Tournament was held at Coney Island on August 15th – 18th of that year, and by all accounts it was by far the largest ever held in history. The *New York Times*, which would normally have little interest in these matters, even ran a story about it:

"The 21st Annual Convention of the New York State Firemen's Association, which begins here on Tuesday and ends Friday, will be the most remarkable gathering of the fire laddies ever held in this country. It is confidently expected by the Executive Committee that when the grand parade occurs on Friday afternoon at least 25,000 firemen will be in line."[clvii]

It was truly a monster of an event. Even New York State Governor Roswell P. Flower and his staff attended: "The Governor has taken a lively interest in this matter, as he has in all affairs appertaining to the public welfare, and he comes here solely for the purpose of studying the work of the various companies represented."

And there were a *ton* of companies represented. A full 57 different companies from New York State – including groups from Seacliff, Rockville Centre, East Islip, Long Island City, Greenport, Patchogue and Sag Harbor on Long Island – were in attendance, along with crews from New Jersey, Connecticut, Massachusetts, Pennsylvania, Rhode Island, Vermont and even Virginia, for a total of 76 different fire companies.

There were predictions of 200,000 visitors each day, which led the *Brooklyn Eagle* to comment: "There was some fear that the thousands of visitors who poured in from all parts of the state would tear the blessed isle up by the roots and transplant it in Lake Ontario, but McNabe's words, 'All we ask of you is that when you have finished with it you will let us have the town back in as good order as you got it,' was too strong an appeal to the honor and the generosity of the firemen to be resisted."[clviii] (The reader will be asked to make note of this quote, as the same appeal will be made again, 54 years later, under similar, but more serious, circumstances.)

Whether or not the crowds were as large as they were feared to be was never confirmed, but there is no question they were huge nonetheless:

"The crowd was enormous. It poured in rivers of humanity through all the streets, lanes, alleys, courts and halls of the place. The walls were blocked solidly, and the beach was black for miles with moving masses. Accommodations were crowded up to the highest notch, and little board shanties on the Bowery were charging Manhattan prices for a cot under a flight of stairs or in a coal hole."[clix]

Which was apparently still not enough!

"As the day advanced the town became decorated with statuary. Sleeping men were the most common of the figures. They were of all sizes, ages and dresses and they slept in all attitudes. Some were curled up on the sidewalks or on the sand, reposing as peacefully as the babes in the woods. Others sat on chairs and slept with heads hanging sideways at an angle that threatened an immediate fall. One genius got himself jammed in between the signs in front of the firemen's headquarters and there he slept, standing, all the afternoon. Another man, after getting intoxicated, must needs [sic] climb the steps of the firemen's headquarters and sit at the top. He flopped about for a time and finally took a header down the whole flight. It is quite a long staircase, but he reached the bottom at last and, addressing one of the witnesses of the fall, said:

'S'ere, Cap'n. Fun's fun, but you do zat 'gin ni bust your snoot, see.' Then he went to sleep again."[clx]

Then, after the convention meeting:

"When evening came hundreds of exhilarated men felt that they were just frittering away their time making individual displays of condition. So they gathered like the Paris students, with no leaders and no organization, simply flocking together and parading with jovial faces and very uncertain steps. The bands which lay around in scores could not stand this. They got up and fell in line, playing every man for himself whatsoever tune he listed. The racket was enough to raise the red Indians who once ate clams on the beach – cornet and saxophone, trombone and drum, horn and trumpet, bassoon and kettledrum. Each had a different air and a different tune. The noise was so hideous and terrifying that Feltman's Carousel, which is the champion disturber of the peace on the island, shut right down and showed signs of nervous prostration. By the time midnight came the island glowed with red fire, and the noise was like that with which New York greets the New Year."[clxi]

As lengthy and elaborate as the media descriptions of the convention were, accounts of the corresponding tournament that year were exactly, and oddly, the opposite: practically nonexistent. The *Eagle*, normally verbose in its descriptions of the events, makes no mention of them in its article, and the *New York Times* merely listed the six contests scheduled and the cash prizes being offered for top finishers. The FASNY Secretaries had traditionally been stingy up until that point when describing the state tournaments, so that was not unexpected, but what happened less than two months later was: at a subsequent FASNY meeting on October 4th, 1893, the Executive Committee adopted the following resolution:

> "WHEREAS, the local organization of the city or village where the annual sessions of this association are held offer certain cash and special prizes for competitive contests of various kinds, emanating such

rules and conditions to govern those contests that they think will best subserve each local interest; therefore

RESOLVED, that the Secretary of this association be instructed to eliminate from the printed proceedings of the annual session of this Convention all reference to such contests or parades as are usually held at the time or immediately following the close of such annual meetings of this organization."

It is unclear just *why* the Executive Committee felt that reporting on firematic contests that had cash (or other) prizes offered by the tournament hosts was an issue that should not be made public; the minutes that describe that meeting are sparse enough in and of themselves so as to make the whole matter rather puzzling. What it effectively meant, however, was that from that point forward FASNY would no longer include *any* information about the tournaments held at their conventions, making the compilation of historical data in this regard a real challenge.

Tournaments on Long Island, however, continued to be well documented, even as they grew and the sport matured. More and more teams were competing at each SCVFA tournament: while the 1890 Islip tournament saw 22 teams competing, the 9th Annual Tournament in Southampton in 1894 saw 33 different teams on the roster. Perhaps learning from past experience, grandstands were set up for the very first time for spectators, as an increase in teams meant an increase in the size of the crowd as well. They were wise to have done so. Regarding the trains carrying both teams and spectators to the tournament, the *Sag Harbor Express*, in its September 6th, 1894 edition, wrote:

"...the Phoenix Hook and Ladder, the Gazelle Hose, and the Hound Engine companies of this place, embarked on the 9:30 special, carrying many excursionists with them, but as there were but three passenger cars and a caboose, four box cars were added, but that was not enough to take the people from this place, to say nothing of those who wished to get on at Bridge-Hampton and Water Mill, so, by authority from Long Island City, two more specials were run.

"The first special train from this place was the first excursion train to arrive, after which excursion trains from Greenport, Hicksville, Huntington, Hempstead, Patchogue and Babylon, making nine special trains in all.

"Besides the large number who went to Southampton by private conveyances, the trains from this place took 622, which number was largely augmented by those who got on at Bridge-Hampton and Water Mill, and all the trains which came from the west were not only long ones, but densely packed with people."[clxii]

Also at Southampton, and for the very first time, a Special Hose Contest was offered, which was won by Bay Shore. In it, contestants ran 100 yards, attached one end of a hose to a hydrant, then laid out 150 feet of hose, broke a coupling, attached a nozzle and flowed water through it. Newspaper descriptions of the contest are not very specific beyond that, but this may well have been the very first running of what went on to become what is called the Efficiency Contest, another contest that is still run today as part of "The Big 8." The Bayport Fire Department, in their history, describes a rather inauspicious start to that contest – at least for them:

"In September of that year at Southampton, a special prize was offered for a similar [wet hose] contest. We went for it. So did eight others, a long list for those days. We were the last to run. Because of flukes and the poorly-drained track, there was a pond around the nozzle end. We made a good run and hydrant connection, but our nozzle man, Dan Gillette, slipped and fell in the pond...of course, Dan should have swum ashore, put the nozzle on and earned time, but we had not yet learned the maxim, now our motto, 'never quit on a fluke.'"[clxiii]

What is interesting, though, is that a contest that today is considered a variation on the Efficiency Contest (one that is not part of today's "Big 8" of motorized contests, but which is still run during Old

Suffolk County Volunteer Firemen's Association

NINTH ANNUAL TOURNAMENT

SOUTHAMPTON, N. Y.

This is to Certify, that ______ Company of ______

won ______ Prize of ______ Dollars in competition of their Class at the Ninth Annual Tournament of the Suffolk County Volunteer Firemen's Association held at Southampton, N. Y., September 5th, 1894.

Signed,

Judges.

Benj. H. Bishop, Chief.

Nathan C. Howell, Assistant Chief.

Fashioned and Juniors tournaments) was actually run for the first time one month *before* the Efficiency Contest at Southampton. The "Efficiency Replacement Contest", as it is now called, was first run at the inaugural Islip Town Firemen's Association tournament on June 20th of that year, and prescribed that firefighters "...run 50 yards, lay 200 feet of hose, break couplings, screw on pipe, put water on target, then exchange end lengths of hose and put water on target again. Time will be taken when water strikes target a second time."[clxiv]

It should be said, for sake of historical accuracy, that the descriptions of these new contests provided by the local newspapers may or may not be completely thorough, and are somewhat open to interpretation. For example, it is assumed that the Hose Contests – "Efficiency" and "Efficiency Replacement" Contests by today's standards – of 1894 were run on foot, since no mention of a hose cart is made within any of those descriptions. However, at the 2nd Annual Islip Town tournament the following year, the Hose Contest there was described with the sentence "...Then take out the middle section of hose and replace it with a new section from same hose cart; put water on target."[clxv] Whether this was the same contest run the year before, or a completely new contest using a hose cart – known today as Cart Hose Replacement – is unclear.

The addition of the Islip Town tournament to the list of firematic competitions being held was a further indication of how the sport was continuing to grow and change. In 1895 Bellport firefighters began hosting their own "Fair & Festival", followed 1896 by the creation of a Queen's County Firemen's

The Riverhead Fire Department tournament team, ca. 1890's

Tournament by the Freeport Fire Department, which was held in Freeport on June 16th. That year also saw the addition of a Whitestone Fire Department tournament on July 21st, and a Far Rockaway Annual Tournament and Parade on September 17th.

The Queens County tournament appears to have been a unique event. Perhaps as an outgrowth of it, on February 10th of the following year, the Town of Hempstead Volunteer Firemen's Association was formed at Freeport, when "...twenty-five out of the thirty-six companies in the town attended a preliminary meeting...George Wallace was elected President for a term of one year. For the past eight years the question of forming an association has been constantly agitated by different companies in the town, but no action was ever taken in regard to the matter until the Freeport department called a meeting on February 10th, as everybody thought the object would be a failure."[clxvi] On June 16th, the first Annual Town of Hempstead Tournament was held.

Chapter 8: The Advent of Northern and Western, and the Golden Age of Tournaments

Although Long Island had by this time a trend of more and more "legitimate" tournaments popping up left and right, that trend was not limited to the island. In 1895, in northern New York, the Jefferson County Firemen's Association was formed. (That association would later go on to become the current Firefighter's Association of Northern New York, one of the four Areas represented in the modern New York State Volunteer Firemen's Parade and Drill Team Captain's Association.) Their first convention – and tournament – was held in Adams, N.Y. On May 29th of that year, the citizens of Adams met in their concert hall, "…all favoring the convention and pledging themselves to do what they could do to make it a success."[clxvii] The convention itself was held on September 18th and 19th, and got off to a nervous start: "The weather prophets prophesied fair skies, but anxious firemen and others who had made extensive preparations and who were awake before dawn on Wednesday, turned uneasily in their beds as they heard the patter of the rain drops falling thick and fast, and mentally repeated the old couplet, 'Rain, rain, go away, come again some other day.' However, after nicely laying the dust and slightly confusing the colors in some of Tuesday's decorations, the rain ceased and at an early hour the citizens were astir putting on the finishing touches until the town was decorated with bunting and gaily waving flags as never before in all its history."[clxviii]

It is true that the newspaper's description of the day starts off sounding like that of every and any other convention and/or tournament of the era, but it is also apparent from these same descriptions that the northern conventions – and those that took place in the central and western parts of the state as well – were not the same, so to speak, as those that took place on Long Island. Notwithstanding that the crowds were probably smaller (the Adams convention drew 8,500), the upstate conventions and tournaments were then – as they are still to this day – much more friendly, casual affairs than those held "down south". An example of this can be taken from a roster of the events that took place in Adams over the course of the two days that the convention was in session. In contrast to the gatherings on the island, which were pretty much limited to activities that were firematic in nature, the first day of the Adams convention included a Bicycle Race, a Potato Race, a Three-Legged Race, a Wheelbarrow Race, a Biscuit Race, an Obstacle Race, two kinds of foot races, a

Band Contest, and even a circus, ending with a big fireworks display. The program for the second day was much more familiar: a firemen's Foot Race, a Dry Hose Race, a Straightaway Race (perhaps the same as a Hub & Hub race), a Prize Drill, and the parade. In a now-becoming-familiar fashion, "…There was a long delay in the dry hose race, perhaps unavoidable under the circumstances, but it made the crowd very impatient" and "…a protest was entered against the Carthage team running because they had picked their men from the department instead of one company, as the rules required. The protest being sustained the Carthage boys did not run, greatly to the disappointment of the crowd, for they probably would have made good time."[clxix] Nevertheless, as the same article concluded, "This first convention of Jefferson County firemen has clearly placed the association beyond the experimental stage and its permanency is assured."

Back on Long Island, where tournaments were more developed, individual contests being held at the tournaments started to become more refined. In 1895, at the 11th Annual SCVFA Tournament in Sayville, "...the Bay Shore Hose Co. came to the tournament with a small wagon in which the hose was laid, instead of being pulled off a reel; thus making a step toward "modern" racing techniques. It was ridiculed and called a 'goat cart' and 'toy express wagon', but the laughing came to a halt when they beat their nearest competitors by 4 seconds."[clxx] Record times for the Hook & Ladder contest were also being established by Sag Harbor: "The 'Old Hay Rack' gained a quarter of a second on its record at Sayville and took first prize. They went rattling down the course at lightning speed, a good ladder was made but as it hit the striking board, one arm snapped in two, but the intrepid "Tommie" Quail flew up the rungs of the unbroken side undaunted and scored the victory."[clxxi]

Even the often-stodgy *New York Times* seemed to get caught up in the excitement when describing the day:

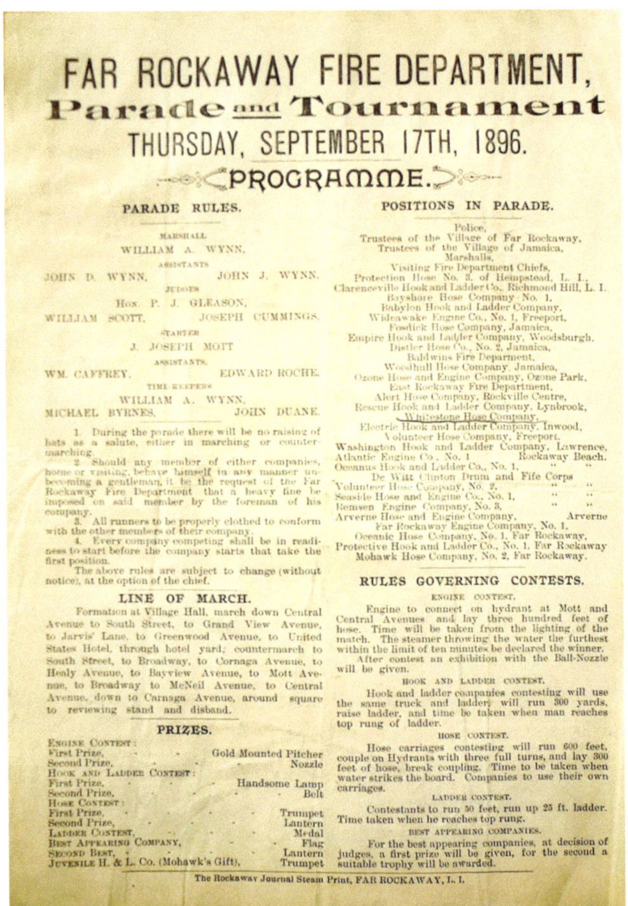

FAR ROCKAWAY FIRE DEPARTMENT,
Parade and Tournament
THURSDAY, SEPTEMBER 17TH, 1896.
PROGRAMME.

PARADE RULES.

MARSHALL
WILLIAM A. WYNN.
ASSISTANTS
JOHN D. WYNN. JOHN J. WYNN.
JUDGES
HON. P. J. GLEASON,
WILLIAM SCOTT. JOSEPH CUMMINGS.
STARTER
J. JOSEPH MOTT
ASSISTANTS.
WM. CAFFREY. EDWARD ROCHE.
TIME KEEPERS
WILLIAM A. WYNN,
MICHAEL BYRNES. JOHN DUANE.

1. During the parade there will be no raising of hats as a salute, either in marching or counter-marching.

2. Should any member of either companies, home or visiting, behave himself in any manner unbecoming a gentleman, it be the request of the Far Rockaway Fire Department that a heavy fine be imposed on said member by the foreman of his company.

3. All runners to be properly clothed to conform with the other members of their company.

4. Every company competing shall be in readiness to start before the company starts that take the first position.

The above rules are subject to change (without notice), at the option of the chief.

LINE OF MARCH.

Formation at Village Hall, march down Central Avenue to South Street, to Grand View Avenue, to Jarvis' Lane, to Greenwood Avenue, to United States Hotel, through hotel yard; countermarch to South Street, to Broadway, to Cornaga Avenue, to Healy Avenue, to Bayview Avenue, to Mott Avenue, to Broadway to McNeil Avenue, to Central Avenue, down to Carnaga Avenue, around square to reviewing stand and disband.

PRIZES.

ENGINE CONTEST:
First Prize, - - Gold Mounted Pitcher
Second Prize, - - - Nozzle
HOOK AND LADDER CONTEST:
First Prize, - - - Handsome Lamp
Second Prize, - - - - - Belt
HOSE CONTEST:
First Prize, - - - - Trumpet
Second Prize, - - - - Lantern
LADDER CONTEST, - - - Medal
BEST APPEARING COMPANY, - - Flag
SECOND BEST, - - - - Lantern
JUVENILE H. & L. Co. (Mohawk's Gift), Trumpet

POSITIONS IN PARADE.

Police,
Trustees of the Village of Far Rockaway,
Trustees of the Village of Jamaica,
Marshalls,
Visiting Fire Department Chiefs,
Protection Hose No. 3. of Hempstead, L. I.,
Clarenceville Hook and Ladder Co., Richmond Hill, L. I.
Bayshore Hose Company No. 1,
Babylon Hook and Ladder Company,
Wideawake Engine Co., No. 1, Freeport,
Fosdick Hose Company, Jamaica,
Empire Hook and Ladder Company, Woodsburgh,
Distler Hose Co., No. 2, Jamaica,
Baldwins Fire Deparment,
Woodhull Hose Company, Jamaica,
Ozone Hose and Engine Company, Ozone Park,
East Rockaway Fire Department,
Alert Hose Company, Rockville Centre,
Rescue Hook and Ladder Company, Lynbrook,
Whitestone Hose Company,
Electric Hook and Ladder Company, Inwood,
Volunteer Hose Company, Freeport,
Washington Hook and Ladder Company, Lawrence,
Atlantic Engine Co., No. 1 Rockaway Beach.
Oceanus Hook and Ladder Co., No. 1, " "
De Witt Clinton Drum and Fife Corps
Volunteer Hose Company, No. 2, " "
Seaside Hose and Engine Co., No. 1, " "
Remsen Engine Company, No. 3, " "
Arverne Hose and Engine Company, Arverne
Far Rockaway Engine Company, No. 1,
Oceanic Hose Company, No. 1, Far Rockaway,
Protective Hook and Ladder Co., No. 1, Far Rockaway
Mohawk Hose Company, No. 2, Far Rockaway.

RULES GOVERNING CONTESTS.

ENGINE CONTEST.

Engine to connect on hydrant at Mott and Central Avenues and lay three hundred feet of hose. Time will be taken from the lighting of the match. The steamer throwing the water the furthest within the limit of ten minutes be declared the winner.

After contest an exhibition with the Ball-Nozzle will be given.

HOOK AND LADDER CONTEST.

Hook and ladder companies contesting will use the same truck and ladder; will run 300 yards, raise ladder, and time be taken when man reaches top rung of ladder.

HOSE CONTEST.

Hose carriages contesting will run 600 feet, couple on Hydrants with three full turns, and lay 300 feet of hose, break coupling. Time to be taken when water strikes the board. Companies to use their own carriages.

LADDER CONTEST.

Contestants to run 50 feet, run up 25 ft. ladder. Time taken when he reaches top rung.

BEST APPEARING COMPANIES.

For the best appearing companies, at decision of judges, a first prize will be given, for the second a suitable trophy will be awarded.

The Rockaway Journal Steam Print, FAR ROCKAWAY, L. I.

"Some of the volunteer firemen of Suffolk County are fat, some thin, some tall, others are stout. But nobody who witnessed the tournament went away without being convinced that whether the laddies be thick or thin, short or long, they are all as limber as India rubber and as tough as hickory. They climbed ladders as nimbly as cats, they ran with their engines and hose carts like a streak of lightning, and when they pointed the nozzles of their hose toward the sky it was evident the sun was in great danger of having his spots washed off in double-quick time. Next to Christmas, the annual firemen's tournament is the greatest event on Long Island, and one look at Sayville was enough to convince the most skeptical that it will hold second place on the calendar of festivals for years to come."

Sag Harbor's Thomas "Tommie" Quail was but one of many firefighters of the day who were quickly gaining popularity in both the media and with the public as "stars" of the sport. In stark contrast to the early part of the century, when the firemen were considered nothing but hoodlums worthy of little respect, by the 1890's firemen, and their tournaments, were considered the height of respectability and civic pride. Said a Greenport newspaper of their team, "The firemen's annual tournaments of Suffolk County are unquestionably excellent incentives to make the

A ladder contest at the 11th annual SCVFA tournament at Sayville in 1895

volunteer firemen of the island look well to their laurels and keep in constant practice as fire fighters if they would wish to rank among the leaders in any of the different villages each year." This may be one reason why West Sayville, at their meeting on June 19th, 1896 regarding the 3rd Annual Islip Town Tournament at Bayport, "…it was voted to travel in style to the town competition, using stagecoaches for transportation."[clxxii]

This phenomenon of praise for tournaments was not limited to New York State. In fact, the extreme popularity of the tournaments even contributed to the benefit of race relations in the south, as indicated by a passage found on the internet regarding the history of black firefighters in Columbia, South Carolina:

"One of the things that helped bolster the popularity of African-American drivers during the 'volunteer days' was their performance in the firemen's tournaments. Most of the tournaments consisted of a three-day program of foot races, reel contests and fire engine races. Many people attended the events, [when] during the 1880's firemen were more popular than baseball players.

"Although African-Americans competed in predominantly white tournaments, no black companies were allowed to compete against white companies. The tournaments that black companies held were just as successful as the tournaments held by their white counterparts. Thousands of spectators came out to see them compete. The stands were filled with black and white spectators who cheered the great skills of the firemen. It was not an unfamiliar sight to see white and black faces in the crowd of an African-American competition. The tournaments provided a rare opportunity for black and white people to come together, free from racial tensions of the time."[clxxiii]

The above statement that "during the 1880's firemen were more popular than baseball players" may seem exaggerated, but in truth it is supported by research of local media coverage of the day. At the 12th Annual SCVFA Tournament that was held in Riverhead on September 6th, 1897, no less than *sixteen* different newspapers covered the event. So large – and apparently important - was their presence that a special "judges stand" was constructed that year to include them. This "stand" was a wooden structure that was built to span one end of the course, and was used for contestants to throw a ladder up against (as opposed to a local storefront, previously), or climb up onto to fill a bucket of water, and as a high, well-placed spot for judges to view the action. There is photographic evidence that these "judges stands" – known today at each track as the "arch", existed as early as 1895, but the one built at Riverhead was different: "…the first 'three-deck' judges stand was noted. The separate decks were for the use of the tournament officials, special guests and the newspaper men, respectively."[clxxiv]

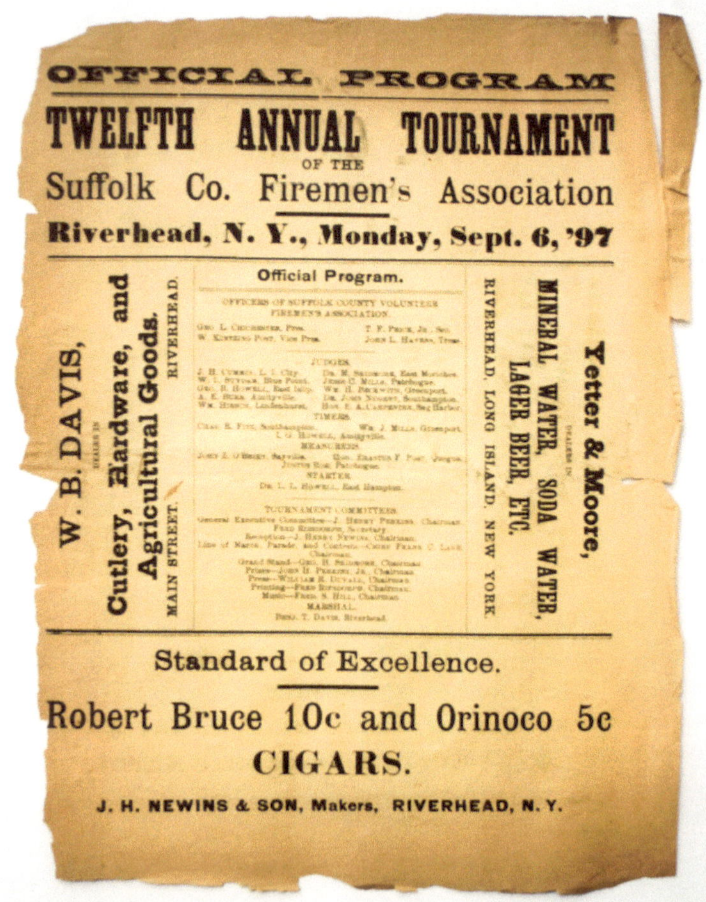
OFFICIAL PROGRAM

TWELFTH ANNUAL TOURNAMENT

OF THE

Suffolk Co. Firemen's Association

Riverhead, N. Y., Monday, Sept. 6, '97

W. B. DAVIS, DEALER IN Cutlery, Hardware, and Agricultural Goods. MAIN STREET, RIVERHEAD.

Official Program.

Yetter & Moore, DEALERS IN MINERAL WATER, SODA WATER, LAGER BEER, ETC. RIVERHEAD, LONG ISLAND, NEW YORK.

Standard of Excellence.

Robert Bruce 10c and Orinoco 5c CIGARS.

J. H. NEWINS & SON, Makers, RIVERHEAD, N. Y.

The huge, festive event that year was not without fodder for the press, at the very least in terms of mishaps. Firefighting can be a dangerous business, at least at the scene of a fire, but it may have been this year that recorded the first fatality associated with the sport of firematic drill team competition. "A letter was received from Sag Harbor," wrote the *Riverhead News*, "announcing that out of respect for Robert Quail, a member of the Phoenix Hook and Ladder Company, who met his death last week while practicing for the ladder contest, no delegates from that place would be on hand."[clxxv] It didn't end there. During the tournament, "...one of Greenport's men, Theodore Howard, suffered a mishap while running over the course, being nearly thrown to the ground by an accidental collision with a dog and a deputy sheriff's horse. Howard pluckily continued toward the goal with his play pipe and saw it safely screwed on, after which he fainted, but soon afterward revived."[clxxvi] And finally, on an even more frightening note: "A sad feature of the day was the injury suffered by John Pollack, the lad who scaled the ladder for Greenport in the juvenile hook and ladder contest. Just as he grasped the top round he fell and struck his back and head in the hard roadway. He was picked up in an unconscious condition and carried into Dr. Raynor's office, where he was attended to by Dr. Raynor and several other physicians. Later in the day young Pollack was removed to the residence of Mr. Thomas Greeves on Roanoke Avenue, and everything possible was done for his relief. His mother was sent for and came to Riverhead Monday evening. As a result of the medical examination it was determined that the lad had suffered a slight concussion of the brain, but no bones were broken.

"Young Pollack is steadily improving and his early restoration to health is anticipated.

"The painful injury received by Mr. Pollack and the previous accident in Sag Harbor by which Mr. Quail lost his life suggest the need of a change in the hook and ladder contests. It is argued that practical work does not require a man to scale a ladder before it is placed against a building, and hence it should not be made part of the tournament exercises. Probably a motion will be made at the next meeting of the association providing that in the future the ladder scaler shall not begin to mount until the ladder is in position against the backstop."[clxxvii] The following year, that issue was indeed raised.

Further west, in and towards New York City, the years 1898 and 1899 brought significant changes to the layout of local government in the area, starting with the official formation of the City of Greater New York in 1898. As part of this reorganization, the City of Jamaica would now be part of New York City. On September 16th of 1897, in a somewhat poignant celebration in anticipation of the coming merger, a "last parade" was held by the Jamaica Fire Department as an independent institution. The *Brooklyn Daily Eagle* reported that close to 1,000 firefighters marched, and that (at the time), "They [Jamaica H&L Co. #1] are, from all that can be ascertained, the only fire company in the state that have been doing active, continuous fire duty for over one hundred years, and are therefore the Centennial company."[clxxviii] The following year, in 1899, the part of Queens County that was not annexed to New York City, consisting of the two towns of North Hempstead and Oyster Bay and most of the town of Hempstead (excluded was the Rockaway Peninsula, which did join Greater New York), was constituted as the new Nassau County.

According to Robert Holzman, in spite the fact that there was now a City of Greater New York, and that the fire service in many communities (such as Jamaica in Queens, and those in Brooklyn) were now run under the authority of the New York City Fire Department, "It was a long time before the spirit of inter-company competition disappeared in the paid fire departments, for too many of the men were ex-volunteers who had grown up in that sort of tradition. One throwback to the competitive days of the volunteers was revealed by Battalion Chief Conley: in 1898, when he joined Engine Company No. 213, there was a great rivalry between that crew and Engine Company No. 229. By reason of a peculiarity in the electric fire alarm telegraph circuit, 213 got the alarms some seconds before 229 did, and the former company would gallop to an intersection of streets which 229 would have to pass. As soon as 229 dashed by, 213 would come up from behind, and as 213's horses had had time to catch their second wind, Conley's boys would overtake the other company, in the old tradition of 'passing' and disgracing a rival en route to a fire."[clxxix]

The closing years of the nineteenth century also brought new growth and changes to the sport of firematic competition. As one might have gathered by now, a *lot* was going on during this time period, and this trend continued well into the first part of the twentieth century.

The Theresa Fire Department tournament hose team, 1898

It is quite possible that the year 1898 holds the distinction of being the first year that the very first "Juniors" competition was held, seperate from that of the adults. The description is rather short and vague, but the *Brooklyn Daily Eagle* refers to a "Juvenile Tournament" that took place in Rockaway Beach on March 21st, 1898, during which first prize was "a large iced cake...and a smaller cake for the second prize." Hempstead Town, Islip Town and Rockaway Beach all held their Annual Tournaments, which were all that much more well-attended due to the apparent sudden popularity of a transportation device called a "Bicycle." On June 8th, at the Hempstead Town gathering, "...Good judges estimated that between eight and ten thousand people saw, or tried to see, the tournament contests. Where the immense crowd came from was a mystery. The bicycle is probably responsible for the two or three thousand who otherwise cannot be accounted for. Every place was stowed with wheels and the conversation of the riders showed they came from Brooklyn on the west, Patchogue on the east, and all intervening points."[clxxx] The Suffolk County Volunteer Firemen's Association that year took recommendations for, and ultimately made changes to, its Annual Tournament, which went into effect that year at their gathering in Huntington on September 14th. Among "numerous rule changes" that took place was a reduction of running distance in events to 150 yards (from 300 yards), and another attempt to eliminate "...all danger from the ladder contests by ordering that no contestant shall touch his ladder until it is securely placed against the structure."

(The latter issue was no doubt brought up as a result of the death of Sag Harbor's Robert Quail, and the subsequent injury of junior competitor John Pollack at the Riverhead tournament the previous year. However altruistic the intent of that rule change was, it obviously didn't stand up to the test of time: in today's B-Ladder and C-Ladder contests the best ladder climbers are often halfway up the ladder before it strikes the arch!)

In northern New York State, in spite of it being only its fourth gathering, the Jefferson County Firemen's Association's convention was, by 1898, gaining popularity at the same rate as its counterparts down south. True to form, the Alexandria Bay meeting included many contests not seen on Long Island, including a Walking of a Greased Pole, Sack Race, a Rowing Race, a Steam Yacht Race, a Sail Yacht Race and a 100-Yard Dash, but there was no difference at all when it came to the crowds attending, and their behavior:

"This year Alexandria Bay entertained the firemen, or rather the firemen took possession of that village and ran things as they pleased...a large number of the visiting firemen arrived at the Bay Wednesday, and that night was not one of peaceful rest for the inhabitants of the village. Someone slept on the bell rope and every time he turned over the bell rang for several minutes, and the only reason they did not turn over every half hour was because they turned over every 20 minutes. As would be expected in a crowd of that size, a considerable quantity of malt and spirituous liquors was consumed, but it is to the credit of the volunteer firemen of Jefferson county that there was no breach of the peace on their part, and everyone was good natured and intent on only having a good time."[clxxxi]

Their convention the following year in Watertown, but for the introduction of a Shoe Box Race, was similar. It wasn't so much the firemen that didn't behave themselves as it was the visiting public:

"The firemen are having a good time, and thus far are having it in a way that could offend no one. There was considerable drunkenness and a great deal of unnecessary and disagreeable noise on the streets last night, but not one of the visiting firemen was observed participating in these orgies. They were conducted by a gang of Watertown loafers who eagerly seize upon the slightest excuse to make others wonder why they were born."[clxxxii]

Of greater significance in 1899, however, was the formation of the Western New York Volunteer Firemen's Association, another organization that exists to this day.

Although there is no hard evidence upon which to base a claim, it's a safe bet to say that the volunteer firefighters in the western part of New York State by that time certainly knew about the different firemen's associations – including FASNY – and their respective conventions that were taking place throughout the rest of the state. A history of the Western Association states that a Mr. Joseph M. Hoffman of the Active Hose Company in Lockport, N.Y. organized the first informal meetings of western firemen, "...with the idea of forming an organization for the purpose of promoting closer friendships among the various fire companies and providing assistance in improving their respective operations in the Volunteer Fire Service." On November 12th, 1899, at a formal meeting, hosted in the Active Hose Company carriage house by Mr. Hoffman, the Western Association was formally recognized. The history states that at that meeting thirty-two different fire companies from four counties (Erie, Monroe, Niagara and Orleans) were represented, and that it was decided that their first convention would be held in Lockport on June 13th of the following year.

Based upon newspaper accounts, however, the first day of the First Annual Convention of the Western New York Volunteer Firemen's Association actually took place one day earlier, on June 12th of 1900, during which the business portion of the convention took place, with the parade and contests taking place on the second day, the 13th. Perhaps because it was a new kind of event to that area, there appears to have been not-quite-as-much-interest in the proceedings by the press as was typical in other parts of the state: the local paper covered the events of the day in only a single column, led with the rather dry announcement, "The first annual convention of the Western New York Volunteer Firemen's Association has been held in Lockport...The visitors were unanimous in the verdict that the convention was a decided success and that the Lockport firemen did nobly at entertaining."[clxxxiii] The story goes on to dryly describe the contests ("The events which followed were very good. A summary follows"), which included the usual Hose Race, Hub-and-Hub Race, Firemen's Foot Race, Wheelbarrow Race, Prize Drill (for marching) and Special Parade Prizes. Two new contests were included there, however, that had not been previously mentioned in other tournaments: a Fat Men's Race, where there was no description at all given, and a Bunkers Race, which was not very clearly described either except to say "Distance one-fourth mile. Four men to each cart. Men wore rubber boots."

One unusual item that did appear in the article described an occurrence that took place that day before the start of the contests that was, by today's standards, politically incorrect to the point of being racist in nature:

"C.L. Nicholls, secretary of the Niagara County Agricultural Society, was on the Grounds, and suggested to Foreman Joseph W. Turner of the DeWitts, overseer of events, that a cake walk[4] on the platform would be good amusement of the crowd while waiting for the events to begin. Acting on the suggestion Mr. Turner persuaded the colored 'mascots' R.J. Berry of Hart Hose, Albion, and Harry Watkins of Dye Hose, Albion to 'do a turn.' They were vociferously applauded."[clxxxiv]

Whether or not it was the result of, or influence from, the non-firematic contests that were being held in addition to the firematic ones in the northern and western parts of the state, these alternative types of contests briefly showed up at tournaments down on Long Island as well. An 1899 tournament held by the Woodhaven Fire Department included a 100-yard dash, and a tournament the following year in Rockaway Beach included one also, as well as a Tug of War and a Greased Pole Climbing Contest. An interesting note to the former was that "...the contests were arranged in two distinct divisions, one for Woodhaven department, and the other for visiting companies. This was done to bring in competition the local department [sic], as none of the companies had ever been entered in a tournament contest, and it was thought that they would be too heavily handicapped if the visitors were pitted against them."[clxxxv] The other interesting side note from Long Island tournaments that year involves the absence of the Greenport from the 14th Annual SCVFA Tournament in Bay Shore, the reason for which remains a complete mystery: "The only circumstances to mar the pleasure of the day was the enforced absence of the Greenport Fire Department and universal sympathy was expressed for the members of that department in their misfortune."[clxxxvi] Why an "enforced" absence would be associated with a "misfortune" is quite puzzling, but this author could not find any other references to it, so it seems we will never know.

The year 1900 was definitely a banner year for the sport of firematic competition. Besides the creation of the Western New York association and a boom in tournament activity in other parts of upstate New York, tournaments on Long Island were reaching never-before-seen heights in popularity. At the 4th Annual Hempstead Town Parade and Tournament on June 26th, the *Brooklyn Daily Eagle* reported that *20,000* people turned out to see the firefighters compete, and that "...an arch, the widest yet built, profusely decorated with the national colors and bunting, and surmounted by two heroic figures of Liberty, spanned Central Avenue in front of the quarters of Washington Hook and Ladder Company. The grand and reviewing stands reached along each side of the arch and were black with people. The Headquarters in Kelly's Hotel were connected by telephone with the arch." At the 15th Annual SCVFA Tournament in Lindenhurst on September 12th, the *Eagle*, in an advance article, further indicated how expansive the events had become: "...not only will Suffolk send its thousands of firemen and visitors, but the sister counties of Nassau and Queens have promised to send large delegations...Not one of the sixty-six companies in the Suffolk County association will lack representation next Wednesday, and nearly all of them will compete in the afternoon's contests." They went on to say, "It is expected that at least five thousand firemen and visitors will be in town Wednesday and arrangements have been made to feed that number of people."[clxxxvii]

---

[4]A Cakewalk is a traditional African American form of music and dance which originated among slaves in the Southern United States. The form was originally known as the chalk line walk; it takes its name from competitions slaveholders sometimes held, in which they offered slices of hoecake as prizes for the best dancers.

ANNUAL PARADE OF THE HEMPSTEAD FIREMEN.

The Finest Turnout in the History of the Town Department.

20,000 PERSONS SEE THE SHOW

Lawrence, in Holiday Garb, Builds a Magnificent Arch in Honor of the Fire Fighters.

(Special to the Eagle.)

Lawrence, L. I., June 26—The fourth annual parade and tournament of the Town of Hempstead Volunteer Firemen's Association took place to-day, and twenty thousand or more came from far and near to see the spectacle and witness the contests of the fire laddies. The village is gayly decorated and everybody is in holiday attire. Early in the morning carriages and farm wagons arrived from outside villages, each bringing its quota of sightseers. Hundreds more came by way of the trains. Everywhere the gallant fire laddies were prominent, and the clanging of bells was heard in every direction as the various fire companies arrived and fell into line.

Along the curbs carriages and wagons were drawn up and the sidewalks were lined with enthusiastic crowds, which cheered each company as it passed by.

An arch, the widest yet built, profusely decorated with the national colors and bunting, and surmounted by two heroic figures of Liberty, spanned Central avenue in front of the quarters of Washington Hook and Ladder Company. The grand and reviewing stands reached along each side of the arch and were black with people. The headquarters in Kelly's Hotel were connected by telephone with the arch.

FIREMEN'S ARCH OVER CENTRAL AVENUE, LAWRENCE.

Rowland H. Mayland,
Grand Marshal of the Parade and President of the Hempstead Firemen's Association.

As the different companies arrived they were met by the committee and assigned to their places and preparations for the start of the parade were made.

Carman R. Lush,
Vice President Town of Hempstead Fireman's Association.

Thirty-four companies and many bands and drum corps were in the line of the parade, the information of which was as follows:

Thomas A. McWhinney,
Chief of the Lawrence Fire Department.

Grand Marshal Roland H. Mayland.
Aids to Grand Marshal.
FIRST DIVISION.
Marshal and aids.
Band.
Freeport Fire Department, R. H. Mayland, chief.
Excelsior Hook and Ladder Company,
C. W. Bedell, foreman.
Ever Ready Hose Company,
A. B. Wallace, foreman.
Vigilant Hose Company, Edward Finch, foreman.
Bay View View Hose Company,
A. P. Lewis, foreman.
Wide Awake Engine Company (steamer),
C. F. Baldwin, foreman.
Baldwins Fire Department, John B. Paton, chief.
Baldwins Engine and Hose Company,
George W. Miller, foreman.
Baldwins Hook and Ladder Company,
Charles Coakley, foreman.
SECOND DIVISION.
Marshal and aids.
Drum, Fife and Bugle Corps.
Rescue Hook and Ladder Company, Lynbrook,
Robert Moods, foreman.
Band.
Inwood Fire Department, E. Zernike, chief,
Electric Hook and Ladder Company,
Valentine Smith, foreman.
Inwood Hose Company.
Empire Chemical Engine Company, Woodmere,
G. H. Schiffmacher, foreman.
Band.
East Rockaway Fire Department,
H. Floyd-Johnson, chief.
Protection Hook and Ladder Company,
Elmore Bedell, foreman.
Vigilant Engine Company (band),
Walter Southard, foreman.
Floral Park Hook and Ladder Company,
S. F. Klein, foreman.
THIRD DIVISION.
Marshal and aids.
Band.
Rockville Center Fire Department,
Simon Stiner, chief.
Eureka Hook and Ladder Company,
S. T. Wright, foreman.
Live Oak Engine Company (band),
E. McGuinness, foreman.
Drum Corps.
Defender Hose Company,
Edward Aldom, foreman.
Alert Hose Company, W. L. Smith, foreman.
Fife and Drum Corps.
Nassau Hook and Ladder Company, Valley Stream, E. H. Gelder, foreman.
Advance Hook and Ladder Company, Bellmore,
T. G. Smith, foreman.
FOURTH DIVISION.
Marshal and aids.
Rockaway Beach Fire Department,
John A. Carle, chief.
Band.
Arverne Hose and Engine Company (steamer),

William Stoffel,
Chairman Prize Committee.

Robert C. Kreuscher, foreman.
Atlantic Engine Company (steamer),
Andrew Churchill, foreman.
Oceanus Hook and Ladder Company,
Albert Ringk, foreman.
Volunteer Hose Company,
Theodore Bennett, foreman.
Reissen Chemical Engine Company,
George Grispenkerl, foreman.
Fife and Drum Corps.
New Hyde Park Hook and Ladder Company,
W. B. Rayner, foreman.
Band.
Hempstead Fire Department,
Robert Vandewater, chief.
Protection Hose Company, John Box, foreman.
Victory Hose Company,
W. R. Mollineaux, foreman.
Union Engine Company (steamer),
W. J. Coles, foreman.
Lawrence-Cedarhurst Fire Department,
Thomas A. McWhinney, chief.
Lawrence Juveniles.
Washington Hook and Ladder Company, Junior.
Band.
Lawrence Engine Company (steamer),
H. Williams, foreman.
Lawrence-Cedarhurst Hose Company,
C. J. Bauer, foreman.
Washington Hook and Ladder Company,
J. J. Phillippi, foreman.

After all the companies had reported to Grand Marshal Roland H. Mayland, the order to start was given shortly after 10:30 o'clock A. M. The line of march was as follows: From Craft and Central avenues, to Cedarhurst avenue, to Broadway, to Lord avenue, to Bay View avenue, to Lawrence avenue, to Central avenue, to Turnpike, to Broadway, to Craft avenue, to Central avenue, to reviewing stand and counter march to starting point, where the companies were dismissed for dinner. The route of the parade was about two and a half miles long, all over good macadam roads.

William B. Osterhout,
Official Measurer.

The committee in charge of the arrangements saw that the hundreds of firemen were well fed and after luncheon the firemen prepared their apparatus for the contests, which commenced at 1:30 o'clock. The hook and ladder contest was the most hotly contested event on the programme and the steamer contest this afternoon will also be a very interesting one. Union of Hempstead, Wide Awake of Freeport, Arverne and Lawrence engines will fight it out.

The contests on the programme for this afternoon are a hook and ladder, hose, steamer, chemical and hand engine, ladder climbing and juvenile hook and ladder.

The contests will be governed by the following regulations:

Hook and ladder—Trucks to run 200 yards, raise twenty-five foot ladder to arch spanning course, and fireman to ascend to top; time to be called when man grasps top rung of ladder and ladder against arch; rung to be held until time is called. Limited to twenty-one men.

Ladder climbing—Fireman to run fifty feet, scale thirty foot ladder placed against arch, time to be called when man grasps top rung; rung to be held until time is called.

Hose contest—Hose carts, carriages or wagons to run 200 yards to hydrant, attach to hydrant, lay 100 feet of hose, break coupling, screw on pipe; time to be called when water strikes back stop, 125 feet from hydrant. Each company to carry not less than 200 feet of regulation hose. Pipe to be carried from starting point. Limited to fifteen men.

Steamers—Steamers to play through 150 feet of hose in one stream. Each steamer will be allowed eighteen minutes in all to make test; time to be taken from first appearance of smoke from the smoke stack; cold water in the boiler. Hose to be attached to ordinary working outlet. First class steamer to use 1½ inch nozzle, second class steamer to use 1⅜ inch nozzle, third class steamer to use 1¼ inch nozzle, fourth class steamer to use 1⅛ inch nozzle and fifth class steamer to use 1 inch nozzle. The award of prize shall be on horizontal distance water is thrown, measured nozzle of pipe placed on ground previous to water entering hose.

Hand engines—Engines to run 300 feet, place suction in tank, lay 100 feet of hose, break coupling, screw on pipe and put water on back stop 150 feet from engine. No water allowed in engine box or hose. Time to be called when water strikes back stop.

The prizes are as follows:

Steamer contest—First prize, Brooklyn Eagle trophy, History of Brooklyn; second prize, silver cup, Dr. H. H. Kane; third prize, clock, from George W. Smith.

Hose contest—First prize, trophy, hose spanners, American Fire Engine Company; second prize, trumpet, J. L. Lawrence; third prize, lamp, J. H. Rauland & Co.

Hook and ladder contest—First prize, tilting pitcher, Smith, Gray & Co.; second prize, cup, Long Island Railroad, Joseph H. Cummins; third prize, desk, Smith Cox.

Hand engine—First prize, lamp, Nassau Recorder; second prize, lantern, J. S. Remsen & Co.

Ladder contest—First prize, comb, brush and platter, C. R. Lush; second prize, ink stand, George Roeckels; third prize, one box of cigars, Frank Blake; fourth prize, one box of cigars, Austin Nichols.

Chemical engine—Steel horse collars, made by Fabric Hose Company and presented by Charles L. Phipps.

Juvenile hook and ladder contest—Rug from John Wanamaker.

Portable hand engine—First prize, Brooklyn Times, clock.

Following are the officers of the association and the committees in charge of the tournament:

President, R. H. Mayland; first vice president, C. R. Lush; second vice president, G. H. Schiffmacher; secretary, F. S. Snedecor; treasurer, H. R. Smith; statistician, S. T. Wright.

Executive committee—T. A. McWhinney, O. H. Henckell, J. T. Hall, E. J. Gayzer, E. Weber.

Parade committee—T. A. McWhinney, Emil Zernike, F. W. Russell, C. Christensen, R. J. Wood.

Tournament committee—H. R. Smith, E. Weber, G. H. Schiffmacher.

Prize committee—W. Stoffel, J. Middleton, C. H. Schiffmacher, S. T. Wright.

Auditing committee—J. B. Paton, F. Lowden, T. Bennett.

Press committee—G. Haukner, E. B. Holland, E. P. Southard.

Reception committee—J. J. Campbell, sr.; J. L. Lawrence, W. Seaz, T. A. McWhinney, A. Englehard, S. Wood.

Badge committee—A. B. Wallace.

The officials of the parade and tournament are:

Judges—William H. Cornell, ex-Battalion Chief Brooklyn Fire Department; A. Robinson, M. D., of Bayville; H. H. Kane, M. D., of New York.

Archer B. Wallace,
Chairman Badge Committee.

Starters—Joseph H. Cummin of Long Island City; Joseph H. Cowning of Brooklyn.

Timers—William Harris of Brooklyn, Eugene Willetts of East Williston, George Roeckel of Foster's Meadow.

Official measurer—William B. Osterhout, C. E.

Hiram R. Smith,
Chairman Tournament Committee.

INVESTING IN MEADOW LANDS.

East Rockaway, L. I., June 26—A deed for a parcel of meadow land near Thurston's Creek has been placed on record. It is from Martin V. Wood and wife of Hempstead to Joseph S. Auerbach of Cedarhurst. The consideration is $3,250. The land is on the south side of Main street and runs to the bay. It is nearly all salt meadows, with frontages on two creeks which adjoin the Rural Farm.

A BOY HELD IN $2,000.

Flushing, L. I., June 26—Richard Hall, 16 years old, of Whitestone, was before Magistrate Connorton in the Flushing Police court yesterday on a complaint of Catharine Newman, who testified that on Friday evening she sent her 6 year old daughter, Wanetta, on an errand to the grocery, and that when a short distance from her home the girl was waylaid by young Hall. He was held in $2,000 bail for examination Thursday.

SIDE PATH BUILDING.

Bay Shore, L. I., June 26—The cycle path between Babylon and Bay Shore, which is to pass in front of the residence of Mrs. Ryan, on West Main street, is being rapidly constructed. Mrs. Ryan has taken an appeal to the appellate division of the Supreme court against the decision permitting the construction of the path in front of her residence.

AN ANTI-BAKER MOVEMENT.

Yaphank, L. I., June 26—A movement to remove Keeper Jonathan Baker of the Suffolk County Almshouse is said to be on foot but as most of the talk of removal comes from Democrats, Keeper Baker's friends say that nothing will come of the matter.

VIGILANT HOSE COMPANY OF FREEPORT, CHAMPIONS OF 1898 AND 1899.

The scale of this event was also reflected in descriptions of other ancillary aspects of the tournament. The Long Island Railroad provided special excursion trains at discounted rates, and for seemingly the first time more attention was spent on the track condition on the day of the contests. The *Eagle* wrote, "The roads committee has placed the roads upon which the parade will be held and the contests run in first class condition, and the expert opinions of the racing course is that it is the most level and finest ever offered the firemen…Every effort will be made to keep the course clear and a board fence has been erected alongside the course. All the 'dry' races will be run first and the 'wet' contests next." Regarding the "wet" contests, "In lieu of a water works system for the 'wet' contests and engine trials a high water tower and pressure tank has been placed in position. The tank has a capacity of 175 gallons and the guaranteed pressure will be 80 pounds. The steamer of the local engine company will keep the tank filled. In the engine contests requiring the throwing of water the contestants will get their supply from the big fire wells with which the village is supplied. The water supply will be more than up to the standard." This is another "first" for the sport that seems to have taken place at that tournament: an interest in and concern about maintaining a consistent water supply, an issue which still plagues modern competition today. (If hydrant pressure varies, and is not the same for each team as they each compete, each team's time would be affected unfairly.)

Other "firsts" at the Lindenhurst tournament that year involved the awarding of Second and Third Prizes for contests: "In several of the events three prizes are offered, something entirely new for the association"[clxxxviii] and Re-Runs: "In no contest shall any allowance be made for any accident, mistake or non-performance of the full requirements in any particular. But if any accident is caused by outside interference, or by the breaking down of anything furnished by the tournament committee, the company so hindered will be permitted another trial. Engines shall be allowed three minutes to replace a length of bursted hose."[clxxxix]

Another "first" contest took place at a different tournament on Long Island that year as well, at Woodhaven's annual gathering. "A feature of the contests this afternoon will be the Horse Race," wrote the *Brooklyn Daily Eagle*. "For the first time in a local tournament horses will be used in this race. The animals will be led to the tongues or shafts of the wagons and hitched and then run 300 yards. On the run the firemen will wear helmets and rubber coats. Time will be taken when the water strikes a backstop, the same as in ordinary hose contests."

Incidentally, Northport, Huntington, Cold Spring Harbor and Lindenhurst were the winners of the day.

Long Island and New York State notwithstanding, the sport was booming everywhere else as well. Regarding the traditional hand-pump "musters" that were still taking place throughout the rest of New England, Dixon reports that during the years 1900 – 1909, "History will show this decade to be the high point of the sport in regards to the number of musters held, and the number of engines competing. Not only was there an average of 35 musters per year, averaging 10 handtubs each, but there were 8 musters with 45 or more competitors, culminating with the largest muster ever held. It was on August 20th, 1908, at Lowell, Mass. When 63 handtubs vied for 20 prizes totaling $1,275.00."[cxc] It was *so* popular, apparently, that on May 30th of 1900 "A house in Arlington, RI burned to the ground while all the firemen were attending a muster in Natick, RI" and that on October 12th, "At a muster in Manchester, NH, a pumper for the Baw Beese of Gardner, Mass. dropped dead after pumping the first stream of that engine."[cxci] ("News accounts of this muster reported that it had the largest number of drunks of any muster held to date. Many of those involved attempted to cover up the fact by wearing ribbons reading 'STILL SOBER.' While this was a noble gesture, no one was fooled.")

And Firematic Competition Fever, apparently, wasn't just limited to the United States…although Americans were very, very good at it. On July 26th, 1900, the *Brooklyn Daily Eagle* wrote:

TO ASTONISH PARIS: Kansas City Team To Seek
Foreign Drill Honors in French Capital

"When the Anchor Line steamer Furnessia sails for Scotland on Saturday, she carries the team of firemen which will represent the United States in the Firemen's Tournament to be held in Paris August 11 – 12. The firemen, all members of the paid department of Kansas City, are at the Hotel Marlborough, Manhattan, in charge of Chief George C. Hale of the department, and accompanied by fourteen well to do citizens of Kansas City who hope to see them duplicate the triumphs of 1893, when they made the British firemen acknowledge themselves novices in the art of fire fighting.

"Chief Hale says his men will secure nearly all the honors at Paris. They are all athletes, and have been training for the tournament tests for a year and a half. They are taking with them a fire engine, some Pompier scaling ladders, jumping nets, etc. and three horses.

"It is with the horses that the Kansas City men expect to astonish their opponents. These have been trained to answer the bugle and will, at its call, run to the harness and take their places to the pole of the engine. They need no fire house to show off their accomplishments, having been taught to run in from the street to the bugler. They can also go through the regular fire drill in the engine house and Chief Hale has with him a skeleton house of the American model that can be erected in a short time.

"The cost of the expedition is being borne by the city, the Kansas City Common Council having voted $15,000 for the purpose.

"The team with the party accompanying it will land at Greenock, Scotland, and then take the train to London via Glasgow, thence to France. At the close of the tournament, Chief Hale said, he will give exhibitions of the ability of his men in London and other English cities. He expects to reach New York on the return trip about October 1st."[cxcii]

As the article indicated, trips abroad to show off firefighting skills were nothing new. Holzeman reports, "International exhibits of fire-fighting techniques began in Paris in 1867, and American companies from that year on could be counted upon to win more than their share of the prizes."[cxciii] In a specific reference to the above-mentioned story, he relates that during that particular visit the United States again did quite well: "At the International Fire Congress in Paris in 1900, for example, the Kansas City Fire Department team of fourteen men won over eight thousand contestants from twenty nations."[cxciv]

Back in the United States, on Long Island, "tournament fever" continued to manifest itself in many ways, as indicated by a newspaper article that told the tale of two fellows in Port Jefferson who were so impassioned that they actually came to blows in 1901 while arguing over a coming event:

"A fearful feud has been brought to a happy finish here and once more the official bass drum looks down upon empty benches...Once more the dahlias lift up their heads and look the town in the face. As of old, the sparrows twitter. Port Jefferson has taken off its coat and gone back to work, for the war between the Squire Squires and Captain Bayles is over.

"The prospective jurymen left hay fields. September corn, potato lots and other important spots, just to attend something that 'was all goldurn foolishness, b'gosh."...Still, while it lasted, the village enjoyed more excitement than it has any time since the 5:05 train came in at 4:06.

"It came at a meeting of the fire tournament committee, held the other night in the Engine House. For some reason, Captain Bayles has been opposed to the plan to hold the annual firemen's tournament here in September. What his reasons was nobody seemed to know yesterday. The tournament will be held any how next month and the citizens are glad.

"'Durned 'f I can see,' remarked one, whose face was partially concealed by a growth of whiskers, in inverse relation to the Kansas corn crop, 'durned 'f I can see why 'twas Cap. Bayles didn't want thet there tour'ment. Why, 'twould be the greatest thing in th' world f'r th' Port. Thet's what it would. Last year me an' m' wife; she an' I went to th' tour'ment over at Lindenhurst and b'gosh, I tell yer 'twuz worth goin' ter see. Wouldn't hev missed it fur a good deal. Didn't see nuthin' 'cause the dust wuz all so fired thick, but what we did see, I tell yer, wuz purty darn fine.'

"That represents public sentiment in the town regarding the coming firemen's tournament. And now the worst must be told. It has been kept back as long as possible, but it must come out at last. Squire Squires, the Postmaster, slapped Chief Bayles of the fire department across the face and the old brass drum on the back platform of the fire engine house, where the tournament committee met, fairly throbbed with agitation.

"'I tell you,' said Squire Squires, 'that the New London fire company is coming here as guests of the Greenport outfit. They will sail from New London across to Greenport and come to Port Jefferson with the company from that town.'

"'No, no,' contradicted Captain Bayles, while all paused to listen, with breath securely baited, 'they are not coming that way. Listen to me. The New London company is coming on its own hook by way of Bridgeport and the Port Jefferson steamer.'

"The whole room was wrapped in silence, painful and momentous. The rivals had met at last, and it was a weighty question over which they quarreled. Was the New London company coming by way of Greenport or was it not? That was the query. Even the legs of the piano shook.

"Quire Squires strode across the room. He approached the seat of Captain Bayles.

"'You're a _________ _________.'

"It doesn't sound nice, but that is what he said. Captain Bayles retaliated by going the Postmaster one better and saying damn real loud. Then Squire Squires struck him in the face, while a shudder ran through the throng. Sherlock Holmes would have called it "an exclamation of horror.'

"'I'm done,' remarked Captain Bayles as he left the room. 'I'm through with fire tournaments.'

"The next day he appeared before Justice Wheeler and swore out a warrant for Squires' arrest, charging him with assault."[cxcv]

The Eagle's description of the SCVFA's 16th Annual Tournament that did take place in Port Jefferson that year never did describe which way the New London crew arrived, but it did say that the Long Island Sound boats ran special excursions that day, and, much like the Hempstead and Rockaway tournaments of 1898, "...There has been a steady stream of bicyclists from the south shore across the cycle path from Patchogue and Squire Norton of Wheelman's Rest of Selden says that 'B'gosh, it is the biggest crowd of folks that have crossed that ere path from "Pat" to "Port" in many a day.'"[cxcvi]

In yet another "first", it appears that the phenomenon of firematic tournaments was responsible for the advent of a modern technology that revolutionized the way people communicated and received their news in that part of the world. Regarding the Port Jefferson tournament, the *Port Jefferson Echo* wrote: "The *Brooklyn Times* was on sale here at 6:45 p.m., containing seven columns of halftone cuts and reading matter, the copy for which was wired. It was a feat in journalism never before equaled on Long Island."[cxcvii] They also went on to say, "The perfect order maintained and the absence of disorderly conduct of any kind was generally remarked."

The latter is worthy of mention only insofar as to illustrate one point: in spite of how large and popular the tournaments were at the time – and the Port Jefferson feud notwithstanding - the *gestalt* of "the gay 90's" seems to have continued on past the turn of the century, with both the firefighters and the public in an overall good mood, where politeness and "good fun" prevailed. This can be seen in descriptions of other

THE BROOKLYN DAILY EAGLE. NEW YORK, WEDNESDAY, SEPTEMBER 11, 1901.

SUFFOLK COUNTY FIREMEN'S ANNUAL TOURNAMENT.

(Special to the Eagle.)

PORT JEFFERSON, L. I., September 11—That tournament day of 1901 will live in the memory of many Long Islanders is evident from the thousands of visitors who are to-day crowding the streets and byways of this picturesque harbor town. It is the sixteenth annual affair of Suffolk County Volunteer Firemen's Association and that it is far ahead of previous years' tournaments, is the universal verdict. Many of the visitors are making their first visit to the village and they have been extended the freedom of the town.

It is estimated that there are at least 1,000 firemen, who in their vari-colored uniforms, intermingled with the bright dresses of the hundreds of women admirers on the field, form a pretty picture.

The great throng began to arrive early, special trains bringing the visitors from all sections, many of the northsiders coming by boat and those from nearby villages driving in. There has been a steady stream of bicyclists from the south shore across the cycle path from Patchogue and Squire Norton of Wheelmen's Rest at Selden says that, "Begosh, it is the biggest crowd of folks that have crossed that ere path from 'Pat.' to 'Port' in many a day."

That Port Jefferson is the "Rome" to which all roads lead to-day is evidenced in the cavalcade of rigs of all sorts and descriptions that have been pouring into town since early morning. The middle island farmers were up before sunrise this morning and are on hand with watermelons and a reserved site along the road, with their families, prepared to fully enjoy the big day.

The Long Island Railroad special trains arrived in good season and brought their freight of firemen in time for the parade, which was a little late in starting. The line of march and formation of companies, as already given in the Eagle, was followed, the Lindenhurst department, who entertained the association last year, being the guests of honor to-day and leading the procession. The Port Jefferson department brought up the rear. The line of march is profusely decorated, every business place and residence being covered with flags and bunting.

Among the visiting delegations are the mayors of New London and Bridgeport, with 100 firemen from New London and a large delegation from Bridgeport. The Connecticut lads were roundly cheered in the parade, as were the different companies encouraged by the plaudits of their followers along the line.

The parade started at 12:15. The right of line was at the old well on Main street, passing down Main street to Hotel Square, and then down Jones street to the harbor, where a countermarch was made back to Hotel Square, to Main street, to Water street, to the judges' stand on the beach, up Barnum avenue to Main street and then disbanded. The line of firemen as it passed the judges' stand was as follows:

Grand marshal, John H. Lawrence. Aides, Dr. T. J. Carrier, Robert Lawrence, Charles V. Platt, Wallace Adams, on horseback.
Wheeler & Wilson Brass Band, twenty pieces.
Officers of County Association: President, W. K. Post; treasurer, A. L. Field; secretary, J. F. Flugrath.
Band.
Sag Harbor Hook and Ladder.
Gazelle Hose of Sag Harbor.
Sag Harbor Juvenile.
Band.
Huntington Hose.
Huntington Hook and Ladder.
Protection Engine, Huntington.
Huntington Juvenile Hook and Ladder.
Babylon Hook and Ladder.
Argyle Hose, Babylon.
Storm Engine, Babylon.
Phoenix Hose, Babylon.
Babylon Juvenile Hook and Ladder.
Center Moriches Hook and Ladder.
Patchogue Engine and Exempt Firemen's Association, with Old Honey Bee.
Union Hook and Ladder, Patchogue.
Vanguard Hose, Patchogue.
Euclid Hose, Patchogue.
Konomoc Hose, New London, Conn., 60 men, led by Chief Stammis.
Delegation Port Jefferson Department as escorts.
Band.
Lindenhurst Department, Union Hook and Ladder, 30 men.
Liberty Hose.
Band.
Bellport Hook and Ladder Company.
Band.
Northport Hose No. 1 and Hook and Ladder No. 1
Centerport Hook and Ladder.
Blue Point Hook and Ladder.
Sayville Hook and Ladder.
Resolute Hose, Sayville.
Sayville Hose No. 1.
Sayville Cornet Band.
Bohemia Hook and Ladder.
Band.
Amityville Hook and Ladder Company.
Amityville Hose No. 1.

The parade was concluded at 1 o'clock, after which the juvenile contests were started.

Huntington engine was decorated with a large picture of President McKinley that was loudly cheered.

Among the conspicuous figures in line was A. A. Fraser of Oakdale, one of New York's early firefighters, and L. H. Clowes of Hempstead, foreman of his local hook and ladder company, who is completing fifty years of fire duty.

The big grand stand, seating 2,500, that lines the course on both sides was filled when the juvenile hook and ladder companies began their contest at 1 o'clock. The greater part of the stand is covered and the price of seats does not go above a dollar. The roads over which the parade passed and the contests are to be run are in excellent condition and fast time is assured in the races. The firemen's enclosure, alongside the grandstand, which insures a clear track, is a new and most agreeable feature.

The principal interest is centered in the hook and ladder contest, number one, and also hose contest, number one. In the former the association trophy and in the latter the Massapequa cup are held by Northport Hook and Ladder and Hose Companies, and every effort will be made to retain it, while equally as strong will be the endeavors of their good natured rivals to wrest it from them. Much rivalry also exists in all the other contests. The appearance of old Honey Bee engine in the engine contest is looked for as being the impetus for record smashing. Honey Bee has a record of 186 feet in contest, but has beaten that by over ten feet in trial practice. All the engines will compete to-day on a general scale, no limit being placed on size of cylinder. Chief Simpson and former Chief Edwin Bailey have been unable to place a single bet on Honey Bee's throw of water.

The reception and entertainment committees have their headquarters at the Townsend House, where the sixty special guests of the local department, including the officials of the day, officials of the association and the officials of Suffolk County, were dined and wined at noon. Press headquarters was at the rooms of the Port Jefferson Social Club.

The members of the Port Jefferson Fire Department who act as hosts to-day are as follows:

Hook and Ladder Company—John E. Overton, A. T. Norton, Joel J. Schryver, J. H. Davis, W. P. Leek, H. T. Bayles, Virgil Ritch, Albert Walker, S. Dreyer, W. H. Tooker, Clarence Lee, O. B. Davis, Holmes S. Lee, John W. Brown, Charles A. Bayles, John I. Miller, John H. Lawrence, Isaac E. Jones, Charles A. Conk, Ebenezer Gildersleeve, Louis Koegel, Lewis H. Brewster, Charles S. Brewster, Charles Terrell, Robert Walker, Benjamin Jarvis, L. B. Homan, M. V. B. Brewster, Alonzo Shute, W. H. Craven, Adolph Tillotson, James Ivens, Olcot O. Davis, H. M. Randall, James E. Bayles, George Brice, James A. Smith, William H. West, W. E. Ryder, Charles A. Squires, Frederick A. Hall, Nelson W. Davis, Joseph Colsh, J. H. Kinner, William H. Kinner, Lester U. Davis, L. A. Tuthill, Louis A. Moger, Charles V. Platt, Robert Dayton, Charles D. Kinner, David Fullwood, Daniel Wheeler, Milton D. Hall, Judson L. Darling, John T. Mather, Charles E. Hulse, Frederick E. Raynor, Frank A. Kline, Ralph B. Dayton.

Rescue Hose Company—Henry West, A. N. Randall, W. H. Robins, A. F. Davis, jr.; Joseph Kinner, Charles E. Tooker, F. B. Robinson, John E. Overton, Herman Floyd, Arthur Parks, William Pope, Thomas Nelson, Robert Lawrence, B. H. Reeve, Frank Robbins, Lewis O. Davis, Howard Hounslen, Frost Hawkins, Frank Darling, Myron E. Overton, Robert Conk, George Brewster, A. T. Norton, jr.; Wilson Ritch, John Gaffga, Walter Scott, Eugene Smith, George Tobiason, Wallace Adams, C. E. Ivens, George Odell, Ralph Wheeler, Robert Brown, Glendon Brown, Albert Cover, Bryant Harned, Bert Dewsick, Harold Dewsick, Charles Dreyer, Daniel Terrell, Gilbert Davis, William Sypher, jr.; Albert Dayton, Charles Clark, Ernest Culver, Charles S. Marsh, John Orth, Henry Jarvis, Eugene Terrell, Clifford Griswold, George Blydenburg, Leslie Homan, William Atkins, David Adee, Ray Dickerson, Ray Randall, Arthur Adams, Elbert Gaffga, George Gaffga, Louis G. Hulse, C. E. Silbon, Byron Leek.

Birdsall, Tarrytown, N. Y.; Cornelius A. Christie, Jersey City; C. D. Chapman, Jamaica, L. I.; John N. Frazer, East Islip; John C. Merritt.

a general grocery store here, yesterday made an assignment to Henry W. Arthur, repre-

OBITUARY—LONG ISLAND.

COURT OF SPECIAL SESSIONS.

Committee, presided. He attacked the present city administration and charged the

tournaments across the state as well, including the 2nd Annual Western Convention that was held in Albion at the end of July in 1901. In spite of having over 5,000 visitors in town to watch over 1,000 firemen march and compete, "…the crowds were good-natured and peaceable. Only a few arrests were made, and those were all sharks and swindlers who came here from the cities with the expectation of reaping profits by thievery."[cxcviii] The *Orleans American* wrote, "A.V. Lord, chief of police, was a busy man during the convention, especially on the last day. He kept order. The public have confidence in his ability and the pickpockets and other toughs on the street had confidence in his club – when he used it."[cxcix] However, "…the usual hilarity and demoralization

which is inseparable from such an event was witnessed on the streets in the evening 'after the ball was over,' but there was less of it than would naturally be expected with the multitude of strangers who flocked here."[cc] In light of these descriptions it would seem that public drunkenness – on the part of either the public or the firefighters themselves – was more a source of amusement at the time than it was a cause for concern or complaint. This fact is further indicated by a story in the North Tonawanda *Evening News* about the Western NY convention that took place there in 1902, which included the following: "A race not on the program occurred during one of the waits at the track yesterday afternoon and furnished much amusement for the crowd. Three men who were feeling good inconsequence of a liberal amount of liquid refreshments started a race of their own. One of them started to fall about 25 yards from the point where he finally plowed the ground with his nose. He kept leaning farther and farther forward until when about parallel with the ground he made a final lunge, dug his head into the track and doubled up like a jackknife." As liberal as the attitude towards intoxication was then, however, the coming years would bring more and more problems associated with it, to the point where major changes in the sport would be made as a direct result.

For the time being, however, everyone was happy, and the tournaments just kept getting bigger and bigger. 1902 saw the SCVFA celebrate their 17th Annual Convention and Tournament in Riverhead, and it was made all the more grand as a result of it being held in conjunction with the Suffolk County Agricultural Society's 50th Anniversary Fair. Nicknamed the "Firemen and Farmers" event, the event was held at the fairgrounds, with running contests on the racetrack and hand- and steam engine competitions taking place simultaneously in the infield, which was also used for the various team camps: "A novelty for the firemen this year is the camp inclosure (sic) in the center of the track, many of the companies erecting tents for

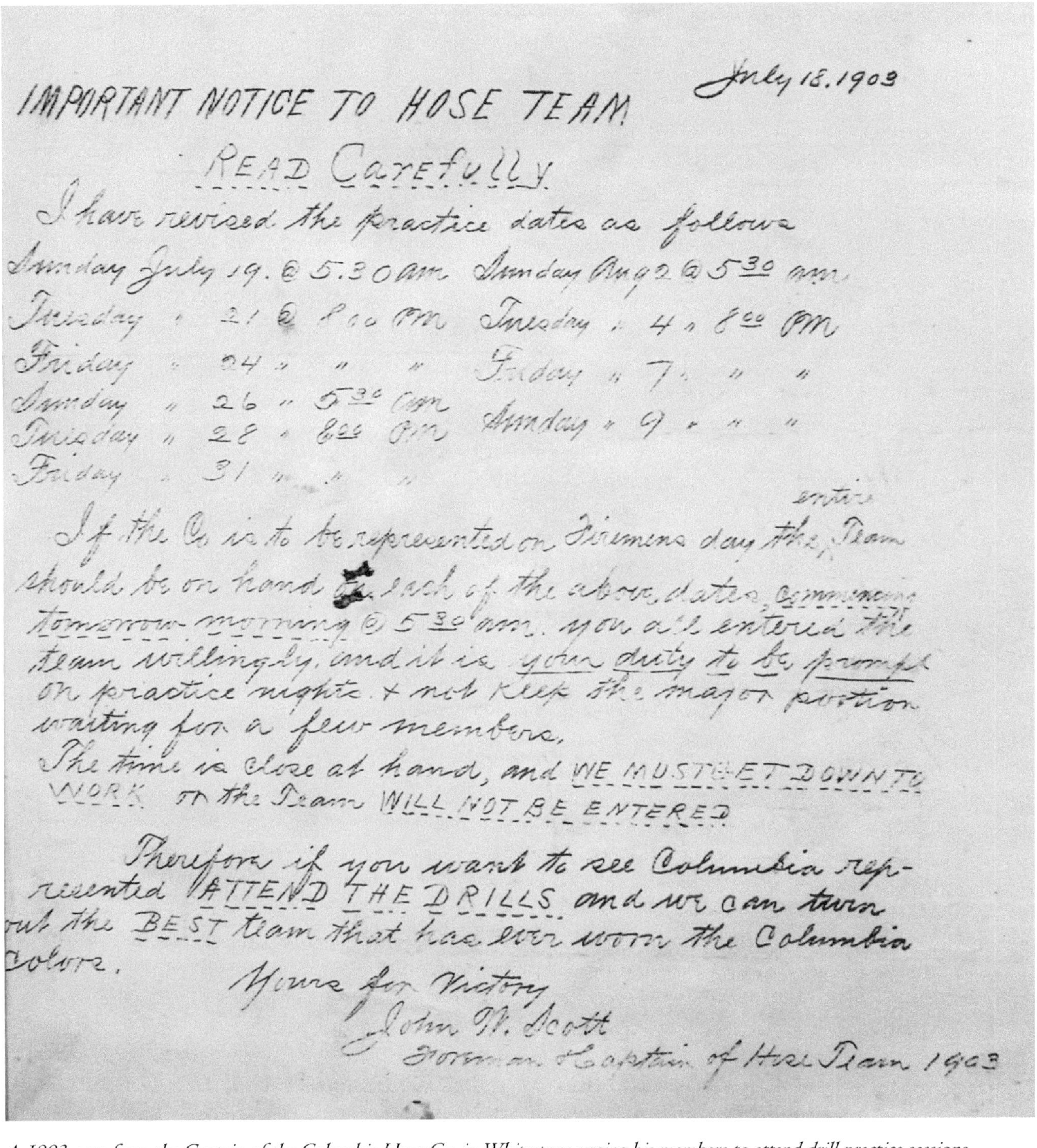
IMPORTANT NOTICE TO HOSE TEAM July 18. 1903

READ Carefully

I have revised the practice dates as follows

Sunday July 19. @ 5.30 am Sunday Aug 2 @ 5 30 am
Tuesday " 21 @ 8 00 PM Tuesday " 4 " 8 00 PM
Friday " 24 " " " Friday " 7 " " "
Sunday " 26 " 5 30 am Sunday " 9 " " "
Tuesday " 28 " 8 00 PM
Friday " 31 " " "

If the Co is to be represented on Firemens day the entire Team should be on hand each of the above dates, commencing tomorrow morning @ 5 30 am. you all entered the team willingly, and it is your duty to be prompt on practice nights & not keep the major portion waiting for a few members.
The time is close at hand, and WE MUST GET DOWN TO WORK or the Team WILL NOT BE ENTERED

Therefore if you want to see Columbia represented ATTEND THE DRILLS and we can turn out the BEST team that has ever worn the Columbia colors.

Yours for Victory
John N. Scott
Foreman Captain of Hose Team 1903

A 1903 note from the Captain of the Columbia Hose Co. in Whitestone urging his members to attend drill practice sessions

headquarters, and the field has taken on a military appearance with the white canvas tents scattered around."[cci] The *Riverhead News* wrote: "The roomy fair grounds, with all the necessary paraphernalia for such an event, was one of the pleasing features of the affair for the firemen. Most of them unhesitatingly said that they had never attended a tournament with more comfort. The contests were exceedingly well managed, and run off with a dispatch that was highly gratifying to the spectators. The track was in excellent condition, and being broad and well fenced in to keep the crowd from interfering with the contestants, provided a track without equal in

The 18th Annual SCVFA Parade and Tournament at Patchogue in 1903

any village. This, too, came in for its share of the general approval of the firemen."[ccii] They went on to say, "The 'croakers' who some months ago predicted that the tournament would be a flat failure, if held under the conditions that it was, must be sorely disappointed at the utter failure of this prophesy."

According to Flugrath, however, the day was not completely without some flukes:

"The only disappointment was the failure of the improvised water supply in the wet contests, cauased by some one tampering with the pipes. In both dry hook-and-ladder events, Sag Harbor met with mishaps. In one of the events the shackles fastening the traces to the truck broke and the runners sped down the whole stretch of track before they discovered they did not have the truck with them. Their feelings were not improved when the band struck up, 'Aint 'Dat a Measly Shame.'"[cciii]

It seems that in these first few years following the turn of the century, the sport of firematic competition began to reach its peak of popularity. One indication of this, in addition to all of the media coverage and descriptions, was the appearance of various accounts of different contests in the form of poetry, as evidenced by this verse that appeared in an unnamed Long Island newspaper (possibly the *Flushing Daily Times*) in 1903:

HOW IT HAPPENED

Columbia's Bard Tells The Story
As Follows:

The day is passed and gone for good,
'Tis history now, I know;
But none the less, we did our best
And were beaten through and through.

Right on the start, we broke our heart
In the first race that we run
The Engine fellars laughed and grinned
Perhaps they thought it fun.

The way it happened, I will tell you
All in this short space.
A pretty bird was asked to be
The Starter in the race.

A bird he was, a bird for fair
Of all the birds I've seen
I never heard of one before
That could beat the likes of him.

They put a flag in his right hand
And told him what to say
Then placed him down the avenue wide
To start us on the way.

And then again the judges all
Told him just what to do
To start us off, then drop the flag
For the timers three, you know.

Our cart all ready on the scratch
And team strung out in line
All ready for the word to catch
And do the trick up fine.

The bird then chirped the starting words:
"Are you ready," "Get Ready," "Set," "Go,"
We rose as if we all were one
And down that street we flew.

Faster, faster grew the pace
Each step grew faster still
Big Kefer hauled off all the hose
Spitz slapped her on to fill.

Then Ike swung 'round that wrench of his
As only I.B. can
From out the hydrant, through the hose
The water then it ran.

Meanwhile, both Cooley and O'Rourke
Had been busy too a bit
The coupling broke and nozzle on
The crowd near had a fit.

But came the water from the spout
In seconds seventeen
We proved that day we realized
Our gallant captain's dream.

Now just imagine how we felt
After making such a run
We almost hugged ourselves with joy
So good we felt, by gum.

But ah, alas, our joy was soon
Replaced by such a pain
For the bird forgot to drop the flag
And we had to run again.

Then all disheartened, we replaced
The hose upon our cart
And took her down the road again
To make another start.

Now every man was out of wind
And as nervous as could be
O'Rourke was shaking like a leaf
And so was Ed. Cool-ey.

Our confidence was on the wane
And speed had lost its edge
But duty said to try again
And try again we did.

You know the rest, you saw it all
Just how the fluke was made
How Spitzy got a nasty fall
And Rooky crossed the thread.

But never mind, it's proud we are
Although we did not win
You'll find us on the scratch next year
To try the thing again.

Chapter 9: Enter the Auto

As the sport of firematic competition reached its zenith, however, so too did it begin to show signs of getting a bit worn around the edges, most noticeably in things having to do with the parades. At the Queens Borough Volunteer Firemen's Parade and Tournament in Jamaica on September 1st, 1902, although there were anywhere from 12,000 to 15,000 onlookers along the line of march, it was a *six mile* march, which caused no small amount of grumbling among the firemen. Two weeks later, on the 16th, when the SCVFA held their tournament in Riverhead, the line of march was thankfully made shorter: "The parade route was very short, and that fact was appreciated by the firemen, who have in the past been taken for miles around the villages for no apparent purpose other than to tire them out."[cciv] But although the event was an overall success, it seems that the planning of it was not without some controversy. The "croakers" that the *Riverhead News* article referred to as predicting the event to be a failure were probably those teams that weren't especially keen on sharing their convention with the Agricultural Society. According to the *Brooklyn Daily Eagle*, Babylon voted to attend, but "...not a great deal of enthusiasm was manifested", Port Jefferson did not attend because they were having a new truck built, and Lindenhurst chose not to attend at all. "The firemen here were very much opposed to holding the tournament at the fair grounds and fought the proposition vigorously when it was discussed at Brentwood in June."[ccv] And, in spite of the fact that "...for the first time in the history of the [SCV] firemen's association the tournament is under its own exclusive management, and its executive committee, ably captained by President W. Kintzing Post of Bayport, has labored faithfully to fulfill the manifold duties that such an event involves," there were problems with the water during the Hose Contest #2: "In this contest the water pressure seemed to be very faulty, and several companies contested, but their time was not taken."[ccvi]

The Suffolk firemen didn't know how well they had it, seeing as they were probably more well-organized than any of the other firematic associations around; other tournaments suffered worse fates. The tournaments were now becoming perhaps a bit *too* big, a bit too unwieldy to control. It is true that these same accusations had been made before, as early as the 1890's, but now more and more references to problems at the

The Montauk Hose Co. tournament team of Sag Harbor, 1903

events started to appear. The same newspaper article that included the poem at the end of the last chapter also stated that that same tournament[5] "...was not finished because it was planned on too big a scale":

"The firemen's celebration yesterday, which began most auspiciously, terminated unsatisfactorily. It was not the fault of committee or individuals, other than that they planned an affair of such magnitude that it was impossible to handle it. Everything was on an immense scale. The trouble was caused primarily by the parade. It was so large, required so much time to get under way, and the line of march so long that it consumed the best part of the day, leaving little time for the contests, of which there were more than enough for a whole day's sport without anything else. As a result not one of the contests was finished, a circumstance that gave rise to a great deal of dissatisfaction.

"The contests were begun at 4:20, more than two hours late, and at 6:40 the tournament was declared off and postponed until next Saturday afternoon. A great deal of dissatisfaction was expressed. Many of the spectators and some of the contestants had come long distances to witness and take part in the tournament, and to thus see everything end in a 'fizzle' was a source of regret. Very little system was evident. The start was delayed a long time on account of the loss of a hose wrench which somebody had neglected to get until the last minute.

"Very little provision had been made for the enormous crowd of people who were present. The crowd opposite the grand stand was halfway out across the course most of the time. It required all of the efforts of the mounted police to keep the crowd back."[ccvii]

Either in spite of, or as a result of all this, other changes continued to be made in the sport. At the 18th Annual SCVFA Drill held in Patchogue on September 9th, 1903, the first hose contest for juniors was held. As

[5] Because this information was gathered from a newspaper clipping that had been cut and pasted into a scrapbook and was missing several key parts (such as a headline), it is unknown which tournament this actually was.

Firemen march to the train station in Patchogue in preparation to board a train to a tournament in 1904

described by Flugrath, "The first hose event for the juniors was held at the Patchogue tournament of 1903. Greenport had formed two juvenile hose companies, the Red Birds and the Stars, and it was on the suggestion of Henry A. Reeves, of that department, that the contest was placed on the program. The Greenport lads were the only contestants at Patchogue. The Stars, which scored 23-1/5 seconds, lacked three full threads on their nozzle[6] and were ruled out, and the prize was given to the Red Birds, whose time was 23-3/5 seconds."[ccviii] Two years later, at the SCVFA's 20th tournament, "...The largest turnout of juveniles in the parade was in Babylon on September 13th, 1905, when 11 companies of 195 boys were in line."[ccix] A month or so after the 1903 tournament, on October 14th, 1903, the SCVFA voted to join the Southern New York Firefighters Association, and that same year the firefighters of newly-formed Nassau County joined together to form the Nassau County Firemen's Association, which still holds its own tournament each year to this day.

In contrast, and rather inexplicably, records show that the firemen's association that would later become the Firemen's Association of Northern New York held no conventions at all for the years 1903 through 1905. No indication could be found as to why this was the case.

What is much more significant, however, is that the year 1903 also saw what would eventually become by far the largest change ever to affect the sport of firematic competition: the introduction of the Motor Car to the fire service. On October 3rd, 1903, Niagara Engine Co. #1 of the New London Fire Department (in Connecticut) took delivery of the first motorized fire truck,[ccx] while the New York City Fire Department

[6] At the time, one of the rules in hose contests required that the nozzle be attached to the coupling with at least three full turns, or "threads", for the run to be considered valid.

Advertisements from the Official Program of the SCVFA's 20th Anniversary tournament in Babylon, September 13th, 1905

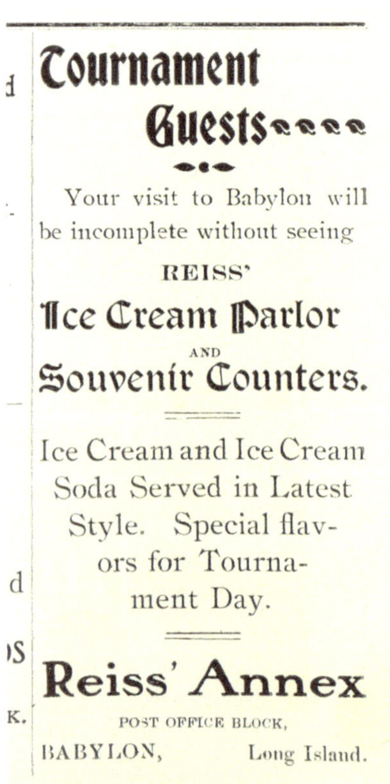

followed suit in 1906 with the order of a four cylinder Know-Springfield hose cart.[ccxi] Although it would be a few years before motorized apparatus would appear at firematic tournaments, the seeds of change had been sown.

North and west of Long Island, things were moving along as well. An interesting item was found in the archives of the *Niagara Falls Gazette*, regarding the parade held during the Western New York firemen's convention in 1905, involving what seems to have been an off-again/on-again relationship between American and Canadian firefighters, who would often march and compete together when conventions were held there:

"As the Canadian contingent passed up Falls Street many were led to remark that the feeling between the firemen of the two sides of the river was different than was the case several years ago. At the time, there was a decided break and relations were severed as a result of an incident at Hamilton. The old Belleview company under Gayton Swan went to Hamilton, Ont., to participate in a parade. While marching down the street a policeman ran out and informed Foreman Swan that he could not carry the American Flag through the streets. The flag was furled, the company fell out of line, their apparatus was immediately loaded on a car and the company lost no time in returning home. There was much bitterness as a result and matters were not changed until the local soldiers departed from the city at the time of the Spanish-American war when the firemen from Niagara Falls, Ont., came across the bridge to give the boys in blue a royal send-off. Since that day the best of feeling has been maintained between the firefighters of both sides of the river."[ccxii]

The Western Convention of 1906, held in Brockport, was held in equal esteem, with *The Brockport Republic* reporting an estimated crowd of over 15,000 attending.

Farther to the south and east, 1906 saw the firefighters of Westchester County, just north of New York City, form their own association. As a page from their written history describes, "…in the fall of 1905, upon returning from the Annual Convention of the Volunteer Firemen's Association of the State of New York, held in August in Glen Falls, N.Y., the delegates of the departments of this area, being disappointed in the lack of recognition of the Westchester County Fire Departments in the election to office in the Executive Staff of the State Association, met in Harrison at the invitation of Assistant Chief Frank Coxe. This meeting laid the ground work for further meetings, which culminated in an organizational meeting held in July 1906, in the then village of White Plains. At this meeting of 79 former delegates to the State Convention, a Constitution and By-Laws were adopted, and the first election of officers held." The Westchester association, from that point on, held an Annual Field Day, Firemanic Tournament and Family Picnic, for its members and their families.

As during those days there were more and more firemen's associations being formed as time went on, the formation of one more local association may not seem like any big deal, but what is noteworthy about the formation of the Westchester Firemen's Association in particular is the circumstances under which it was formed, as described above. Many years later, when the organization and running of firematic tournaments on the state level by FASNY was taken over by the New York State Volunteer Firemen's Parade and Drill Team Captain's Association, participants from Westchester County – and indeed the entire lower Hudson Valley – in the new governing association were strangely absent, a trend which continues even to this day. Although it is complete speculation to suggest that their lack of participation in modern competition is a holdover from a grudge held back in those days, it's an interesting possibility.

As far as the FASNY conventions and tournaments were concerned, in spite of the "sour grapes" feelings that were felt by the Westchester teams, by 1908 even more delegates and teams from Long Island were starting to attend, albeit mostly those from the western part of the island. At their 36th Annual Convention in Oswego, groups from Sea Cliff, Jamaica, Flushing, Far Rockaway, Maspeth, Long Island City, Freeport, Rockville Center and Woodmere made the trip north, but they still represented only a tiny fraction of the FASNY membership and attendees, which continued to grow each year. Even so, in spite of *that* fact, and also as a result of the fact that firematic tournaments were at the zenith of their popularity, it is almost humorous by now to note that the leaders of FASNY appeared to once again be struggling with the issue of the "worthiness" of parades and tournaments at their conventions. It seems that the problems that Long Island's firemen had had to deal with having to do with hosting Bigger and Better tournaments had now reached the other parts of the state, as indicated by an entry in the Secretary's report of the minutes of their convention in Poughkeepsie in 1909:

Topic No. 11:
SHOULD WE DISPENSE WITH THE PARADE, TOURNAMENT, ETC.

> There is a feeling among some of the members of the State Firemen's Association, that at our annual conventions the tournament and parade features should be eliminated, and nothing but the business meeting held.
>
> In my opinion it would not be a wise course to pursue. The holding of a tournament and parade is entirely in the hands of the local committee of the city taking the convention, and it is a

An unnamed tournament in Greenport, September 17th, 1907

matter over which we should have no control. All that we could do would be to refuse to hold a convention in that city that would eliminate the tournament and parade features.

What has started this feeling is that for the last five years cities have not been bidding very actively for the convention on account of the excessive cost of holding it. Each city tries to out-do the former one holding the convention, with the result that every year the cost of entertainment has increased. We could cut down the item of badges in the cost to the local committee, a matter of about $1000.00, by adopting and furnishing each member of the Association a badge of suitable size and design, such as the Association has repeatedly been asked to do, and so save that cost to the city taking the convention.

Since the inception of the State Association and the holding of the first convention, in 1872, the annual conventions of this Association have been looked forward to by the volunteer firemen throughout the state with enthusiasm. Those who remember the earlier conventions of the Association as compared with the conventions of recent years must admit they were pretty tame affairs.

The tournament and parade features are what have built the Association up. The very object of the Association is for the purpose of improvement, by exchange of views, and stimulating competition among the companies of the State.

The desire to excel has created a rivalry among the companies of the State in the matter of apparatus, uniforms, drilling, etc., and so created an interest in the annual conventions, and by doing so they have joined the State Association, and taken an active interest in its conventions.

- Let us hope the day will never come when the present program will be laid aside for a mere business meeting of delegates.

Wm. Fronfield, Life Member

On the old-fashioned Handtub Muster side of things, that branch of the sport was enjoying an unequalled popularity during that time as well…however more dangerous it appeared to be compared to the more modern tournaments. Dixon writes that the largest handtub muster ever held took place in Lowell, Mass. on August 20th, 1908: "Anticipating a large turnout of participants accompanying the 63 handtubs, along with an equal number of spectators, the muster sponsors encouraged the American Red Cross to provide a medical tent on the grounds, staffed by four doctors and two orderlies. Twenty-two cases were attended to throughout the day, which included an amputated finger, a fractured ankle, fractured ribs, one case of angina, several instances of heat exhaustion, a few lacerations, and one man who stubbed his toe on a tent rope, fell to the ground and shoved a tent peg up his nose, causing a severe fracture."[ccxiii]

Back in the modern-day world, references to that newfangled invention called the "automobile" start to appear a bit more in historical references. A July 30th, 1908 newspaper account in the *Orleans American* of the Western New York convention in Albion includes the paragraph that states "The automobile trip starting from the Orleans House at 7:30 o'clock, around the village pleased every delegate, and they had an opportunity of seeing Albion at its best. The red fire illumination was a fine feature and added much to the trip."[ccxiv] At their convention the following year in Dansville, the *Dansville Express* reported that on the second day of the convention, "…before six o'clock loads of people came in town from all directions, some in rigs and others in autos…"[ccxv] and at the 1909 St. Lawrence and Jefferson County Firemen's Association convention in Potsdam, the local paper wrote "…No teams or automobiles will be admitted to the Fair Grounds."[ccxvi]

Of much greater import, however, is what might be one of the very first mentions of an automobile being involved in a firematic parade. At that same Potsdam convention, in its report describing the order of march of the parade on the second day, the *Potsdam Herald-Reporter* wrote, "…Potsdam Band and Fire Department; the Fire Team driven by Arthur Jenner and resplendent in decorations of red, white and blue; the Fire Engine; then autos, hacks, private carriages, etc."[ccxvii]

Ditzel makes a brief reference to a novel contest that took place in 1909 in Birmingham, Alabama between a horse-drawn steamer and a motorized Seagrave apparatus, although he doesn't say who won. For some reason steamers were not all that prevalent on Long Island; whereas there are numerous photographs and illustrations of them responding in the larger cities, such as New York, Flugrath writes, "Steam engines have been few in Suffolk County. Up to 1907 there were but five in the county, viz., Greenport, two; Riverhead, Shelter Island Heights and Lindenhurst, one each. All but Shelter Island Heights had been in competition at the various tournaments up to 1910, the last actual contest for these engines including Torrent, of Greenport, and Washington, of Riverhead, at the Sayville tournament that year."[ccxviii]

The newspaper account of the Western convention in Dansville in 1909 provides some interesting, if not humorous, reading. It was reported that during the opening addresses, a B.G. Foss, Esq., in extending the Mayor's welcome to the delegates, stated that "...when the committee met to arrange for [the convention] coming here a gentleman from a nearby town told what terrible things confronted them if they came to Dansville, no railroads to get in, they could not get out because of the hills that surrounded the village, and he advised his friends to 'bid goodbye to God, as we are going to Dansville.' Mr. Foss said it was safe to predict that when the delegates come to go home a spirit that comes from the heart and soul

The Protection Hose Co. of Hempstead, 1909

will prompt them to express another sentiment, namely: Thank God that we went to Dansville!"[ccxix] Elsewhere in the story is a description of the banquet that was held for the delegates that same evening: "A banquet was given to the delegates in the Heckman opera house Tuesday night and about 350 sat down to the long and well filled tables that had been arranged and enjoyed the splendid viands. Mrs. A.B. Lindsay had charge of the affair and everything passed off delightfully. While enjoying the cigars a novelty was introduced that afforded amusement for all. Prof. Max Huber the hypnotist gave an exhibition of his wonderful powers; G. Frank Spencer and F.G. Austin gave character sketches in costume, Herbert Kidd made some wise sayings that took the crowd. Mr. Maybee sang and a couple of young fellows did a few stunts in fancy clog dancing. The visitors pronounced the banquet and entertainment very satisfying."[ccxx]

As was usual for tournaments in that part of the state, the holding – and reporting – of the firematic contests was not of as much importance as it was on Long Island. The usual contests were held: Prize Drill, Hub-and-Hub Race, Hose Race, Hook and Ladder Race and a Tug-of-War, but not much was written about them except to say that "...the crowd came on Main Street and filled it completely, the crowd swaying back and forth, the bands playing, and the funny fellows marching up and down and making hideous noises. It was an orderly lot of persons and we have yet to hear of any disturbance or accident, all of which speaks well for the crowd and for the excellent manner in which the arrangements were carried out."[ccxxi]

Even if it is true that the Northern firemen were the very first to include an automobile in a parade, the crews from Long Island were still the vanguard of the sport of firematic competition. As the first decade of the new millennium saw advances in the science of firefighting, in the form of new methods that could be used to extinguish fires by using a combination of chemicals instead of water, it should not be surprising that in 1909 a new firematic contest, called the Chemical Contest, was introduced at the SCVFA tournament in Southampton. That tournament itself was noteworthy mainly for that fact, but history also notes that during the day "…a cloudburst swept that section. The storm caused a freight wreck on the Long Island Railroad that delayed the leaving of the special excursion trains for several hours."[ccxxii]

The quote goes to illustrate that as necessary as the railroads were to the sport of firematic competition, they were not infallible. There's a humorous anecdote from Dixon regarding the aftermath of a New England muster on September 20th, 1910: "The most reliable means of transportation at this time was the train. But even they were prone to occasional miscues. While the crew of the Hancock V.F.A. of Brockton, MA anxiously waited at the train depot in Rochester, NH for the arrival of their engine, they were notified that it had been shipped to Rochester, NY."[ccxxiii]

Railroad mishaps notwithstanding, it seems that by this time, albeit slowly, the disparities in style and popularity between Long Island events and those that were being held in other parts of New York State were starting to diminish. Of the three major areas – Northern New York, Western New York, and Long Island – Northern events were still quite provincial in nature, but the Western galas were gradually catching up to those on Long Island in terms of their size and sophistication. The Western convention in Lockport on July 26th - 28th, 1910, is a good example, with the *Lockport Union-Sun* devoting practically its entire front page of two issues to coverage of the event.

"Recipients of the most kingly reception in the history of the association of Western New York Volunteer Firemen," they wrote, "the fire laddies yesterday showed their appreciation of the city of Lockport's grand entertainment by presenting the most magnificent parade Lockport has ever seen or can expect to witness again in this generation. Thousands of men and hundreds of musicians matched in formation before the thundering thousands of Lockportians, and a mass of visitors the likes of which Lockport has never seen."[ccxxiv] Over 15,000 people were reported to have attended, a figure not unlike what one might find at a Long Island celebration. Two weeks later, at FASNY's 38th Convention in Watertown, the Nassau County Firemen's Association members were represented for the first time, followed less than a year later by the SCVFA, which voted to join FASNY on June 13th, 1911, electing James W. Eaton of West Islip as its first Delegate.[ccxxv]

And Long Island returned the favor…at least for a moment. At the SCVFA tournament in Amityville that same year, Henry Blake, a member of the Quogue-Westhampton Beach FD was selected as the "handsomest" fireman in the parade by the local tournament committee. The awarding of

THE WEATHER.
Washington, July 27.
For Western New York—Showers this afternoon, generally fair tonight and Thursday.

LOCKPORT UNION-SUN.

RAINY DAYS
Are just as good for using Union-Sun want ads.

THE ONLY NEWSPAPER IN THE CITY OF LOCKPORT HAVING A WIRE TELEGRAPH SERVICE.

FOUNDED 1821. LOCKPORT, N. Y., WEDNESDAY. EVENING, JULY 27, 1910. EIGHTY-NINTH YEAR

FIREMENS GREAT PARADE MOVES THROUGH CHEERING THOUSANDS

TEN THOUSAND VISITORS ARE IN THE CITY TODAY

Dozens of Bands Fill the Air With Music and Firemen Are Enjoying Themselves as Never Before.

Through the cheering masses of thousands of people the greatest firemen's parade in the history of the Western New York Firemen's Association and possibly in the history of the State of New York or firemen's gatherings of other states moved this afternoon. Ten thousand visitors flocked into the city this morning from all corners of Western New York. The Central specials carried hundreds and the trolley lines were burdened to the limit. Automobiles came in droves and the streets are a mass of happy, fun seeking humanity.

The firemen own Lockport today. They have been given the key to the city and most of them are in an orderly manner taking advantage of our hospitality. Some of them seemed to forget themselves. But on the whole the visiting boys are having a good time without treading on anyone's feelings.

Their various races, etc., took place this morning on Chestnut Street. None of the local companies were allowed to compete. The results will be published tomorrow.

The visiting fire companies began to arrive early this morning and by 10.30 the streets were gay with their giddy uniforms. The boys cheered to the echo when they reached Main St. and gazed on the grand illuminations. They were taken to the houses of the various fire companies where they will be entertained this evening after the parade.

The formation of the parade was a big job because of the great number of visiting companies but Grand Marshal Samuel Cunningham had the matter well in hand before and was aided by competent marshals.

The big feature of the day outside of the parade will be the massed bands following the parade. These will assemble on the Big Bridge under the direction of John W. Bolton of the 74th Regiment, Buffalo, and will move from the Big Bridge to Elm Street at 5 o'clock, playing "Onward Christian Soldiers."

Following this the prize drill will take place on the Big Bridge.

This evening the sensational free acts, band concerts, etc., will be given. The midways will put on extra shows. [illegible] show at the corner of Walnut and Locust Streets, there will be a female wrestling match this evening between Mable Dixon, who defeated the famous Cora Livingston, and Clara Dean. It is expected that a big crowd will witness it.

The fireworks display this evening will be from the tower of the Lockport Light, Heat & Power Company's plant on Elm Street and will no doubt be the most beautiful of the week, considering the vantage point.

LOCKPORT FIRE DEPARTMENT.

First Row—Chief Engineer Cunningham and Assistants Few and Cameron, Washington Chemical and Hydrant Hose Co. Second Row—Mayor Minard, Fire Commissioners Graham, Oliver, McGlynn, Murphey on running board; Protection Hose & Ladder Co., DeWitt Clinton Hose Co., Active Hose Co.

THURSDAY PROGRAMME.

10.15 a. m.—O. K. Stuart on high wire at Opera House Square.
11.20 a. m.—Sangster's Shows open.
12.00 m.—Dixie United Shows open.
1.30 p. m.—Industrial Floats assemble at Division Headquarters.
2.00 p. m.—Floats leave for headquarters.
2.30 p. m.—Parade starts.
4.30 p. m.—Sensational Smithson in leap of death, Big Bridge.
6.00 p. m. to 8 p. m.—Band concerts.
8.00 p. m.—Grand electrical illumination.
8.00 p. m.—O. K. Stuart on high wire.
9.00 p. m.—Grand display of fireworks on Big Bridge.
10.30 p. m.—Sensational Smithson in leap of death.

TODAY'S PROGRAMME.

5:00 p. m.—Firemen's drill contest on the Big Bridge.
6:00 p. m.—Hand engine contest on Lift Bridge, Lower Town.
6 p. m. to 8 p. m.—Band concerts.
8:00 p. m.—Grand electrical illumination.
8:00 p. m.—O. K. Stuart on high wire.
9:00 p. m.—Grand display of fireworks.
10:30 p. m.—Sensational Smithson in leap of death.

DARKTOWN BRIGADE WINS GROTESQUE PARADE PRIZE

Lasts Nights Crowd on Streets Taxed Every Available Inch-- The Was Fun Aplenty.

The Grotesque Parade last night gave the people a little diversion in the line of parades. It was just a comglomeration of ridiculousness and the people had spasms watching the antics of the ragtime artists as they went by in the most ludricrous outfits that the human mind ever conceived.

The Darktown Fire Brigade of Niagara Falls won the first prize of $50. This bunch of artists got their start at the Falls Carnival and when Lockport's was planned they voted to come here. Fred C. Stricker is Grand Marshal of the Darktown Brigade. H. McMillan was assistant.

The men were dressed in regular red short and duck trousers, but their black face makeup and their paraphernalia were laughable in the extreme. It would take a sketch artist of the funniest kind to depict the steam engine and the truck. The Darktowns had about 50 men in the line and made a great showing.

The parade was headed by Grand Marshal Harry B. Suthers on horseback behind the mounted police. Then came the City Band and a sketch made up as Uncle Sam. The Phoney Police Force of Niagara Falls which is a part of the Darktown Fire Brigade Company was next in line. There were two platoons of these cops with clown makeups and each smoking a broken stemmed clay pipe.

Marshal Daniel Kline and Chairman A. T. Chilcott of the Parade Committee rode at the head of the second division, consisting of the funny trucks, etc., of the Falls contingent.

The North Ends won second prize in the parade. The boys of that club had some of the most original makeups ever perpetrated. If they did not win the first prize they gave the Falls crowd a run for the money. Second prize was $10. Third prize went to Miegel's German Band of Swormsville. The band was made up in a most provoking style and rode on a big farm wagon. The music the band attempted was enough to drive the perfectly sane to a bughouse. Seated on a throne in the center of the wagon was the big Fat Mayor of Swormsville in a make up like King Dodo.

Richard Daley was marshall for one division and in this were the Cyclones with a lot of screams in funny clothes. A bride and groom in an awful looking outfit that beggars description formed part of the division and got a thousand laughs.

The prizes for the individuals went to the following: The Mule and Sulky, first prize, $10; Cyclone Lady Lifters, second prize, $5, and the "My Float," third prize, $1, and the Cyclone Bathing Girl fourth prize, $1.

The parade moved from West Ave. to Main to East Ave. to Lowertown, where the fireworks display was given last night. The attendance at them was not large, the crowd preferring to remain in the lighted lane of Main Street.

The big sensational free acts were given. O. K. Stuart on the high wire performed, but some of his slow comedy ought to be chopped, as the people get tired waiting for him to get down to business.

Springville For Firemen in 1911

Gets Convention--Mr. Melody of Dansville Elected President-- That Home Proposition.

The annual convention of the Western New York Firemen's Association came to a close yesterday afternoon in the Hodge Opera House with the election of officers as follows:

President D. J. Melody, Dansville; vice president, E. N. Thurber, Springville; secretary, Charles F. Foley, Lockport; treasurer, C. Royce Sawyer, Albion; trustee, for three years, Charles A. Kandt, Lockport.

Delegate to state convention—John R. Mahaney of this city, retiring president. Messrs. Foley, Kandt and Sawyer were re-elected. They have served the association for several terms in their respective capacities and are valuable aids to it.

A resolution was adopted calling on state superintendent of insurance William H. Hotchkiss to enforce the law governing the tax on mutual fire insurance companies that the volunteer firement may get the financial returns due them. The mutual companies were recently put under the same head as other foreign companies in the state acting without agents and they are subject to the two per cent tax. This tax goes to the firemen.

Another resolution was adopted yesterday after a long discussion against making the line of march of any parade at any convention over three miles in length. In some of the small places where the convention has been held the local officers have dragged the firemen over every street of the city, taking all the afternoon and half the evening to it. The firemen voted to stop this sort of thing. The Lockport line of march is a trifle over two miles.

Springville put up a hard fight for the 1911 convention of the association and won out. Rochester was after it. John P. Kislingbury of the Rochester Exempts fought for the Flower City. He retired finally in favor of Springville on the understanding that Rochester would have first choice for the convention in 1912. Attorney David Wilcox of Springfield put up the fight for his town.

Lockport got a great hand from the convention delegates last night. The boys were wild about the decorations, illuminations and otherwise grand reception accorded them. "We never saw anything like it, Foley," they yelled to the secretary. Lockport was cheered and resolutions were adopted thanking the Old Home Week General Committee, the local firemen and the citizens. Lockport never could expect to get the boosting that it was recipient of from firemen yesterday. Some of these delegates were men of prominence in their sections of state and they will take back to their homes the greatest praise of the city.

Among other resolutions presented and passed was one referring the matter of a contemplated home for aged firemen in Western New York back to the committee to communicate with all companies in the association and ascertain which one of the propost-

(Continued on Page Five.)

THE INDUSTRIAL PARADE MOVES IN SEVEN DIVISIONS TOMORROW

The Industrial Day parade promises to be one of the biggest features of Old Home Week. Hundreds of handsome floats have been prepared for the parade, which will move promptly at 2:30 o'clock.

Owing to the length of some of the floats it will be impossible to countermarch on East avenue and the line of march has been changed. The parade will turn over Cave street to Walnut to Washburn thence to High.

The formation will be as follows:

Hawley to West avenue to Main, to Cave, to Walnut, to Washburn, to High, to Pine, to Genesee, up Buffalo to Big Bridge and disband.

Mounted Police.
Grand Marshal, J. H. Ellers.
Aides—Cyrus D. Ormiston, Robert Hall, Albert M. Little, Richard Williamson.

FIRST DIVISION.

Builders Trade Unions will form on Prospect Street, right resting on West Ave.

Marshal, Albert Nott.
Aides—Wm. C. Rhodes, Daniel O'Brien, Thomas McDonald, J. Downey.

The various unions of the city of Lockport that are not marching with the manufacturers' floats will march in this division.

SECOND DIVISION.

This division will form on Elm street, right resting on Main street, no later than 2 P. M., and from there march down Main, up Niagara street to Prospect street with the right resting on Park Ave. The following are included in this division.

Marshal, Geo. Castle.
Aides—Bert A. Lewis, Harry Gardner.
Band.
Corson Mfg. Co's. float.
American District Steam Co.
Lockport Cotton Batting Co.
Williams Brothers.
Parsons Drug Co.
Taylor & Reynolds.
J. K. Perry.
C. Dole.
Lockport Ice and Cold Storage Co.
Home Baking Co.

THIRD DIVISION.

This division will form on Prospect street, right resting on Niagara street.

Marshal, Harry Ransom.
Aides—Geo. A. Welsh, Sheldon Cook, Edward Fritton.
Band.
Niagara Textile Co's Float.
Lockport Light, Heat & Power Co.
D. R. Sillesky & Co.
Niagara Laundry.
Home Telephone Co.
John Young.
McCarthy's Nursery.
Jenss Brothers.
Gerner Brothers.

FOURTH DIVISION.

This division will form on Ontario street, right resting on Prospect street.

Marshal, Frank G. Oldham.
Aides—Frank Woodruff, W. D. Smith.
Band.
Oliver Brothers.
Western Block Co.
Thompson Milling Co.
Citizens Dairy.
Cocker Mfg. Co.
Noble Bakery.
Review Publishing Co.
E. P. Marvin.
Lock City Brewing Co.
J. W. Turner Box & Lumber Co.

FIFTH DIVISION.

This division will form on Mill street, right resting on Clinton street, ready to march at 3:30.

The United Box Board Co., Niagara Paper Mills, Lockport Paper Company and Niagara Cooperage Co., will march to corner of Clinton and Gooding streets, and wait for Westerman & Co., Gould's Electric Smelting Co., Whitmore & Co. From there they will march to Grand, to Church, to Niagara street. Up Niagara street to Prospect street, right resting on Prospect street.

Marshal, Arthur Brim.
Aides—A. F. Sellick, Wm. Case, Raymond Ehlers, Frederick Callahan.
Band.
United Boxboard Co.
Niagara Paper Mills.
Westerman & Co.
Lockport Paper Co.
United Indurated Fibre Co.
Electric Smelting & Aluminum Co.
Niagara Cooperage Co.
C. H. Whitmore & Co.
Downe's Brewery.

SIXTH DIVISION.

This Division will form on Niagara street, right resting on Hawley street.

Marshal, E. B. Omstead.
Aides—A. J. Aldrith, L. E. Merritt, J. M. Woodruff.
Band.
Lockport Textile Co.
Covert Mfg. Co.
Lockport Glass Co.
American Motor Truck Co.
Federal Milling Co.
Mansfield, the Florist.
Lockport Ice Co.
Richmond Mfg. Co.

SEVENTH DIVISION.

Merchants Division. This division will form on Park Ave, resting on Prospect street.

Marshal, A. Williams.
Aides—Seldon Woodward, Dr. Blackley.
Band.
Burtt Brothers.
Shearston Mfg. Co.
Ferrin Coal Co.
R. Urtel.
Pillsbury Flour Mills.
And other Merchants.

SATURDAY IS A HOLIDAY

Mayor's Office, Lockport, N. Y.
PROCLAMATION.

To the Citizens of Lockport:

The Old Home Week celebration now being held in our city marks one of the greatest events in its history, and I deem it fitting that we should recognize the same by declaring a Civic Holiday.

Therefore, I, George H. Minard, Mayor of the City of Lockport, N. Y., do hereby declare that Saturday, July 30th, 1910, the closing day of this memorable event, be a Civic Holiday within this city, and respectfully request of all our citizens to duly observe the same as a holiday.

Witness my hand and seal of the City of Lockport, N. Y., this 27th day of July, 1910.

GEORGE H. MINARD,
Mayor.

(Seal.)

VETERANS ATTENTION.

The members of Sprout Post, G.A.R., an dall veterans of the Civil War, who will take part in the parade Military Day are requested to meet at Grand Army Hall Friday at 1 p. m. Automobiles will be there in readiness.

No Person Under 18 Can Run Autos

Albany, July 27.—No person under 18 years old will be permitted to operate an automobile after August 1st unless accompanied by a licensed chauffeur or the owner of the machine.

Secretary Koenig has received numerous inquiries as to whether a person under 18, who is the owner, can operate a motor vehicle under the Callan law upon the public highways of the state.

Attorney General O'Malley, in an opinion today holds that one of the purposes of the new law was to protect the public from accidents, which are due in many instances to the operation of automobiles by careless, incompetent and inexperienced operators, and that the legislative intent is plain that a person should be at least 18 years old before he has judgment, experience and discretion sufficient to qualify him to drive a car.

Russia's copper production is increasing, amounting to 10,306, 14,534 and 16,591 tons in 1906, 1907 and 1908, respectively. Deposits of copper in the Caucasus present vast opportunities, and enormous beds of coprolites exist in central Russia and in the Dniester basin, the quantity in the latter alone being estimated as high as 27,000,000 tons, says a consular report.

OLD HOMERS—REGISTER AT HEADQUARTERS. OPERA HOUSE

The arch and grandstands under construction prior to a tournament in Mineola, 1910

these kinds of prizes for various and sundry distinctions at parades was a common occurrence up north and west, but not so much on Long Island; in fact it is significant because it is the only time in the history of the SCVFA that that award was given.[ccxxvi]

It is hard to say for sure, but it is possible that the reason for this might be that parades – and tournaments – were taken just that much more seriously by the fire departments on Long Island, as compared to their brethren in other parts of the state; yet another trend that continues to this day; at times being one of the chief complaints from northern and western teams regarding their Long Island counterparts. It could be said that whereas on Long Island, in spite of a strong sense of camaraderie, the spirit of competition amongst the teams ran a bit higher, as opposed to those firemen from North and West…who simply wanted to have some fun.

This is indicated as well by a front-page story about a particular parade at the Lockport convention on its first day. At the time it was no doubt considered silly and somewhat harmless, but by today's standards it would be considered horrifically and inexcusably politically incorrect: the Grotesque Parade.

"The Grotesque Parade last night gave the people a little diversion in the line of parades. It was just a conglomeration of ridiculousness and the people had spasms watching the antics of the ragtime artists as they went by in the most ludicrous outfits that the human mind ever conceived.

The Protection Hose Co. of Hempstead runs the Cart Hose contest at a tournament in 1911

"The Darktown Fire Brigade of Niagara Falls won the first prize of $50…The men were dressed in regular red short [sic] and duck trousers, but their black face makeup and their paraphernalia were laughable in the extreme."[ccxxvii]

And although it was not nearly as politically incorrect, there is evidence that the northern firemen at the time were also all about having a good time, as indicated during their convention in Carthage on June 5th, 1912:

"Several practical jokes were played by the firemen during the convention which caused considerable merriment among the crowd.

"It is reported that late Wednesday evening or early Thursday morning, some of the firemen put one of their number in a rough box, loaded the box on to an empty wagon and drew it around town. Several bands of them performed at different times and marched up and down the street, using squawkers and horns for music. One of the firemen, dressed up in feminine attire, was the center of attraction yesterday. The firemen were, however, without exception, orderly."[ccxxviii]

But the big story that year was the Carthage-Sackets Harbor rivalry, which no doubt provided a good deal of fun for all as well:

"Many people, no doubt, noted yesterday morning that the large banner which had been strung from the cupola of the firemen's hall to one of the highest branches across the street had disappeared during the night. This banner was about 25 feet long and 15 feet wide. Upon it was printed in an artistic manner the picture of firemen among the flames, and in the center was a fireman upon a ladder with his trumpet to his lips. Upon the banner were also painted the words: 'Welcome, Firemen to Carthage. The town is yours.' It seems there is a history connected with this banner, which very few of the younger members of the Carthage fire

West Sayville runs the Cart Hose contest, ca. early 1900's

department knew anything about. Some two months ago while a meeting of the firemen was being held and plans were being made for the convention, the question of a banner was brought up and the janitor said that an old banner was stored in the cupboards of the firemen's hall. The banner was brought before the meeting and spread out upon the floor. Upon the banner were the words, 'Welcome, Firemen to Sackets Harbor. The Town is yours.' Some of the older firemen cast knowing looks at each other as the flag was unrolled. It seems that 16 years ago, this flag had been appropriated by the Carthage firemen, the convention being held that year at Sackets Harbor. Just who the firemen were who secured this souvenir is not generally known, but not all of them are living at present.

"Two pieces of cloth were sewn over the words 'Sackets Harbor' and 'Carthage' was painted on to take the place of those words. Considerable difficulty was had in stringing this large flag over the street and it had no more than been put into position when a strong breeze ripped it. It was taken down again, reinforced by a backing of cloth and once more flung to the breeze. It is reported that the Sackets Harbor firemen had no sooner turned the corner at Mechanic street than some of their older men recognized the old flag and a plot was at once formed for carrying the banner back home again. The plot seemed to work all right, for the policemen who were on duty stated that early in the morning they noticed the flag was there and within ten minutes after it had disappeared."[ccxxix]

Once again, however, the advent of the automobile was evident in western New York with a front-page story in the Union-Sun on July 27th, 1910 ("No Person Under 18 Can Run Autos") and again on the 28th ("Great Auto Parade Will Take Place Friday Night"). However, all did not necessarily receive these new

The Greenport tournament team, 1912

machines with open arms. From that same issue, within a description of the evening's festivities after the parade:

"The evening's entertainment of big sensational free acts, band concerts, fireworks, etc., pleased the vast concourse of people. The streets were jammed for hours, but while some were noisy none were rough. The automobiles formed the most objectionable feature of the evening. They should be barred from the streets during the carnival hours. One machine, owned by a Lockport man, made 40 trips exactly, back and forth the length of Main Street through that dense mass of people last night. There seems to be no necessity for this and the danger is great. The police claim they can do nothing so long as the autos are allowed on the street."[ccxxx] As part of the description of the northern firemen's convention in Watertown in 1913, the Watertown Daily Times wrote, "The third accident of the day happened when a man by the name of Henderson and said to be from Theresa went into the ditch with a Ford automobile after having its front tire blow out. It is said by those who witnessed the accident that the automobile turned completely around while on the highway and went into the ditch, throwing its occupant out."[ccxxxi]

Just exactly when automobiles came to the fire service on Long Island isn't very clear. One might assume that being so close to New York City they would have appeared sooner than they would have upstate, but there is no solid evidence of this happening prior to 1911, where Fred Flugrath, in his Suffolk County Volunteer Firemen's Blue Book, states that the Steamer Contest was amended to include gasoline engines, which would imply that motorized apparatus of some sort were most likely being used for the first time within a firematic competition.[ccxxxii] At the SCVFA's 27th Annual Parade and Tournament in Sag Harbor on August 28th, 1912, historian Charles Zigrosser refers to motorized apparatus appearing in the parade, but oddly enough the August 29th issue of the Sag Harbor Express makes no mention of automobiles at all. One might presume that it would be a pretty big deal, and worth mentioning, if they had been part of the line. That same article does recount one of the contests as being the "Steam and Gasoline Engine Contest", so it's likely that there were automobiles in attendance; whether or not they competed, or actually appeared in the parade for that matter, is unclear.

What is definitely clear is that motorized apparatus both paraded and competed the following year, at the SCVFA's 28th Annual Parade and Tournament in Bay Shore on August 27th, 1913:

"According to the official figures there were 1,459 men and juveniles in the line, 60 companies, all but two or three of which brought their apparatus. There were four automobile apparatus in the parade, and about a dozen bands and drum corps."[ccxxxiii]

"Lindenhurst had a walkover in the gasoline engine contest. It had no competitors, and owing to difficulty with its suction hose made a short throw of 92 feet."[ccxxxiv]

What's interesting is that the same article, in the "Results" section of the story, states that the record for that event was held by Lindenhurst with a throw of 187'10". That fact, combined with the fact that they were the only team entered in that contest, implies that not only had that contest been held in prior years, but that there is a good chance that Lindenhurst was the very first team to compete with motorized type of apparatus – at least in Suffolk County.

The 1914 SCVFA tournament in Riverhead brought more mentions of the automobile ("This was the first time that motor apparatus has appeared in the line," wrote the Riverhead News. "Patchogue had two beautiful pieces of apparatus; Southampton had four. It attracted great attention."[ccxxxv]) and up north there was mention of what may have been one of the very first responses by automobile to a fire by volunteer firemen:

"Automobiles saved the farmhouse and hay barns of J. A. Harper near Lisbon Monday afternoon when fire destroyed a granary, hog pen and a smoke house on the farm. Summoned by telephone farmers from nearby rushed to the scene of the fire in their cars and managed to get it under control.

"When the fire was discovered Mr. Harper tried to reach central to summon assistance on the telephone. His line was in trouble, but he did succeed into getting into communication with Dan Akin's house, not far away. Mr. Akin summoned seven automobiles and pretty soon firefighters were whizzing over the country roads to the rescue of the imperiled buildings. It is the best piece of fire fighting that has been done in many a day and last night Mr. Akin was being complimented on the way he suited the word to the action."[ccxxxvi]

It was during the following year, however, that a true benchmark was made in the sport of modern-day firematic drill team competition. On September 1st, 1915, at the 30th Annual SCVFA Parade and Tournament that was held in Bayport, the Motor Contest was introduced. As distinguished from the Gasoline Engine Contest, wherein a motorized apparatus was used to simply pump water, the Motor Contest was the very first to involve a motorized vehicle actually running down the track:

"The motor contest introduced at the tournament for the first time was of prime interest. It was a race of 345 feet from still engine start to hydrant, 150 feet of hose to lay and stream to be played on backstop. Penataquit Bay Shore won. The VanGuards claimed the best time but were ruled out because their coupling broke before the hose was all laid. This was an accident in which Bobbie Roe took a long ride on the seat of his pants. There were two other flukes. Dauntless of Amityville got water too quick and could not get the nozzle on, and Halesite made a great run only to find that they had no nozzle with them all."[ccxxxvii] (The South Side Signal wrote regarding the latter, "The Halesite company made a good run but the nozzle man forgot

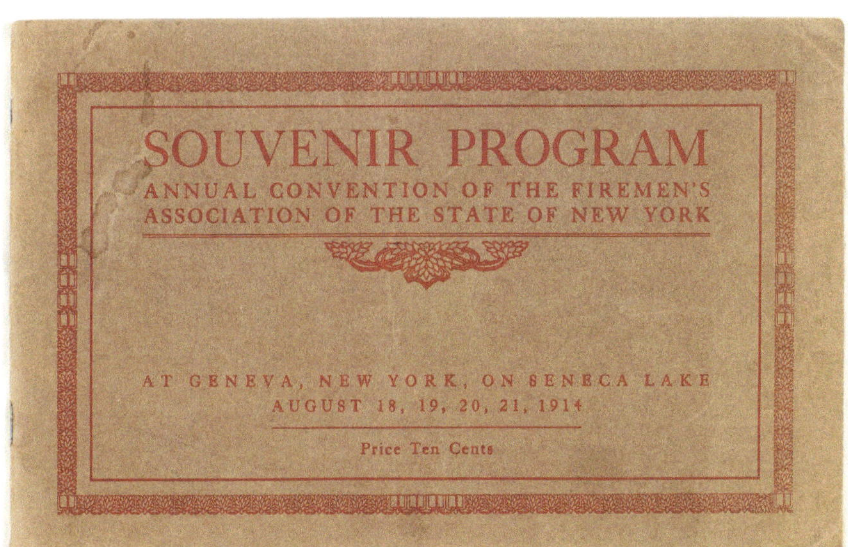

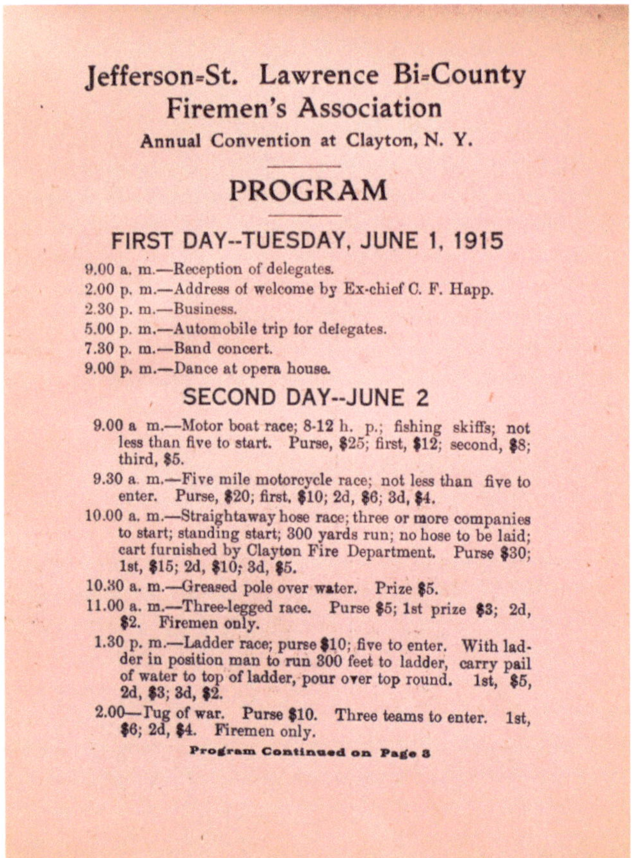

Jefferson-St. Lawrence Bi-County
Firemen's Association
Annual Convention at Clayton, N. Y.

PROGRAM

FIRST DAY--TUESDAY, JUNE 1, 1915

9.00 a. m.—Reception of delegates.
2.00 p. m.—Address of welcome by Ex-chief C. F. Happ.
2.30 p. m.—Business.
5.00 p. m.—Automobile trip for delegates.
7.30 p. m.—Band concert.
9.00 p. m.—Dance at opera house.

SECOND DAY--JUNE 2

9.00 a m.—Motor boat race; 8-12 h. p.; fishing skiffs; not less than five to start. Purse, $25; first, $12; second, $8; third, $5.
9.30 a. m.—Five mile motorcycle race; not less than five to enter. Purse, $20; first, $10; 2d, $6; 3d, $4.
10.00 a. m.—Straightaway hose race; three or more companies to start; standing start; 300 yards run; no hose to be laid; cart furnished by Clayton Fire Department. Purse $30; 1st, $15; 2d, $10; 3d, $5.
10.30 a. m.—Greased pole over water. Prize $5.
11.00 a. m.—Three-legged race. Purse $5; 1st prize $3; 2d, $2. Firemen only.
1.30 p. m.—Ladder race; purse $10; five to enter. With ladder in position man to run 300 feet to ladder, carry pail of water to top of ladder, pour over top round. 1st, $5, 2d, $3; 3d, $2.
2.00—Tug of war. Purse $10. Three teams to enter. 1st, $6; 2d, $4. Firemen only.

Program Continued on Page 3

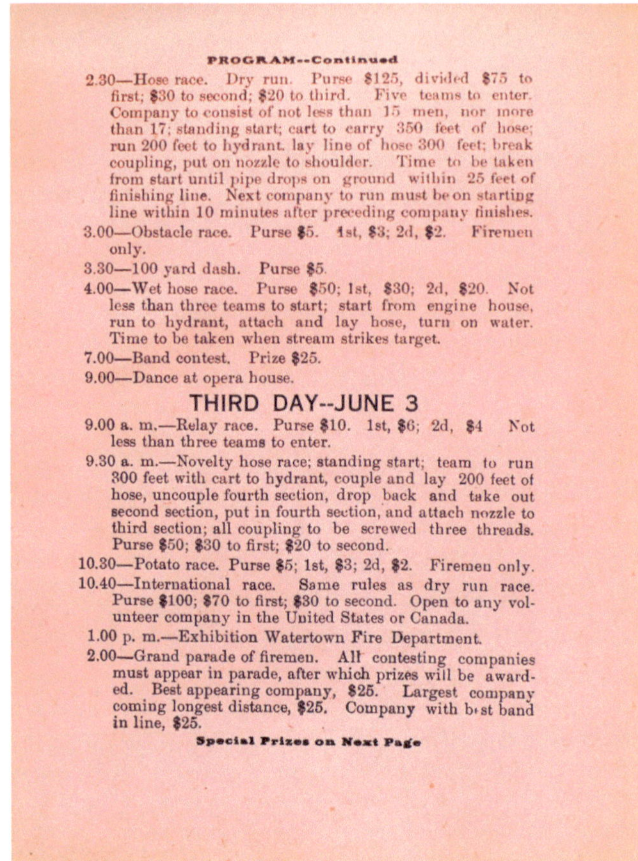

PROGRAM--Continued

2.30—Hose race. Dry run. Purse $125, divided $75 to first; $30 to second; $20 to third. Five teams to enter. Company to consist of not less than 15 men, nor more than 17; standing start; cart to carry 350 feet of hose; run 200 feet to hydrant. lay line of hose 300 feet; break coupling, put on nozzle to shoulder. Time to be taken from start until pipe drops on ground within 25 feet of finishing line. Next company to run must be on starting line within 10 minutes after preceding company finishes.
3.00—Obstacle race. Purse $5. 1st, $3; 2d, $2. Firemen only.
3.30—100 yard dash. Purse $5.
4.00—Wet hose race. Purse $50; 1st, $30; 2d, $20. Not less than three teams to start; start from engine house, run to hydrant, attach and lay hose, turn on water. Time to be taken when stream strikes target.
7.00—Band contest. Prize $25.
9.00—Dance at opera house.

THIRD DAY--JUNE 3

9.00 a. m.—Relay race. Purse $10. 1st, $6; 2d, $4 Not less than three teams to enter.
9.30 a. m.—Novelty hose race; standing start; team to run 300 feet with cart to hydrant, couple and lay 200 feet of hose, uncouple fourth section, drop back and take out second section, put in fourth section, and attach nozzle to third section; all coupling to be screwed three threads. Purse $50; $30 to first; $20 to second.
10.30—Potato race. Purse $5; 1st, $3; 2d, $2. Firemen only.
10.40—International race. Same rules as dry run race. Purse $100; $70 to first; $30 to second. Open to any volunteer company in the United States or Canada.
1.00 p. m.—Exhibition Watertown Fire Department.
2.00—Grand parade of firemen. All contesting companies must appear in parade, after which prizes will be awarded. Best appearing company, $25. Largest company coming longest distance, $25. Company with best band in line, $25.

Special Prizes on Next Page

to take the nozzle out of the cart, and he looked foolish when the crowd laughed at his predicament."[ccxxxviii]) The introduction of this contest pointed the direction in which all future tournaments in the state would grow, eventually making New York State unique from all other states in the country – and indeed the world – in terms of firematic competition.

By all accounts however, that tournament would have been one for the books anyway, regardless of the introduction of the Motor Contest. First was the size of the event, and its venue:

"There were 1,226 firemen and musicians in the parade…18 automobile apparatus, 13 ordinary hook and ladder trucks, 9 hose carts, 6 chemical engines, 3 hand engines and 3 juvenile companies."[ccxxxix]

"The grandstands were very substantial structures, and so designed as to give every occupant an excellent view of the whole course. The Cove property, where the running track was laid out and the stands erected, was admirably adapted and the location was a picturesque one, giving a fine view of the bay. It also afforded ample parking space for hundreds of automobiles and for the erection of booths for refreshment vendors and the various fakirs and side shows which always assemble at this gala occasion."[ccxl]

Then there was the choice of some unusual apparatus by Southampton:

"The department used its 'cesspool pump' and its old Agawam machine in both these [Special Engine and regular Hand Engine] events, and the unusual combination presented by the Southampton fire laddies caused considerable comment."[ccxli]

A record was broken in the Individual Ladder Contest, won - with a new technique - by Huntington with a time of 5-3/5 seconds:

"This was the only event in which a record was broken, and the winner and the second man introduced a new form of climbing, taking two rungs of the ladder at a time in the climb."[ccxlii]

Sag Harbor's burned hose cart after the Bayport tournament after the SCVFA 1915 tournament in Bayport

Other contests included Juvenile Hook & Ladder, Hook & Ladder Contests #1 and #2, Hose Contests #1 and #2, a Department Contest, the Hand Engine and Special Hand Engine Contests, and Chemical Engine and Steam Engine Contests. The West Sayville team ended up winning the day by taking home ten of the forty prizes being offered, but it was the team from Sag Harbor that ultimately won the hearts of the crowd by taking First Place in the Dry Hose Contest, after overcoming almost insurmountable odds.

It was – and to this day still is – extremely rare to see any sort of malice within the sport of firematic competition. Thus, it was a great shock to everyone when, as it sat atop a railroad flatcar the night before the Bayport tournament, Sag Harbor's Montauk Hose team hose cart was set on fire by unknown persons, in an obvious attempt to sabotage them before the next day's contest. In an amazing display of teamwork and determination, the team not only rebuilt the cart in time for the contest, but won the contest with a Personal Best time.

"Several departments offered the Montauks the use of their carts, but the Sag Harbor men hustled together other hose and repaired their cart as best possible..."[ccxliii]

"That the Montauk [Hose Co.] running team was able to win the dry hose contest after its apparatus had been destroyed by fire, was considered nothing less than remarkable. In the parade the burned cart attracted much attention and the Sag Harbor boys were loudly cheered when they made the best time in the contest, using the charred and dilapidated cart."[ccxliv]

North and west of Long Island, the conventions and tournaments of that period remained more or less the same in terms of their tenor and content. Western firemen ran a Hose Race, a Hub and Hub Race and a Hook and Ladder Race as well as a Special Canadian Prize (a Hook and Ladder Race for Canadian Teams only), while Northern contests routinely included a Dry Hose Race, a 100-yard Dash, an Obstacle Race, a Relay Race, a Novelty Race, a Straightaway Hose Race, a Three-Legged Race, an Egg Race and a Ladder Race. One interesting contest, however, was a Wet Hose Race that was held at the Bi-County ("Northern") convention at Clayton in 1915, as it involved – perhaps for the first time – the use of a live-fire target. As the Watertown Daily Times described it:

"The last event of the afternoon was the wet hose race. This race constituted running from the Clayton fire department station, through one or two streets and extinguishing a barrel elevated on a couple of stakes and ignited with gasoline. The race was very late to start. It was scheduled for 4 p.m. and did not begin until nearly 6. The only team entered was the battalion from Morristown. The Morristown firemen came down the street in fine style, but the man at the hydrant had taken off both caps and the water, following the lines of least resistance, flowed through the open cap instead of through the hose. Before the mistake could be remedied the fire had died of its own accord. Because of the poor showing of the first organization due to the mistake of the man at the hydrant, the Sackets Harbor battalion agreed to run merely for exhibition, very graciously surrendering first money to the Morristown men."[ccxlv]

The pranks and hijinks seen in the past at these conventions were still prevalent, including the turning in of multiple false-alarms to the point where the alarm system was turned off altogether (Postdam, 1914) as well as Lockport exempts swiping a dummy dressed in an exempt's uniform from the roof of the Niagara Falls headquarters (Niagara Falls, 1915).

But the world was now starting to undergo some great changes. One of the first, as already described, was the invention and swift rise to popularity of the automobile and other motorized vehicles. For the fire service it was a boon as well, but newspaper accounts of the day make it clear that it was still somewhat of a novelty, and something that would take some getting-used-to, as indicated by a clipping accompanying a story about the 1915 convention in Clayton:

"Frank Hall, the local motorcycle daredevil, nearly broke up the firemen's convention at Clayton yesterday. He arrived at the convention yesterday morning on his high powered Indian and immediately put more vim into things by tearing about the streets of the village at a rate not in keeping with the speed laws. The streets were crowded and all made a wild scramble to get out of harm's way, as Hall had not notified them of the fact that he was tuning up his machine for the scheduled race there.

"The trial spin was enough to convince those in charge of the motorcycle race that they had got it decidedly wrong on scheduling such an event and this event was immediately stricken from the program."[ccxlvi]

Beyond the rise of the automobile, however, there were bigger changes happening: social unrest was starting to occur. Apparently the size of the gatherings of firefighters at their conventions and tournaments was now being foreseen, and taken advantage of, by groups of all sorts who wished to draw attention to themselves or their causes. One such cause was that of allowing women to participate in the voting process. Flugrath writes: "Marching in the Bay Shore tournament parade of 1913 was a group of a dozen women, enthusiastic leaders for the cause of women's suffrage. The women were dressed in white and carried yellow banners, with yellow ribbons proclaiming, 'votes for women'".[ccxlvii] Another mention is found two years later, at the convention in Clayton: "The suffragists are busy about the city, and have an automobile running about the streets. The car is in charge of Miss Anna Laughlin, and is gaily decorated with suffrage banners."[ccxlviii]

Of far greater significance, though, was the "social unrest" (to put it mildly) that was taking place overseas in Europe, with the assassination of Austrian-Hungarian Archduke Franz Ferdinand in June of 1914, an event that triggered the series of events that became World War I. Although the United States did not officially enter the war until April 6th, 1917, it followed events very closely up until that time due to its allies being involved.

A tournament at Greenport in the early 1900's, possibly the 31st SCVFA tournament in 1916

The war affected all aspects of everyday life, including – understandably – the fire service, which, as somewhat of a paramilitary organization, was closer to being involved than most everyday citizens might have been. The Suffolk County Volunteer Firemen's Association – as well as other firematic associations throughout New York State and the rest of the country – did their part by directing their attentions to the war effort, by putting their conventions and tournaments on hold and promoting recruitment in the armed forces. Numerous references can be found that refer to this, including one in Sag Harbor's history: "...the association devoted its full energies to serving the government during the period of hostilities of the World War..." and in Flugrath's "Blue Book": "One of the [SCVF] Association's most conspicuous activities was its work in assisting the recruiting of volunteers for the land and sea forces of this country when the United States entered the World War. Not only did the Association postpone the holding of its annual tournaments during the years 1917, 1918 and 1919, but it played an important part in the enrollment of the firemen for war service. An intensive campaign was planned and carried through by the 'Quarter Century' Executive Committee, led by President [Waldron Kintzing] Post, to encourage and solicit the firemen to join the country's military and naval units."[ccxlix]

To make matters worse, the year 1916 also saw an outbreak of Infantile Paralysis – also known as Polio – that spread across sections of the northeast United States and New England. This outbreak had a detrimental effect on the fire service as well, at least as far as the hosting of conventions and tournaments was concerned. Dixon, in his history of old-fashioned handtub musters, wrote that it "kept muster attendance down", and FASNY even went so far as to relocate their 44th Annual Convention to Rockaway Beach instead of holding it in White Plains, as they originally had planned: "The epidemic of Infantile Paralysis spreading throughout this section alarmed the health authorities of the City of White Plains, and they, fearing a breakout of this disease in their community, issued an order and served it upon the President of this Association and the Chairman of the Local Committee forbidding the holding of the Convention in White Plains."[ccl]

On Long Island, the 31st Annual SCVFA tournament (the last that would be held until 1919) went ahead as scheduled in Greenport on August 30th, 1916, but in spite of an estimated crowd of between 6,000 and 8,000, attendance "...was materially reduced over former years by the Infantile Paralysis scare."[ccli] An online history of the West Sayville Fire Department states that the tournament was "...diminished by the absence of the juvenile contests. [The] statewide outbreak of Infantile Paralysis not only eliminated the participation of the junior firemen, it also kept many spectators away." The website also states that in addition, the Individual Ladder Contest was discontinued that year, and would not be reinstated as a contest until a statewide reorganization of the drill teams in the 1950's.

The conventions and tournaments during the first year immediately following the end of World War I appear to have been a bit subdued compared to previous years. At the northern convention and tournament held in Massena in 1918, newspaper coverage of the event was not even an entire column long, and in contrast to a program of only two contests ("sporting events") were held – a Hose Reel Race and a baseball game – compared to ten different contests that were held at their 1915 convention. As it was, the hose race was marred by injury:

"The hose reel race resulted in a tie for first place between Clayton and Adams, with Massena third. Canton entered a team but dropped out when one of its men was hurt by falling. This fireman, George Wheelock, stumbled, and the cart passed over his shoulders. It was thought that he was seriously hurt and he was taken to the Hepburn hospital. He was able to return home Friday night. The man opposite Wheelock also fell but he saved himself from injury by grasping the wheel of the cart and throwing it over him."[cclii]

By 1919, however, things seemed to be getting back to normal. Regarding the Western convention in North Tonawanda on July 30th, the Evening News reported that there were 2,000 firemen comprised of 55 companies and 30 bands in line for the parade, and prizes for all the silly parade categories such as Tallest Fireman, Homeliest Fireman, etc. were once again awarded in good order. In true fire service fashion, there was even a dust-up regarding them:

"A prize was to have been awarded to the handsomest man in the parade. There were so many of them that the judges were unable to agree on the winner. It was, therefore, decided to substitute the department having the most complete fire apparatus in line for the prize."[ccliii]

Chapter 10: Changes, Sammis and Standardization

If the sport of firematic competition was "getting back to normal" by the beginning of the 1920's, it seems that the term "normal" was becoming more and more synonymous with "change". With the advent of the automobile, a historical and significant era in the history of tournaments was passing, as fewer and fewer people were using trains to travel to and from the conventions. In fact, at a SCVFA meeting on June 20th, 1920, "…it was reported that the railroad company had discontinued the practice of carrying apparatus free, but would run four special tournament trains with reduced rate fares, for the Lindenhurst tournament of that year, the first after the World War. No special trains were run by the railroad company for the Bay Shore tournament of 1923, which marked the close of that feature. Reduced rates of fare, however, were given by the railroad company that year, but the sale of tickets was small, only 261 being sold at 16 stations on the Montauk division."[ccliv] And then, in 1924, after selling only 306 tickets from 17 stations: "P.H. Woodward, general passenger agent of the Long Island Railroad, in a communication to the [SCVF] Association, wrote, in part, 'I am giving the information which shows that practically everybody goes to these tournaments, fairs, etc. in motors, and that the railroad is not justified in running any special service or giving any special rates for the same. I think this is the last year we will make any special reduction in fares on account of these propositions."[cclv]

That same tournament in 1920 in Lindenhurst was also marked by change. First was its date: although it was a unique exception, it is the only time in the history of the SCVFA that the tournament was held on a date other than in August or September. (The date was chosen to coincide with the celebration of the 50th anniversary of the founding of the village of Breslau, before its name was changed to Lindenhurst.)

In keeping with the theme of change, at the SCVFA tournament the following year in Riverhead on August 31st, 1921, a new contest was again introduced: the Motor Hook and Ladder #2 Contest, which is known today colloquially as the "Dummy Contest". In this contest, a 125-pound dummy is "rescued" from atop the arch and carried down the ladder for time. Once again quoting Flugrath: "The carrying down of a

125-pound 'dummy' on a ladder from atop a structure, while affording much amusement and thrills to the spectators, nevertheless requires considerable skill on the part of the men assigned to the 'rescue' work."[cclvi]

Although it is a complete coincidence, it's fun to speculate whether or not the members of the SCVFA got their idea for the "Dummy Contest" from firemen from Western New York. Even though it's not very likely since both tournaments occurred within a week of each other, the following blurb appeared in the Orleans Republican on the same day the Riverhead tournament took place, while reporting on the convention and tournament in Albion:

"A unique firemen's display worthy of mention, during the parade last Wednesday, was a 'dummy' fireman descending from a window at the home of Jacob Carter on Caroline street, carrying a small 'dummy' child rescued from the supposed burning house. A stream of water, flowing from a hose, was carried by the 'brave fireman.' It certainly created lots of applause by the marchers and others."[cclvii]

Another noticeable change of that time – although arguably more ongoing – was the continued homogenization of the different regions of the state when it came to conventions and competition. In spite of the fact that in 1920 FASNY was reportedly going through a period of decline in membership (their Annual Report of that year stated, "...mention of the fact [was made] that the membership is becoming less and less each year and that special effort was being made to revive interest, collect back dues and recruit new members...") their membership in 1922 included an all-time high of representatives and teams from Long Island. Among others, firefighters from Amityville, Baldwin, Babylon, Bellmore, Elmhurst, Elmont, Freeport, Flushing, Floral Park, Lindenhurst, Long Island City, Massapequa, Rockville Center, Southampton and Westhampton Beach were all listed as having new delegates that year...but it also may have helped that FASNY was celebrating it's 50th Anniversary that year, and that the convention was being held on Long Island, at Freeport. Both the Southern New York Firemen's Association and Nassau County Firemen's Association were listed as member organizations, but – for some strange reason – the SCVFA membership, as a body, was not included. That had changed by the FASNY 1924 convention in Fredonia, which had Suffolk included in its roster. By then East Islip, Greenport, Islip, Long Beach, North Bellmore, New Hyde Park and Roosevelt were listed as "New Members" as well.

On the track, on Long Island, other changes were being made as well. It has been mentioned earlier in this text that for whatever reason, Long Island firemen were just that much more competitive when it came to tournaments as compared to their brethren north and west. For this reason it's a safe bet to say that immediately after the first motorized fire apparatus made its way down the track in the Motor Contest, teams began trying to figure out different ways to make them go faster so as to beat the competition. One must remember that in these very early days of motorized competition, tournament teams were still using what is called "Class A" or "frontline" apparatus. That is, the trucks they were using to compete with on the track were the very same trucks that were being used to respond to actual fire alarms. In the years to follow the sport would grow and diversify to include special classes of racing "trucks" that were used only to compete with on the track, but in the 1920s, in order to compete at a tournament, a team would need to strip all the extra hoses, ladders, firefighting tools, etc. that wouldn't be needed at the tournament off of the truck, drive the vehicle to where the tournament was being held (sometimes quite a distance), compete, and then drive the truck all the way back to the firehouse. This traveling, running hard during the event, and then traveling back again was very hard on the equipment, which is one main reason that other,

The Dummy Contest at the SCVFA's 34th annual tournament in Sag Harbor in 1922

more specialized types of competition vehicles were eventually developed.

Perhaps it is because of their heritage, as hosts of the very first firematic tournament on Long Island, that the firefighters from Patchogue were considered some of the most competitive in the sport. This was illustrated even back in 1921, in a newspaper article reporting on the SCVFA's tournament in Riverhead that year:

"While coming home from the Riverhead firemen's tournament on Wednesday evening the VanGuard Simplex racing truck was held up in Brookhaven by Motorcycle Officer Jansen, who alleged that the machine was traveling at the rate of 45 miles per hour, says the Patchogue Argus. Lorenzo Hendrickson, driver of the car, was handed a 'ticket' for appearance in court. The VanGuards have engaged counsel to fight the case when the summons is returnable before Justice Johnson on Sept. 13th.

"The VanGuards are of the opinion that some of their 'hard luck' in the Riverhead tournament was not so much 'luck' as poor sportsmanship on the part of some enemies mad at last year's tournament. They charge that the rope which broke in the hook and ladder contest had been deliberately cut. In this event, also, their mascot, a snake which had made its headquarters at the VanGuard house for several months, fell from the truck and was killed. In the tournament parade Sherman Reid sat in the VanGuard running cart with the snake about his neck, representing a 'wild man'."[cclviii]

Dixon notes a few changes that went on as well in the early 20's with the Old Fashioned Muster competitions, such as the first recorded muster held on a cement platform (including a built-in drafting tank) that was specifically erected for those kinds of musters, at Quincy, Massachusetts on May 30th, 1921. Mostly, however, things in upper New England remained fairly status quo, with their historical notes of that period being comprised mostly of anecdotes, such as a muster held in Cambridge, Massachusetts later that year that was finished in the dark, with the course lit by paper lanterns, and an amusing entry regarding a muster held on July 12th, 1924, wherein "...When telephone lines crossing the street in Byfield, Massachusetts bothered the streams being played, they were taken down for the day, affecting both phones in town."[cclix]

That status quo of the hand-drawn musters in areas other than New York didn't hold true on Long Island, however. As Flugrath put it:

"With the extension of water supply systems throughout the county and the rapid growth of motor apparatus, the old hand engines gradually disappeared. When the [SCVF] Association was organized there was great enthusiasm among the engine boys and competition was at a high tension for several years. The old fire wells that furnished the water supply for these hand pumpers were slowly but surely giving way to the march of time and progress, and it is doubtful if any of these reminders of the early days are now in existence. The last of these events was listed on the 1920 Lindenhurst program."[cclx] The Steamer Contest, a long-standing fixture of firematic competitions, also had its swan song at that tournament. Long Island firefighters, it seems, were keen to stay at the forefront of technology, and in turn bring those advances to the track as soon as they were able. If the automobile was the wave of the future, so be it! Indeed, it was not long after, at the SCVFA's 36th Annual Parade and Tournament at Bay Shore in September of 1924, it was written that spectators "...witnessed the final withdrawal of hand-drawn machines, with 52 motor apparatus and no hand-drawn carts."[cclxi]

The early 1920's also saw more mention of new firefighter's associations. On April 2nd, 1923, the Sunrise Firemen's Protective Association was formed to serve the interests of firefighters from the eastern part of Long Island; they subsequently held their first tournament the following year in Westhampton Beach. Islip Town firefighters also began holding tournaments again in 1923, and the Huntington Volunteer Firemen's Association was holding their own tournaments at that time as well. It was at one of their tournaments that a rather shameful, and in hindsight somewhat bizarre, occurrence took place.

It has to be reiterated at this point that the world of the 1920's was a different place than it is now; there were different beliefs and mores then, some of which, in light of those of today, can seem rather barbaric or unconscionable. One of those involves a group called the Ku Klux Klan, which in the early 1920's was considered a respectable civic organization. Newspapers of the day refer to them in passing, as they would any other local organization, without any stigma attached to them whatsoever.

It appears that the Klan, as a respected organization, began sponsoring trophies for certain contests at firematic tournaments on Long Island as early as 1923, if not before then. These trophies, in many cases, still exist. Their whereabouts, however, are a closely guarded secret; although they represent an award for a job well done within the history of certain individual fire departments, it is understandable that at the same time they are not something that these same fire departments can be publicly proud of or show off in any way without risking public outcry or censure. So today, then, they rest tucked away, out of sight in attics and storerooms, an embarrassing reminder of a different, less tolerant, era.

At the Huntington VFA tournament that took place on Labor Day of 1924, however, it was business as usual, with the Ku Klux Klan trophy offered to the winner of the tournament. The trouble was, First Place that day ended up being a tie between the Huntington and Halesite teams, and no one was quite sure what to do about the trophy. Newspaper accounts describe a fair amount of discussion, but in the end, in what has to be one of the strangest decisions in the history of firematic competition:

"Huntington, L.I. – A silver cup donated by the Ku Klux Klan of this town, to be awarded to the winner of a tournament held on Labor Day by the Huntington Volunteer Firemen's Association, will be sawn in two and the halves given to two departments which tied for first place, it was decided at a meeting last night in Greenlawn.

"The meeting was attended by the chiefs and three delegates from each of the nine departments which participated. A letter from the Klan was read expressing approval of the decision. The halves of the cup will go to the Huntington and Halesite departments."[cclxii]

Whether or not the Klan had a presence in the northern and western parts of the state during the 1920's is not clear – at least through newspaper accounts – but those same accounts do seem to reinforce an ongoing attitude of their firefighters that perhaps placed their interest in competition second only to simply having a good time at their conventions, as indicated by a Page 1, Column 1 story about the Western convention at Warsaw in 1925:

"This is the morning after the night before of a 'very, very large party,' it being the third night of that party, and Warsaw awoke this morning with a clear head, looked back over the events of one of the biggest and most successful 'parties' Warsaw ever gave, and congratulated the Warsaw Fire Department and its chief on the way the party had been 'pulled off.' It was a wonderfully orderly and well behaved party too, considering that it WAS a party. It was not a prayer meeting nor a White Hope gathering, or anything like that; it was the annual outing of a bunch of the finest chaps to be found in Western New York, chaps who give their time and efforts

The Riverhead Fire Department Juvenile team, 1923

and sacrifice their health and clothes and money, often, in their work of VOLUNTARILY saving property of their neighbors from destruction by fire.

"The writer actually saw, with his own eyes, a few young fellows from out of Warsaw who were gaited a little queerly, in all three days. But it was their day, and I am not bigot enough to insist that everybody shall take his pleasures in the same way I like mine."[cclxiii] (Rather interesting, in light of the fact that Prohibition had been in effect since 1920!)

That the firefighters should want to be in "party mode" during that era, though, is not surprising. Since World War I had ended, things were really looking up. There was a marked increase in economic development in the United States and Europe, as well as "...several inventions and discoveries of far-reaching importance, unprecedented industrial growth and accelerated consumer demand and aspirations, and significant changes in lifestyle."[cclxiv]

It was the Roaring Twenties.

"The spirit of the Roaring Twenties was marked by a general feeling of discontinuity associated with modernity, a break with traditions. Everything seemed to be feasible through modern technology. New technologies, especially automobiles, movies and radio proliferated 'modernity' to a large part of the population."[cclxv]

In terms of the fire service and firematic competitions, the sense of "breaking with tradition" and "modernity" seems to have manifested itself in significant changes in the way tournaments were organized and run, especially during the latter half of the decade. The trend of homongenization of groups from Long Island into FASNY continued at an ever-increasing pace. At their 53rd convention in Mamaroneck in 1925, quite a large number of "new" organizations from the island had joined, prompting the secretary to write in his report: "Fifty new organizations have joined this year, twenty seven being from Suffolk and Nassau counties, where good, hard work was done by brothers Sammis and Howland."

Albert Sammis

Frank A. Howland, of Islip, was President of the SCVFA at the time, and Albert V. Sammis, of Huntington, was Vice President of the association, as well as Trustee of the Fireman's Home in Hudson, New York, a facility built and maintained by FASNY for the care of elderly and infirm firefighters. Although Howland should be acknowledged as doing a good job as the SCVFA President, it was Sammis who should be recognized as being one of the true movers and shakers of the sport of firematic drill team competition, as it was he who was directly responsible for "merging", as it were, the planning, administration and execution of modern-day tournaments on Long Island with FASNY's conventions, for the first time in the history of the sport. Flugrath writes: "It was largely through the efforts of Mr. Sammis that the tournament contest feature was introduced at the State convention gathering in Mamaroneck, on August 21st, 1925, when Mr. Sammis was chairman of the [SCVFA] tournament committee. Mr. Sammis was also chairman of the State tournament committee for several years thereafter."[cclxvi]

Now, one might be thinking at this point, "but FASNY has been holding firematic competitions at their conventions for many years now!" True enough, but one interpretation might be that in 1925, for the first time, the firematic contests held that year were done so separately, and more formally, from the convention activities themselves. In other words, from that point forward the "tournament feature", although both supported and sponsored by FASNY, would be a separate entity unto itself, as opposed to an informal gathering of firemen who wished to compete with each other once the formalities of the convention were over with.

It was also Sammis, together with SCVFA President Howland, who was first responsible for organizing – and formalizing – the sport of firematic drill team competition by creating, again for the very first time in history, a uniform list of firematic contests that would be held at a tournament, along with a set of associated rules to govern them.

"At the [SCVF] Association meeting of October 14th, 1925, President Howland, in his report, recommended that action be taken toward adopting uniform contests and governing rules. Vice President Sammis said his experience as an official at eight recent tournaments had shown the necessity of standardizing the contests with a view of fully determining the island champions in the several events and to establish proper basic records.

"President Howland then named a committee to confer with the Nassau Association, the Islip and Huntington town associations and the Sunrise County Protective Association."[cclxvii] This they did, on November 4th of that year.

Over the course of the winter Sammis worked on the problem with his committee. Then:

"After a series of fifteen sessions of the joint committee, Chairman Sammis reported to a special Association meeting, in Patchogue, May 12th, 1926, that the Nassau Association has previously adapted the list of 21 standard contests. In his report, Chairman Sammis said, in part:

> 'It is the thought of this committee to have no tournament or contest recognized by either one of the major associations unless the local tournament where such affairs are held is governed in every particular by the requirements set forth herein.
>
> <u>The 21 Standardized Contests</u>
>
> A – Ladder Climbing (Dry)
> B – Running Hose Contest #1 (Wet)
> C – Running Hook and Ladder Contest (Dry)
> D – Extension Ladder Contest (Dry)
> E – Motor Hook and Ladder Contest No. 1, Class A (Dry)
> F – Motor Hook and Ladder Contest No. 1, Class B (Dry)
> G – Motor Hook and Ladder Contest No. 1, Class C (Dry)
> H – Efficiency Contest (Wet)
> I – Department Contest (Wet)
> J – Motor Pumping Contest, Class A (Wet)
> K – Motor Pumping Contest, Class B (Wet)
> L – Combination Motor Contest (Wet)
> M – Running Hose Contest No. 2 (Wet)
> N – Motor Hook and Ladder Contest No. 2 (Dry)(Dummy)
> O – Motor Hose Contest No. 1, Class A (Wet)
> P – Motor Hose Contest No. 1, Class B (Wet)
> Q – Motor Hose Contest No. 2 (Wet)
> R – Booster Tank Contest (Wet)
> S – Chemical Contest (Wet)
> T – Motor Bucket Contest (Wet)
> U – Juvenile Hook and Ladder Contest (Dry)'"

Within the motorized contests, the three different classes listed above (A, B, and C) referred to the use of three different kinds of vehicles, as determined by a set of guidelines that had been drawn up. Although this author was unable to locate a rulebook from 1926 after the guidelines were created, it is assumed that those guidelines had not changed much by 1931, when these qualifications were listed in a Nassau County rulebook:

> "In Motor Hook and Ladder contests there shall be four classes, Class A to consist of Regular Factory Built Hook and Ladder Truck carrying 150 feet of ladder on permanent racks in middle of truck and to carry same in contests. Class B, Regular Factory Built Combination Fire Truck. Class C, all other rebuilt pleasure cars. Class D, all cars not propelled by gasoline motors. In Hose contests there shall be three classes. Class A, to be Regular Factory Built Fire apparatus and rebuilt trucks. Class B, to be rebuilt pleasure cars. Class C, all cars not propelled by gasoline motor."

As it turned out, even though it was Sammis, who was Vice President of the Suffolk County association at the time, who helped create the standardized contests, it was the Nassau County association who was the very first to try them out at their tournament in Great Neck on June 16th, 1926.

An unknown tournament in Nassau County in 1926

Apparently they were quite the success, and apparently that word traveled fast, as they were also adopted for use by the Central New York Firemen's Association that same month. According to Flugrath, "Roosevelt tried them on July 5th, Islip had them for the Southern New York tournament on July 17th. Floral Park used them at its annual affair on August 14th, while the State association proved their worth at Alexandria Bay, August 19th."[cclxviii]

On September 1st of that year, the SCVFA held its 38th Annual Parade and Tournament in Riverhead, and in doing so ran most – but not all – of the 21 Standardized Contests. Of the 15 contests that were held, the Riverhead News reported:

"Rapid time was made in most of the contests, but as the rules were standardized this year for the first time the officials did not consider that any record was broken; the records made here will become the initial records for other tournaments to be judged by."[cclxix]

Things, for the most part, went well:

"Of course during the afternoon there were many ludicrous situations when the contestants went skidding about in the mud, often on their backs, making flying leaps from rapidly moving motors to adjust a hose to a hydrant, or shin a ladder – but these were all part of the great 'doings.' And of course at times the fans in the grandstand got wet when a hose went on a rampage, but all such accidents were taken in good spirit."[cclxx]

In a testament to just how popular automobiles had become by that point (Wikipedia states that Henry Ford had sold 15 million Model T's by 1927), the newspaper also reported that "...the Standard Oil

The Port Washington "Rangers" compete in a hose contest at an unknown tournament in the mid-1920's

Company gave free gasoline and oil to all of the motors contesting in the events during the afternoon – over 700 gallons." (One certainly wouldn't see that today!)

If there was anything to complain about that day, it was that the Suffolk County association was perhaps a bit too enthusiastic about their new set of Standardized Contests, in that if you have too much of a good thing it can work against you:

"The contests were snappy to the point of thrilling, and there were lots of them. If any criticism could be made it would be to urge the association to eliminate a part of the games. There were so many events and so many entries in each that it was not until nearly 9 p.m. when the last was finished. Darkness had long since come, of course, and the final race had to be run by the aid of motor lights."[cclxxi]

But this rather minor complaint did little to stem the flow from the floodgates that seemed to have been opened as a result of Suffolk County's decision to try to move the sport forward by formalizing things. It was a Pandora's Box of sorts, or perhaps a Sleeping Dog that had been awakened; now that the subject of "we want to reorganize things" had been effectively raised, it was inevitable that every firematic organization from Niagara Falls to Montauk was going to have their own opinion on just how this should be done.

It should be noted that things moved in that direction pretty quickly at that point. Most likely through the influence of Suffolk County's own Albert Sammis – and the fact that there were now 90 Nassau and 41 Suffolk organizations who were part of it - FASNY was the first association to formally re-address the issue of firematic tournaments, which they did at their 55th Annual Convention at Niagara Falls on August 16th – 19th, 1927, less than a year after the introduction of the Standardized Contests. Sammis, as head of the State Tournament Committee, presented the following report to W. J. Coleman, FASNY President:

FIREMEN'S TOURNAMENTS, THEIR EFFECTS ON FIRE DEPARTMENTS AND COMMUNITIES AT LARGE

It is impossible in a short article to cover every phase of firemanic work as it pertains to the volunteers, but there is one branch which deserves special treatment, and should be brought more clearly before the people of New York State, particularly the taxpayer, and, in fact, everyone who carries fire insurance. It also effects directly the safety of lives of residents everywhere.

This phase of firemanic activity is known as the "Firemen's Tournament," where a practical application of real firemen's work is put in competition on a speed basis where the most efficient body of firemen are awarded prizes for doing a given task more quickly than the competing team.

These contests consist of hose races – both foot and motor, hook and ladder races, both foot and motor, and contests using motor pumpers of various sizes and capacity.

Firemen's tournaments have been held in Nassau and Suffolk counties for many years, and it is what might be termed the "hot-bed" of firemen's contests. The Committees who have stood behind these games for years have given much time and thought in order that all the contests might be brought to as near a practical standard as possible, eliminating to a large extent those things which would not be considered in regular fire service.

The results of these activities have been far reaching in their effect, both on the Fire Departments themselves and the benefits to the various communities, to say nothing of the benefits to the Fire Insurance Companies.

From a Fire Department standpoint the association of the fire companies in friendly rivalry has brought about a better feeling, and the interest becomes more keen in firemanic affairs in general, not only locally, but throughout the state. These tournaments have been the means of building up Fire Departments, bringing into them young men who have become a real asset to their various companies and to the community in which they live. They like the spirit of the thing, and when once you have the young men interested, all things are possible.

The tournaments of Nassau and Suffolk counties have become so large in attendance that it is almost impossible to complete the program in one day, for as many as 65 or 70 entries are secured in a single contest. Just think for a moment what this really means! It means that fire companies in many communities must give a considerable amount of time for two or three months in the year to keen, live practice in order to show their efficiency on Tournament Day. No one can deny but that practice of this sort in handling the tools which they have to use in actual fire service must fit them for the more serious work which may come to them any time, night and day.

Another result of the keen interest of the Fire Departments in tournament work is the demand for better tools to work with, better apparatus and better equipment, until today the territory covered by Nassau and Suffolk counties is unquestionably better equipped than any similar territory throughout the United States, or, in fact, the world.

Just at this point is where the taxpayers and residents of the various communities come in. The Fire Department, or Company, in any community, equipped with the most modern fire fighting apparatus, can not help but be a real asset to that community, and this is true of the equipment. The very fact that there has been a desire on the part of the firemen for apparatus of this sort makes it self-evident that the Fire Department and its personnel must be of high class.

There is, of course, in any Department what we might term 'dead wood," and there always will be, but comparing the Volunteer Fire Department with other organizations, you will find, without question or doubt, that the personnel of the Fire Department is above the average, and added to that the spirit of the Volunteer makes a factor for safety the value of which is inestimable and gives to the residents of the community that sense of security which permits them to turn over on their pillows and go to sleep, knowing full well that the interests of the community at large are being taken care of whenever the alarm of fire is sounded.

Recently there has been some rumors to the effect that the Insurance Companies throughout the state, and outside of the state, as well, have offered some objection or criticism of communities where their apparatus was allowed to be taken out of the district on certain days when Firemen's Tournaments were held, the claim being made that the various communities were left unprotected. This is very untrue and really a false statement of fact, for our experience has been that wherever apparatus has been taken out of the community it has always been replaced by something which would answer the purpose for the short time or, as an alternative, definite arrangements have been made with nearby towns to completely cover the territory where apparatus has been removed to attend one of these firemanic affairs. It might be said that the community fire equipment might be lessened to a certain extent, and yet the community has no claim upon the personnel of its Department, and all of the apparatus in the world would not be of any avail if there were not men at home to operate it. Men go to the city, go to work, to various sections of the country, and oftentimes the firefighting force is very limited, and yet it is a known fact that wherever fire has occurred in a live community it has been amply taken care of regardless of the time when such fire happened.

We have yet to know of a serious loss brought about through the shortage of equipment or manpower on account of the Firemen's Tournaments.

It would seem that the Insurance Companies were very short sighted in bringing criticism to bear on the Volunteers for being interested in something which worked directly to their own benefit, for surely the keeping of losses down to a minimum must be of real financial benefit to the Insurance Companies.

If an effort is made to kill the initiative on the part of the Volunteer it is easy to see what would result once the firemen became uninterested in firemanic affairs. The Fire Department itself would naturally become an uninteresting body to them, and in a short time would be of no value, at which time the only remedy would be the establishment of a paid Department, with a limited number of men, at great expense.

We have endeavored to interest the Fire Departments of New York State to follow the footsteps of Nassau and Suffolk counties, and hope the time may come when every county throughout the great Empire State will have its County Tournaments, and the smaller communities their local affairs, using Standardized Rules, making records, and with a final result that we may look forward to the time when championship races may be held at some point in New York State, taking the fastest times from each county and bringing them together for the great finals.

It would be a wonderful thing to see the racing teams from the eastern and western parts of the State competing for the highest honors in firemanic activities, and while the eastern end of the State might have some advantage over the western at this time, the spirit of the volunteer fireman is the same in both sections, and it will be but a short time when decisions would be so close that it would be hard to determine who were the real champions except by fractions of a second.

It is to be hoped and urged that in the near future every county will take this matter up seriously, and get into this most important of games with so far-reaching effects.

Although in his enthusiasm Sammis goes a bit over the top – it's definitely a bit of a stretch to imply that the overall fortitude and success of a local volunteer fire department is dependent upon them having a well-practiced drill team - one has to tip his hat to him for his efforts to promote the sport, especially in areas of the state north and west of Long Island. Sammis is, in fact, probably the most significant "unsung hero" in the sport of firematic competition, as it was he who was the first, with his pro-tournaments enthusiasm and tireless behind-the-scenes work, to cause the sport to grow from a relatively disconnected group of individual teams and local associations into an organized body of competitors. If not for his efforts, FASNY would probably never have taken over sponsorship of the tournaments, which ultimately led to the formation of the New York State Parade and Drill Team Captain's Association that exists today.

The FASNY convention that took place in Niagara Falls that year that Sammis presented his report was unique in that it was held concurrently and in conjunction with the Western New York association's convention. Ironically, though, in light of such a pro-tournaments atmosphere, there doesn't seem at all to have even been one at that convention: the Niagara Falls Gazette, in its coverage in the August 18th issue, goes into some detail about the parade and its participants, and what prizes were won for Best Appearing, etc., and then goes on only to describe a Prize Drill that was held on Kaltenbach Green, where the LeRoy Chemical Hose Company took first "on very small points of difference." There is no mention of any kind of races – hose, ladder, pumper or otherwise.

Back on Long Island, though, no grass was growing under anyone's feet, as no time was wasted with picking up the ball that Sammis had given. At the October 19th, 1927 meeting of the SCVFA, "...Vice President Robinson, in his annual report, expressed a hope that the first, second and third place winners of the respective contests of the Suffolk and Nassau associations could meet at some central place, to determine the Long Island champions, the winners there to meet up-state winners, selected by the same process of elimination, to decide the State championships."[cclxxii] Eight months later (in somewhat typical fashion) Nassau and Suffolk were still trying to get it together, as on June 27th, 1928, "...Vice President Robinson reported on the proposed inter-county meet, that Nassau had not accepted Suffolk's invitation to hold a series of contests to determine the Long Island champions."[cclxxiii] It wouldn't be until another year had gone by, on June 19th of 1929, that Robinson was finally able to report that as a result of a June 6th meeting with Nassau officials a championship meet had been arranged at the Mineola Fair Grounds.

But even though there may have been snags and

delays in getting the championship tournament organized, the level of competition on Long Island remained at an all time high. In what is probably an indication of this, another, more shameful "first" took place there at the time as well, on August 11th, 1928:

"Floral Park, L.I. – Five persons were seriously injured and a dozen more were cut and bruised this afternoon when a motor fire truck dashed into a crowd lining the course at a tournament in which the fire departments of many Nassau County towns and villages were competing.

"The Roosevelt Fire Department's hook and ladder truck, manned by the 'Brush Rabbits,' as the volunteers are known, and driven by Jack Bier of Roosevelt, was at the head of a runway preparing to speed several hundred yards, stop, raise a ladder and send a man to the top. David Herk, a Floral Park motorcycle policeman, rode down the runway to clear the spectators from the course. The crowd gave way and the truck started. Then Bier, with his machine gathering speed every second, saw that Herk had turned his motorcycle and was riding back directly in the path of the fire wagon.

"He threw on his brakes and swerved his truck to the right to avoid killing the policeman and the heavy truck dashed into the throng. Men, women and children were bowled over before Bier could stop."[cclxxiv]

Meanwhile, northern New York firemen just went about their business of simply enjoying themselves and each other's company. At their 1928 convention in Clayton, Norwood beat Massena in the Running Hose Race, and shenanigans were still king:

"At 5 p.m. 'Dutch' Moulton of the Norwood company, who had been the principal entertainer of the convention, was scheduled as 'Cyclone Pete' to leap from the roof of the Ellis block to the pavement below, without means of safety devices. A crowd of about 400 gathered at the scene and 'Dutch', after announcing his presence, plunged from the top headfirst amid the gasps of the crowd below. On viewing the body it was found only to be a well made dummy cleverly thrown from the building without detection."[cclxxv]

One will note that the only firematic contest to be held there that year was the "Running Hose Race". Indeed, in spite of the by-then-ubiquitous automobile having almost replaced cart and foot contests on Long Island, trucks still had yet to make an appearance on a northern track. It would not be long before that changed, however, as during the convention of the following year the following motion was passed during the meeting:

"No. 5. Resolved – That this association establish a motor hose race to be an annual event wherever the convention is held, to be known as the "Bi-County Association Motor Hose Race." Rules governing the race to be the same as our association rules."[cclxxvi]

And which rules might those be? From the same meeting:

"No. 2. Resolved – That the rules and regulations governing all tournament races of the Jefferson-St. Lawrence Bi-County association be the same as the rules and regulations that have been adopted by and are now being used by the New York State Firemen's association and that any changes made and adopted in state rules will be adopted by our association and that this association have sufficient copies of said rules printed so that all departments in our association may be furnished with them."[cclxxvii] Albert Sammis's influence on the tournaments, it seems, was continuing to spread.

Chapter 11: No Need for Prozac

Sammis's influence continued to be felt on the state level, as well. In a realization of his dream, the first tournament ever held by FASNY in which State Champions were crowned in each firematic contest took place at their convention in Saranac Lake on August 22nd, 1929. As Louis Marks, then Chairman of the New York State Tournament Committee, stated it in his notes: "...due to the fact that this and other tournaments will be held each year in connection with the convention, that the winners of the different races be declared champions of N.Y. State tournament events and that the winners of the point trophy be declared the state champions."[cclxxviii] He goes on to report (perhaps not surprisingly) that, "...It is interesting to note that all of the officials and most of the winners of the various contests were from Long Island, with the exception of the Potsdam Fire Department from St. Lawrence County. This department came in second in the Department Contest."[cclxxix]

So what team holds the honor of being the very first official New York State Tournament Champions? None other than the "Brush Rabbits" of Roosevelt, who won three of the nine contests held – Motor Hook and Ladder #2 (in 43 seconds), Motor Hook and Ladder Class C (in 26-4/5 seconds) and Motor Hose #1, Class B (in 33-2/5 seconds) - and tied in a fourth, Individual Ladder. Williston Park, Syosset and Inwood were the other winners that day. The latter took first in the Motor Pump Contest under circumstances that were perhaps unique, as "...all companies in this contest used the Saranac Lake pumper manned by the operator from Saranac Lake who operated for all the contestants."[cclxxx] (According to Marks, Inwood also came out on top in a water fight that was held on Main Street between six different companies, although no points were awarded.)

It is a bit ironic to note that even while Roosevelt – a team from Long Island - was being crowned State Champions on August 22nd, it wasn't until September 17th that Nassau and Suffolk counties finally got their act together and held an inter-county tournament at the Mineola Fair Grounds to determine who the Long Island Champions would be.

Although it seemed to take forever for the two counties to finally make the arrangements – discussions had been going on since the mid 20's – the idea for such a tournament was actually much older. According to Flugrath, "…This gathering was the development of a plan originally proposed by [SCVFA] President W. K. Post, at a dinner tendered by him to the officers of the Suffolk and Nassau associations at his winter home in Lawrence on April 21st, 1904."[cclxxxi] He goes on to say: "The Committee on Standardized Contests, in its final report on May 12, 1926, through its chairman, Albert V. Sammis, said it had in mind a final tournament where the fastest teams of the counties of Suffolk and Nassau could get together at some central part like the Mineola race track."[cclxxxii]

From the way it was described, the affair seems to have been a misadventure from the start. Apparently there was a fair amount of squabbling over just which contests should be included, and who was eligible to compete. Eventually the joint committee settled on six of the standardized contests, and limited the number of contestants to the first five teams to finish in those contests in the two county tournaments. By coincidence, both Suffolk and Nassau entered 20 companies of 15 departments each.

Unfortunately, the net result of all the bickering was rather inconclusive: while Suffolk took first place in four of the six events, Nassau scored higher in points, beating Suffolk 52 to 38. This fact did not go unnoticed by the tournament committee, as "…Owing to the failure of the executive committee of the two associations to agree upon a program of contests, a revised plan of scoring and listing of the respective county tournament winners for the final championship continuation of the joint affair was abandoned the following year."[cclxxxiii]

The decision to nix the inter-county tournament probably wasn't made until some time later within that year, however, as Flugrath reports that roughly one month later Albert Sammis – who had apparently been wearing his usual rose-colored glasses - was using that tournament to press his argument for a state-wide championship tournament:

"In his annual report at the meeting of October 19, 1929, President Sammis, spurred by the success of the Mineola competition of the previous month, urged that all the individual county winners throughout the State come together in a final struggle for the State-wide championships, similar to the Mineola plan."[cclxxxiv]

Those words were echoed by FASNY Tournament Chairman Louis Marks, in his report to FASNY secretary Fred Davis regarding the results of the state tournament that was held in Nyack the following year:

"Since the Saranac Lake tournament many new companies have corresponded with me and have been given instructions by mail on the ways of running these tournament races and many have come to my home town with their racing cars to get instructions as how to do several of the little stunts pertaining to the races…As Chairman I feel positive that next year, wherever the State affair will be held that many new teams will spring up and in time to come I feel it will be necessary for the different counties to hold tournaments and select the winners of each contest to be sent to the State Tournament to compete for the Championship of the State of New York."[cclxxxv]

Marks and Sammis were no doubt in bed together when it came to promoting the sport, but Marks had statistics on his side to back his/their argument up. Whereas during FASNY's first state championship tournament in 1929 there was an average of only 12 entries per contest, "...the second year there were 53 entries in the tournament, with 80 fire trucks and racing trucks taking part in 12 races with 263 individual starters and 26 officials from all over the state in charge."[cclxxxvi] (According to Marks's notes, the "Brush Rabbits" of Roosevelt again took top honors, but those same notes go on to say that "Due to a dispute regarding the point prize up to this time the prize trophy has not been awarded to any company or department.")

But in spite of the fact that Sammis was one of the greatest influences in the sport of firematic competition, his plan to hold a statewide championship tournament comprised of winners of county tournaments never did become realized. Instead, from that point forward, State Champions were crowned only as winners of a stand-alone state tournament, a practice which continues today, where, for one day only, the

Scenes from the first Inter-County Championship Drill at Mineola in 1929

The Potsdam "Sandstoners", 1930

playing field is leveled for everyone, and every individual or team has a chance to win, regardless of how well or how badly they have been doing during the season up until that point.

This theme, of "leveling the playing field", seems to have been one of the mainstays of firematic competition in more ways than one. Time and again, as world events happened throughout history that brought humanity to either new highs or new lows, firefighter's tournaments – at least of the modern variety - seemed time and again to remain relatively unaffected. A case in point: exactly one week after Sammis extolled his plan for a New York State Championship Tournament, Black Tuesday began the catastrophic fall of the stock market, and the beginning of the Great Depression. One year later, as it reached one of its lowest points, firemen in all parts of the state were still carrying on with their tournaments – and even improving upon them - almost as if nothing had happened. In northern New York, the convention of June 4th – 6th, 1930 at Carthage was augmented by the addition of firefighters from Lewis, Essex and Franklin counties, bringing the estimated total of delegates and visitors to 1,200. And, in what was surely a show of extravagance for the time, the delegates from Clayton – all four of them – arrived via airplane!

On Long Island things were much the same. Rather than things slowing down as a result of the times, they were status quo or better. A February 13th, 1931 entry in the minutes of the Southold Fire Department reads: "Chief N.E.

The Clayton "Calumets" at Alexandria Bay, 1932

Booth spoke in regard to us having a Firemen's Tournament in Southold this summer and he feels as though we will have a record crowd as the western part of the county are very anxious to come here to this part of the county." He was not wrong. The SCVFA's 43rd Annual Parade and Tournament took place in Southold on August 26th, and "...over 10,000 crowded the grandstand and sidewalks, and hundreds sat in their parked cars."[cclxxxvii] For its part, the SCVFA even added two new events to the list of Standardized Contests: the Combination Pump and Hose Contest ("V") was added on June 4th, 1930, and the Juvenile Hose Contest ("W") was added on June 8th, 1932 (although neither was adopted for use by any other organization, so they remained as Suffolk County contests only).

It is hard to say what spared the then-modern-day sport of firematic competition from the effects of the Depression, except that possibly it was saved by its very nature of being contemporary and cutting-edge, and thus more popular, as Albert Sammis portrayed it in his 1927 FASNY report. This argument is perhaps borne out by an entry by Dixon, as he describes how things were in the 1930's for the sport of old fashioned muster competitions:

"This decade was an extremely difficult one for mustering. The Great Depression prohibited any kind of muster sponsorship, and many associations folded up as members found it increasingly difficult to come up with cash to support them. With meeting halls being repossessed, and handtubs being broken up for scrap

metal and firewood, times looked very bleak."[cclxxxviii] Regarding the New England League Championship Muster that was held in Providence, Rhode Island in 1936, he goes on to say: "Also attending this muster was the handtub 'Uncle Sam' of Marlboro, MA which was manned by a crew of boys from CCC camp #197 of Fitchburg, MA. These camps, officially known as Civilian Conservation Corps camps, were prevalent during the depression, giving young men from impoverished areas something to do in return for three square meals and a roof over their head."[cclxxxix]

Just why there was such a disparity in the success of the old and new forms of firematic competition is difficult to say, considering that they were both made up of firemen, but one guess might be that the former involved the use of, well, old equipment and practices, while the latter made use of equipment and techniques that were cutting-edge, and in use at the time. This seemed to generate more enthusiasm towards the modern-day sport. In a nod perhaps to Holzman, former SCVFA President and Chief of Bayport W.K. Post was quoted at the time as saying:

"...Our meetings were not all play, far from it. Our tournaments had more to do in the development and improvement of the fire service in the county, and out of it too, than any other factor. Fire drill, and putting all in order after it, are hard work and interfere with the precious spare time of youth. As a duty only, it is apt to be neglected and has to be enforced with fines. But the sporting spirit of the young American can be depended on for almost anything. Put a tin cup in front of him, to betoken his triumph over a competitor, and he will cheerfully work himself to the bone. The Navy has used that spirit to make its engineering the best in the world."[ccxc]

It almost seems that ironically, as things in the world got worse and worse, the sport of firematic competition just got better and better. By many accounts, July 8th, 1932, marked the lowest point of the Depression, but according to a history of the Bayport Fire Department, the SCVFA's 44th annual tournament a month or so later was a record-breaker: "At the tournament in Bayport in 1932 a record number of 1770 persons were in the Association parade. It was also the highest record of 113 companies representing 43 Departments in the parade, with a record number of 144 pieces of apparatus in line."[ccxci] Even the weather seemed to get into the act: "A heavy storm in the morning and an almost complete eclipse of the sun in the afternoon was a feature...the storm ceased just before the parade started. The sun shone throughout the rest of the day and the eclipse was a spectacular success."[ccxcii]

Equally healthy was FASNY's tournament in Lake Placid on August 19th of that year, where an average of 30 entries was listed per contest, and the Mamaroneck "Tigers" took home championship honors.

In light of who won, it should be noted here that during the 1930's, in contrast to today, Westchester County teams were still very active in the sport as well. A program from that association's 34th tournament on October 8th, 1935 lists 11 different teams

TM: Nitty Carpenter was a hell of a good guy. He could do anything. Good football player. Good on the racing team. In fact, he was a driver on the racing team after Woodie [George Wood]. Mac McCarthy was the first driver, and then Woodie. I'll never forget Woodie with the Laraby. We'd put it-we would start--you'd have to push it, and then you'd run and get in. You stood so far behind it. And then, you had to run and get in the driver's seat. So, Woodie had it down. We had it down. Decided to practice it. All you have to do is run up and stop and push and go into second gear and pull the gas down. It had a gas ... [?] ... on the steering wheel, as well as one on the floor. So, he jumped on. We's all on. Here's Woodie standing on the side, coming down the course, not in the driver's seat. The quickest hook-up you ever seen, perfect time ...
Q: And he's not driving.
TM: And he wasn't in the driver's seat (laughs). We got disqualified because he wasn't in the driver's seat. If he just crawled in the driver's seat on the way down, we wouldn't have been disqualified. So, we got disqualified for that. So, we asked the--went up and asked for the reason, and we told them we wouldn't leave the course. We wouldn't take a-- then they said, "We'll tow the truck off the course" (laughs). We had a beef. And they showed us the rules, and then we had to comply by them. But it was perfect. It was good. You know, when you're racing, you can tell something when it's good. I mean, same as baseball, you could tell if it's going to be a good hit. I mean, you got that feeling.
Q: Everything clicks.
TM: Everything clicks. And it had clicked, yeah. Yeah, I had to laugh that day. But we had a lot of fun. I'll never forget Poughkeepsie. We were up to this tournament. And they had like a carnival in the night time up there. It was quite a college town. There was a lot of young girls up there. And it was easy pickin' if you wanted to. All the girls would ask you up to the college hill, and one would take a walk up to College Hill. So, we went into this--it was Harry Carpenter from the Atlantic's. Nicky Gazzo, myself, Mount Carey, and a couple of others--Ray Driscoll, I think it was. And we went in to see this girlie show. So, we get in and they charged us a dollar. And you see nothing. And then they put a couple of dancers up there all on the--and that was it. Now, they said three dollars to go in the back a little further to see it. So, a couple of 'em went in the back. And they came out, and they said it was a gyp. So, what did they do? They cut the tent ropes, and they put the old guy out of business (laughs).

(From interview with Port Washington firefighter Tom Mahoney, www.pwfdhistory.com)

competing in 10 different events, including both Class A and Class B motorized contests.

Northern tournaments, too, were advancing, albeit not at the same pace as those down south. After introducing a motorized contest for the first time in 1928, the 1932 convention in Alexandria Bay included no less than four motorized contests, including Class C Motor Hook and Ladder, and the 1934 convention in Gouverneur had 20 "squad cars" from 18 different teams entered. In addition, one of those cars even provided a bit more excitement that day than it would have by simply speeding down the track:

"Many thrills were provided at the Thursday tournament races of the Northern New York Volunteer Fire company association in this village. Just before Potsdam had finished in the Class C motor hook and ladder race the Marble City Boys' squad car took fire and box 15 at Gordon and Main street, close to the finish line, was sounded.

"The fire apparatus was compelled to make a detour to reach the scene. The squad car, one of the finest equipped at the meet, was not much damaged. The Marble City boys are not eligible for prize winning as they are the convention hosts, but the car was being kept busy."[ccxciii]

It is quite unusual to have a real alarm of fire occur at and during a tournament, but strangely enough a similar thing had happened on Long Island, less than two years earlier:

"Lindenhurst, L.I. – A demonstration of fire-fighting at the first annual tournament of the Volunteer Firemen's Association of the town of Babylon took a realistic turn here today when a shout of 'Fire' resulted in a race of all the participating contestants to a burning garage. A building large enough for two cars was destroyed, but the firemen saved the home of Frank Holbein about six feet away, when the roof caught fire from spreading flames.

"The last of the contests was under way and the Amityville Fire Department had clinched first place in the tournament when the alarm occurred. The Argyle Hose Company of Babylon, the last contestant, was preparing to give a demonstration of fire-fighting with chemicals when the cry of 'Fire' was heard from the reviewing stand.

"Speeding past the target at which they were preparing to aim, the hose company dashed to the burning garage on South Sixth Street, with the chemical apparatus ready to 'shoot'. The other companies rushed from Meridale Park, where the tournament was being held, and assisted in putting out the fire."[ccxciv]

An alarm of a different nature took place at a Western tournament during that same period as well. At their 1934 tournament held on July 24th at Tonawanda:

"An exhibition First Aid drill by the members of the Hamburg Volunteer Fire Department late yesterday afternoon turned into a real First Aid, when one of the spectators collapsed from heat exhaustion and the volunteers went to his assistance.

"Charles Wagner, 65, of 232 Grove Street, who had been working all day in the hot sun, went to the Tonawanda High School athletic field behind his home, with members of his family, to witness the demonstration. After the mock patient had been completely bandaged and treated and was about to be placed on the stretcher, Mr. Wagner toppled over, his head hitting hard on the gravel.

"He was revived by one of the squad and examined by Dr. J. T. O'Connor, who was in the audience. He gave his opinion that Mr. Wagner had been stricken with heat exhaustion."[ccxcv]

It has already been referred to several times in this text that firefighters can sometimes be a bit of a rowdy lot, and the 1930's were no exception. The enthusiasm of theirs that Sammis often referred to

The Port Washington "Rangers" compete in a ladder contest, ca. 1930's

was, a vast majority of the time, put to good use in a variety of positive ways that benefited the fire service. On the track, one of the ways it manifested itself was through the use of nicknames that the various teams would create for themselves, much like a baseball or football team would do. It would be extremely difficult to pin down just when the first nickname of this sort was used, but whereas earlier where teams had simply run using their traditional company names – the Otter Hose Company of Sag Harbor or the Van Guards of Patchogue, for example – the late 1920's and early 1930's saw a sudden blossoming of the use of team nicknames that were much more competitive-sounding, fun or even frivolous in nature. An illustration of this would be the lineup of teams that were competing at the Northern tournament in 1932: the Dawn Patrol of Dolgeville, the Mountaineers of Tupper Lake, the Panthers of Lake Placid, the Beavers of Beaver Falls, the Wolverines of Lowville, the Night Hawks of Carthage, the Stars of West Carthage, the Rockets of Harrisville, the Power City Pirates of Massena, the Sandstone City Boys from Potsdam, the Fireflies from Herman, the Racketeers from Norwood, the Ramblers from Dexter, the Ponies from Chateau-gay, the Snails from Norfolk, the Wild Geese from Alexandria Bay, the Scamps from Star Lake and the Greyhounds from Clayton.

Team names were often created in some very interesting and unusual ways. Since a number of teams were created in the late 1920's and early 1930's when the public consumption of alcohol was a sensitive issue, some team names were created as a result. The Red Onions of Spencerport, for example, took their name from a local establishment that provided "refreshment" to firemen during Prohibition, as did the Rum Pointers of Roosevelt, after the town garnered the nickname "Rum Point" during the same period due to its alleged number of illegal distilleries. In the same vein, the Guzzlers of East Islip started using their moniker after a comment was made from some East Hampton firefighters about their drinking abilities.

Not all team names were derived from such bawdy circumstances, however. The "Schlagers" of Islip Terrace took their team name from the German word "schlager", which translates as "someone who beats someone else", and the North Massapequa "Gremlins", who began competing just after the second World War, originally called themselves the "Smokey Stovers" but changed their name before the end of their very first season to the "Gremlins" to honor those team members and/or their family members who had just returned from the conflict: "Gremlins" apparently had been a nickname that had been applied to airplane pilots who were known for having been "invincible." And the very first team to host a tournament on Long Island – that same Van Guard Hose Company of Patchogue – changed its name to the "Forty Thieves" in 1914 to emulate the Forty Thieves street gang that was one of the very first to intimidate the citizens of New York City in the early 1800's – perhaps hoping to intimidate their competitors on the track as well.

There are many, many stories – too many to be included in this text – about how these names came to be. It is a tradition in the sport that has been a long-standing, and one that continues to this day as a bit of good fun.

As with any bit of good fun, however, it's all great until it starts to get out of hand, and history has shown that placing competitive, enthusiastic firefighters in a festive, celebratory atmosphere can sometimes lead to problems. Among a long list of pranks and hijinks, the turning in of false alarms during conventions is near the top, to the point where in one case, at the 1914 Northern convention in Potsdam, the city fire alarm system had to be shut down altogether. The 1934 Western convention in Tonawanda suffered from similar woes, but authorities decided to handle the matter a bit differently, as indicated in a feature newspaper story about it:

"Despite the warning issued by Judge Ralph W. Simon, while acting as temporary chairman at the opening business session of the Western New York Volunteer Firemen's convention yesterday, that the sounding of false fire alarms in the Tonawandas will not be tolerated during the convention, five of the fire alarm boxes in Tonawanda were sounded last night. This fact was known only to the police and fire departments, where gongs are located to sound the numbers of boxes pulled.

"That the general public was unaware of the alarms was due to the foresight of the authorities in deciding not to sound the whistles of the pumping stations during the convention, but in the event of fires found to be responsible for the alarms a general call was to be sounded for apparatus. On each alarm sounded last night only one piece of apparatus was sent out.

"The alarms were sounded from boxes in the outskirts of the city. The boxes in the downtown section of Tonawanda have been under almost constant guard by Boy Scouts, since the opening of the convention, which thwarted any attempt to sound alarms from them. During the parade this afternoon the boxes in Tonawanda and North Tonawanda will be guarded by members of the Tonawanda Post, American Legion who have volunteered their services so that the parade can be conducted without interruption from false alarms."[ccxcvi]

As is usually the case with these types of situations, there is a lot of finger-pointing that goes on afterwards as to who is to blame:

"Officers of the association said that they did not believe that visiting firemen were responsible for the false alarms, but that they were perpetrated by miscreants who took advantage of the convention to perpetrate such acts. Police Chief George C. Marohn this morning said that he has every reason to believe that the visiting firemen were responsible for the alarms sounded in Tonawanda. He declared that any person, no matter whether a fireman or other person, will be arrested and summarily dealt with if apprehended."[ccxcvii]

The net result of all this, though, was that whether or not firefighters were responsible for the mischief ended up being beside the point; either way the misbehavior became associated with the conventions, a stigma that in the coming years would play a significant role in how the future of the sport of firematic competition would progress. Having said this, however, it cannot be emphasized strongly enough that this was a relatively minor issue when considering the overall history of the sport, as in the early days of firefighting the volunteers were still greatly loved and appreciated by a large majority of the populace. Attendance of 10,000 or more at parades and tournaments was proof enough of this, and any shenanigans that took place were – at least at that time – somewhat accepted as being part and parcel of the fire service and their gatherings.

Oddly, there is little in the literature of the period to suggest that Long Island conventions and tournaments suffered from the same troubles as Northern and Western events did. There was fun to be had, for sure, but it seems that once again, from the perspective of the Northern and Western teams, the Long Island firefighters took their contests much more seriously than they did. Too seriously, in fact, some would complain, but it was that very criticism that may have indeed been their saving grace; they were just too busy taking their tournaments seriously to play games and get into trouble.

Evidence of this can be found in the Blue Book, where Flugrath indicates that during this same period the SCVFA stayed all-business, keeping its nose to the grindstone by continuing to try to change old tournament rules and create new ones so as to keep the playing field as fair and level as possible for all the teams in light of the continued growth of the sport. At their 47th meeting in Riverhead, on October 19th, 1932:

"Among the recommendations of the officers, was one by Third Vice President Colson, for the unification of the starting and finishing lines for the contests, in order to standardize the distances and help save time. Chief Frank Hubbard, of Bay Shore, on behalf of the Bay Shore department, urged the changing of the tournament date to the first or second Wednesday in July, in order to gain extra daylight for the contests."[ccxcviii]

And again, the following year:

"Speed development has been the mode of recent years, and thus the reason for the special attention given to the so-called 'speed' events for motor apparatus. However, consideration is also given to the smaller departments which do not have the opportunity to invest in expensive apparatus, the aim of the Association being to devise contests with governing rules in order to give every department an opportunity to compete in several contests.

"With that purpose in view, a special committee on revision of contests and rules, with the Executive Committee, held a series of meetings in several sections of the county in the early part of 1933 to obtain the views of the firemen."[ccxcix] (Flugrath did not relate the results of those meetings, as the Blue Book went to press before the results were summarized.)

THE COMPANY WILL APPRECIATE SUGGESTIONS FROM ITS PATRONS CONCERNING ITS SERVICE

1201-S

CLASS OF SERVICE

This is a full-rate Telegram or Cablegram unless its deferred character is indicated by a suitable symbol above or preceding the address.

WESTERN UNION

R. B. WHITE, PRESIDENT — NEWCOMB CARLTON, CHAIRMAN OF THE BOARD — J. C. WILLEVER, FIRST VICE-PRESIDENT

SYMBOLS

DL	Day Letter
SER	Serial
NM	Night Message
NL	Night Letter
CDE	Code Cable
LC	Deferred Cable
NLT	Cable Night Letter
	Ship Radiogram

The filing time shown in the date line on telegrams and day letters is STANDARD TIME at point of origin. Time of receipt is STANDARD TIME at point of destination.

Received at

NBV40 32 GOVT=THE WHITE HOUSE WASHINGTON DC SEP 19 1119A

GEORGE W HILDRETH=

MINUTES IN TRANSIT — FULL-RATE — DAY LETTER

THE PRESIDENT HAS LEARNED WITH ESPECIAL INTEREST OF THE TESTIMONIAL DINNER BEING TENDERED THE RIVERHEAD TOURNAME TEAM STOP HE HAS ASKED THAT HIS CONGRATULATIONS AND BEST WISHES BE EXTENDED TO ALL PARTICIPATING=

STEPHEN EARLY ASSISTANT SECRETARY TO THE PRESIDENT.

1131A

The telegram from the White House received by the Riverhead "Ironmen" congratulating them on their 1935 State Championship

In spite of this, Suffolk firefighters were not completely without a sense of humor. At FASNY's 63rd Convention and Tournament that took place in Greenport on August 23rd, 1935, the Ironmen of Riverhead took State Championship honors, in spite of having re-formed the team only eight months earlier. In honor of their win, the merchants and residents of Riverhead sponsored a "Testimonial Dinner and Entertainment" for the team at the Hotel Henry Perkins. In the spirit of the moment, as a goof, the team sent invitations to the event to President Franklin D. Roosevelt, New York State Governor Herbert H. Lehman, former Governor Alfred Smith and Internal Revenue Collector Alnon G. Rasquin. Surprisingly, they actually received a telegrammed reply back from the President, offering his apologies for not being able to attend and congratulating them on their win!

The Riverhead firemen, it seems, really knew how to celebrate: Two years later, after winning the 1937 Suffolk County tournament, the Riverhead News-Review wrote:

"Joseph F. Walsh's invincible Iron Men of the Riverhead Fire Department added another collection of scalps to their string Wednesday and made themselves county champs.

"With speed and brawn they bowled over nearly everything in sight at the annual Suffolk County firemen's tournament in Port Jefferson, rolling up point after point until they had collected 34 of them, then they called it a day and came home.

"Here they were met by other members of the local department and a procession was started into town, with sirens screeching, fire sirens blowing, motor horns tooting and what have you. And they were met too, by their wives and sweethearts of the Women's Auxiliary at the firehouse, and dutifully proud women as they are where their men folk win laurels, the women had a nice dinner ready for the victors.

"And after dinner the big crowd stayed in the firehouse and whooped it up in merriment until a late hour. Most of the tournament judges, several of the Greenport firemen and others from other sections came here to join in the celebration."[ccc]

Things seemed to calm down a bit at the Northern and Western conventions in 1936, with no mention of any problems associated with false alarms or other misconduct. However, good-natured horseplay and pranks were still in order. According to the Watertown Daily Times, "...One of the fire department bands put on a rehearsal in preparation for this afternoon's parade by marching along the downtown streets at 3 this morning. Howling sirens, clanging bells and automobile horns furnished accompaniment for the musicians in their impromptu concert."[ccci] On the track, at the same drill, "...One squad team member disqualified his team in the bucket contest by leaving a bucket up on the arch and paid the penalty with a grin when his teammates tossed him bodily into a vat of water at the edge of the course. Another racer did the same thing and was doused with a bucket of water."[cccii] Out west it was much the same, at the convention in Rochester:

"Volunteer firemen substituted the picket line for a hose line yesterday.

"A number of the delegates to the Western New York Convention took placards away from WPA white collar strikers yesterday and took up the march around Convention Hall where the strike is in progress. They even carried the slogans into the basement of the hall and gave workers who remained on the clothing distribution project somewhat of a scare.

"At first, the workers thought the strikers were descending on them for hand-to-hand combat. After parading around inside for a while they returned to the street and relinquished the signs.

"Always the gentlemen and equipped to rise to any emergency, the firemen set up an impromptu picketing code. No woman could picket while they were around. When one took a sign, from two to a dozen of the 'volunteers' sprang to her side and took the label during her planned tour of duty.

"'More fun than a three alarm fire,' said one fireman. 'And it's good training for the parade this afternoon.'"[ccciii]

At the same convention, there was also another case of inter-team rivalry and revenge, similar to that between Carthage and Sackets Harbor in 1912 (involving a banner) and Lockport and Niagara Falls in 1915 (involving a dummy in uniform). This time it was in Rochester, between the exempts of Barnard and St. Paul, involving a hose cart:

"According to the St. Paul firefighters it was all bravado on the part of the Barnard men that brought belated justice. Not only did Barnard's exempts steal the cart but yesterday they bragged exceedingly by displaying the cart in the parade of the Western New York Firemen's Association.

"Henry Diechmiller, an old Barnard resident, who now lives in the St. Paul district, constituted himself a committee of one to watch the cart. He pursued it up Main Street in the parade and around to Convention Hall where the Barnard men broke ranks. He watched its transfer under canvas to a vegetable truck and followed the truck to Dewey Avenue.

"Then while Barnard men returned to town to celebrate the convention with visiting firemen, he called his St. Paul friends. To his aid came Walter Hess, Edward Slater, Irving Briggs and George Baist. They too had a truck and now the hose cart is home in the St. Paul section. Its present owners are not disclosing its whereabouts.

"St. Paul firemen said last night they might have let the cart and the possession rest but for that sign displayed over it in the parade

reading 'Still Barnard's Truck.'"[ccciv]

A rather embarrassing explanation of this appeared elsewhere in that same issue:

"Over the wheels of the cart in the parade was the large sign, 'Still Barnard's Truck,' a pun to remind spectators that some months ago the truck was found by federal agents in a Greece building that contained a wildcat still."[cccv]

The tournament there that year was amusing as well, but for the wrong reasons: only one team showed up. As the story reads:

"Point Pleasant 'Pea Pickers' had a field day yesterday. In the absence of competition they captured all five prizes offered in the tournament.

"With their green hose cart and bright new uniforms they reported for duty on South Street prepared to do battle with the best of the association. So they climbed ladders with the speed of monkeys, coupled their hose lines to the hydrants in record time, rallied their forces with a speed and efficiency which drew applause from the veteran spectators and rounded out the demonstration with a first aid routine.

"Then they settled back to see how the five companies from the Buffalo area who had been entered in the event do these things. But the entrants were all busy with parade plans and failed to appear."[cccvi]

In hindsight, the fact that this happened may not have been all that surprising. In spite of ongoing efforts at the time to standardize the rules of the sport and the regulations under which the tournaments were run, most of the work being done in those areas was based on Long Island, and – as we have seen – those kinds of changes weren't happening overnight. In addition, any new policies that were eventually enacted in Nassau or Suffolk took some time to reach the other areas of the state, as much as they may have been willing to go along with them.

As a result, Northern and Western tournaments of the middle to late 1930's were still relatively disorganized affairs compared to those on Long Island. According to Tommy Champion, a member of the Sea Breeze "Indians" team who raced with them in the 1950's, aside from the fact that each team brought one Official, most tournaments were conducted on an informal basis, with no governing body overseeing them, and no real structure or set of rules. "You didn't even know what races you were running until you got there the day of the tournament," he says.

Not that Long Island was much help. According to an article in the Autumn, 1938 issue of The Volunteer Fireman, the house organ of FASNY, Nassau and Suffolk's continuing efforts to create order out of chaos wasn't going so well. After the SCVFA's Golden Anniversary tournament in Smithtown on July 13th, in spite of the fact that "...State and county records toppled like ten-pins..."[cccvii] the following article appeared:

> "This past season an effort has been made to establish a system of so-called State records by the adoption of an extended list of contests, designed to be uniformly used throughout the State, by county and other local group tournaments and at the annual tournament held in connection with the State Firemen's Association convention. While the plan may be idealistic in its intent, yet the several contests being subject to local interpretation in parts and changed ever so slightly so as to meet local desires, it is questionable whether the high ideal of unifying the forms of contests for the purpose of establishing a series of so-called State records can be fully continued.
>
> "In both Nassau and Suffolk Counties, where the spirit of competition exemplified in the respective local tournaments is strongest of anywhere in the State, not all of the contests established by the States Rules Committee were carried out as originally laid down; changes were made in several of them, and also added to the various programs were forms of contests previously used by the respective local groups. Quite naturally, the changing of a single word or even the slightest condition in any contest by any local group at once nullified any prospect of those changed contests from being eligible for a State record.

The Manhasset-Lakeville "Minute Men", 1938

"Along the line of standard State forms of contests is a justified criticism brought forth by one of the best informed and best known of Long Island's volunteer firemen, in that the forms of contests should be simplified to that there would be less confusion of starting points for all contests, dry and wet. Said this critic: "The present track layout under the State rules, and even for our local organizations heretofore, has too many starting lines, which makes the track look like a football gridiron; an effort should be made to reduce the distances to be run to as few as possible or, practically, one starting center. As to the wet contests it is impossible to set a fair standard record due to the difference in water pressure at the various places where the tournaments are held. At one of our Island tournaments the hydrant pressure was around 90 pounds, and at other tournaments the pressure ranged to figures much less. It is impossible to obtain a fair time record for any wet contest where the pressure varies ever so slightly."

"Recalling a bit of Long Island tournament history, a movement was started thirteen years ago to establish a series of standardized contests with a view of fixing permanent State records therefore. A joint committee of representatives from the senior Suffolk County and Nassau County firemen's association, after a series of extended discussions, presented a series of Standardized Contests that were used in the 1926 tournaments of the Nassau and Suffolk associations, the State convention and several other up-State county and local tournaments. Each year thereafter changes were made by the different organizations in these Standardized Contests until finally they were scarcely recognizable from the original 1926 draft."[cccviii]

Chapter 12: A World's Fair, and a World War

Even as the sport struggled along with growing pains, it continued to stay true to its nature as it defied all logic by thriving as the rest of the world went to hell: as on August 18th, 1939 newspaper headlines screamed "2000 Bombers Invade Britain", the previous day saw the sport of firematic competition have one of its proudest, grandest moments in its entire history, as FASNY held its 67th Annual Convention and Tournament in Flushing, Queens – the site of the 1939 World's Fair. That tournament was "…'the most successful ever held in the history of tournaments held by the Firemen's Association' according to Henry N. Buck, the Chairman of the Tournament Committee. 'There were eight events with two hundred and forty starters in the time of seven hours and thirty eight minutes. The contests were run under State Rules and there were no protests. Over 85 beautiful trophies were presented and several new records were established.'"[cccix]

"The annual tournament of the association opened this morning on Blossom Avenue, between Main and Lawrence Streets, Flushing, one-half block from the Flushing entrance to the World Fair," wrote the Long Island Daily Press on the 17th. "A highlight and probably the most spectacular feature of the organization's 67th annual convention, now in progress, the tournament attracted 5,000 spectators before 10:30 a.m. Special grandstands on both sides of Blossom Avenue were filled."

Several State Records were broken that day. As the Victory Chemical Engine and Hose Company of West Hempstead (also known as the "Westerners") took State Champion honors, there were record times set by Franklin Square in the Motor Pump Class A contest and Port Washington in the Efficiency contest. Buck, in his report, glosses over it completely, but a surprising record time was also set in an unexpected category: the length of the drill overall.

"State officials set a new time record in running off the eight contests in which 61 crack racing teams of New York and New Jersey fire departments were entered. The entire tournament took only eight hours whereas in other years it took from 16 to 20 hours.

OFFICIAL PROGRAM
NATIONAL TOURNAMENT · NATIONAL PARADE
FIREMENS ASSOCIATION STATE OF NEW YORK
INCORPORATED
67th Annual CONVENTION
FIREMEN'S ASSOCIATION
STATE OF NEW YORK
FLUSHING . . . . N. Y.
NEW YORK
home of WORLD'S FAIR 1939
TOURNAMENT · AUGUST 17TH, 1939
PARADE · AUGUST 18TH, 1939
25¢
© N.Y.W.F.

"Yesterday's contests were expected to take 13 hours but President John J. McConnell of Flushing attributed the efficient handling of the tournament to the police protection afforded by 100 patrolmen under the command of Inspector Charles P. Mooney and Deputy Inspector John J. Donegan."[cccx]

Unfortunately, another high-water mark of a different sort was also made, as a record number of injuries occurred that day as well:

"Twenty-two persons, an unusually high toll, were injured during the running of the contests. Most of the injured were firemen taking part in the events but the person most seriously hurt was a woman spectator, Mrs. Marie Hertzhe, 35, of 132-18 Maple Avenue, Flushing.

A 1939 tournament in Sea Breeze, at their original track on White Street

"Mrs. Hertzhe, the wife of a New York City fireman, was severely injured trying to dodge a stream of water which poured from an engine pumper during an incompleted contest trial.

"She was sitting on the south grandstand with her husband, Otto, watching the events as the Bernardsville, N.J. team[7] ran down the course during the Motor Pump A contest. The men had coupled the suction hose to the engine in an effort to aim the water-filled hose at the target when the hose suddenly parted from the engine, allowing the pump-driven stream to pour into the stands.

"Hundreds of spectators fled the powerful water pressure but Mrs. Hertzhe was hit squarely by the stream and knocked among the grandstand seats.

"She was removed to Flushing Hospital in a hospital ambulance on duty at the scene and went home after being attended for shock, leg contusions and abrasions of the body."[cccxi]

The "incompleted contest trial" that day was the Motor Pump Contest Class B, and it is fortunate that even more spectators weren't injured in the same manner as Mrs. Hertzhe. Every team on the roster had trouble with it that day, as none were used to or prepared for the higher amount of water pressure that New York City used in their hydrants, compared to the outlying areas where most of the teams were from:

"In the motor pump contest, Class B, the Brentwood Racing Company of Brentwood staged a surprise by capturing this trial in 11-2/5 seconds as the highly-touted teams failed to control the high pressure water in the city hydrants.

"The water pressure accounted for the number of injuries and the failure of the teams to complete their hose contests as they let the streams of water spray the crowd and police and fire officials."[cccxii]

---

[7] From all research and indications, the participation of the team from Bernardsville, New Jersey at the 1939 State Tournament in Flushing marks the only time in the history of firematic competition in New York State that a team from outside the state has competed in a New York State Tournament. According to Bernardsville firefighter Jim Berenger, whose father was on the Bernardsville team, some New Jersey fire departments of the time had similar kinds of teams and would compete against one another on an informal basis, and continued to do so until the early 1960's when safety concerns led to their having to be discontinued.

The West Sayville "Flying Dutchmen" during the C-Hose Contest at a tournament in 1939

All of this was quickly forgotten the next day, when one of the most spectacular parades in the history of the New York State fire service took place. It was truly a day to remember: there were 10,000 firefighters in line, and "...the parade was viewed by more than 40,000 cheering spectators, one of the largest crowds that was ever attracted to a similar event in the history of Flushing."[cccxiii] (FASNY Statistician Henry Buck wrote, "...at noon, when the parade was at its height in the business section of Flushing, the mobs had swelled to 100,000" That's quite a discrepancy in numbers, but either way there is no doubt it was a huge crowd.)

The FASNY convention of the following year was held in Dunkirk, New York, but for reasons lost to time their associated tournament was held, once again, at the World Fair grounds in Flushing, Queens on October 12th, in conjunction with Volunteer Firemen's Day. Very little, if anything was found in writing describing the contests, at least as far as newspaper accounts go. The only description of what happened that year comes from SCVFA Secretary's notes:

"It was a 'nip and tuck' struggle between Bay Shore and East Hampton, Suffolk's high scorers, and West Hempstead, high scorer for Nassau County, until the final time was announced in [the] last contest; but on the runoff of the tie between Bay Shore and East Hampton in 'dumping the buckets' Bay Shore beat its original time, winning the event and the much coveted department championship. West Hempstead was placed third, in that event."

The Clayton "Calumets", 1941

The Suffolk County association continued to move the sport forward in other ways as well. A new contest, the Three Man Ladder Contest, was added at or about that time, and on February 2nd, 1940, the Suffolk County Running Captain's Association was formed, "...to eliminate arguing and the time wasted discussing rules on the track."[cccxiv] The formation of this association made it the first group, organized separately from a firefighter's association, to be created solely to address the organization and needs of firematic competition teams in the history of the sport. The name would eventually be changed to Drill Team Captain's Association, and other areas of the state soon followed suit by creating Associations of their own.

In stark contrast to the Flushing parade of 1939, the parade of the Western firefighters at their 1940 convention in Brockport was the antithesis of "huge". Smaller in comparison to begin with, it got even smaller due to the extremely warm weather. How small it was for sure we'll never know, as the "provincial" (and somewhat humorous) reporting style of the local paper attests:

"The Volunteer Firemen's parade was all that was expected. Many companies, bands, drum corps, etc. were in line. Also, as expected, the temperature played a big part in the event by cutting down the line of march from five and three tenths miles to approximately three and a half miles. It was just too much for those taking part to take.

"This writer missed his cue to give an accurate description by sitting on his side porch waiting, pencil in hand, for the parade to go by. It did, in part, but not the entirety.

The Cutchogue "Panthers", 1940

"Anyway, the whole affair so far has been a huge success and the balance should be likewise.

"Playboys of the various departments have made merry by building large fires on the pavements, guests from far and near have come back to Brockport for the event we are now celebrating.

"As this is being written, horns, sirens and what have you, are being blown, people are shouting, it is difficult to write, however, the occasion is one we will not have again for many years – the last one was, should you recall, in 1911."[ccxv]

Note that the writer refers to how they "...made merry by building large fires on the pavements..." It's not clear when this form of "celebration" started to come into vogue, but there were signs that they were becoming more and more prevalent each year, as indicated by a story in the paper about the Western convention in Dansville the following year:

"Out of a possible thousand firemen in town Tuesday evening to take part in the Firemen's Convention and Parade, a few became over-hilarious to such a degree that they became over anxious to see local fire fighting methods in operation, and are reported to have started several blazes which caused more or less concern and more damage.

"General alarms, sounded about ten thirty o'clock in the evening, one thirty and five o'clock in the morning, brought the Dansville Fire Department into action.

"A barn in the rear of the St. Patrick's Parochial School, the owner of which, according to report, had granted permission to have it torn down or disposed of, the fence blocking off the F. H. Willey vacant lot between the Super Market and the Ferguson Dress Shop and a wagon belonging to the Scovill estate, all went up in smoke. Whether Juveniles or adults were responsible for the fires has not been determined and police are investigating.

"A small group of enthusiastic fire-fighters also started bon fires in Main Street, stopped automobiles passing through the business section and otherwise demonstrated their gaiety during Tuesday afternoon and night."[cccxvi]

The Bellmore "Ballbreakers" during the 1940's

At this point, during the summer of 1941, in spite of America watching anxiously what was going on overseas as World War II raged through Europe, the running of firematic tournaments continued in a more or less status quo manner. One indication of this comes from a local Riverhead paper from that summer's racing season:

"There is one sport on Eastern Long Island, it seems, that even a rain can't wash out – a sport that is popular with a large following, too. It's a firemen's tournament.

"So, in spite of the bad weather on the Fourth, the North Fork Volunteer Firemen's Association went calmly ahead with its determination to hold its annual tournament in spite of the bad weather, and the fair attracted a considerable number of people to the Fair Grounds in Riverhead during the afternoon.

"Riverhead played the host on this occasion, and perhaps the local firemen weren't as polite as hosts are supposed to be, because they won the tournament."[cccxvii]

Then, as we all know, everything changed on December 7 when the Japanese attacked Pear Harbor; the nation was once again at war.

In comparison to the First World War, World War II had a much more significant impact on the fire service, and in turn its conventions and tournaments. The worst hit, perhaps not surprisingly, were the old-fashioned New England musters. Dixon writes: "The outbreak of World War Two saw the continued downward spiral of Firemen's Musters. During the first six years of this decade there was an average of one muster per year, averaging 12 handtubs each."[cccxviii]

Modern-era competitions changed and adapted as well. Although the United States had not yet entered the war, the FASNY tournament at Geneva on August 21st, 1941, included a new contest that was probably promoted – if not inspired - by current events: the First Aid Contest. According to state Statistician Buck, "They were very instructive, and in line with National Defense."[cccxix] Kelmworth, Point Pleasant, Albion and Bellmore finished in that order in that inaugural running.

Although they were new to FASNY tournaments, first aid contests had been taking place on a local level since the mid-1930's, with the contest that took place in Tonawanda at the Western convention in 1934 being just one example. In that case there was no war to think about; the idea behind the contest had come from a simple effort to get firefighters to become better at their job:

"The drill…had been to take a hypothetical problem of a case as follows: At a large fire in a business block, a fireman working on the second floor is overcome by smoke. He is found at the foot of the ladder with the following injuries: Right lower arm, compound fracture, bleeding in spurts; left upper leg, simple fracture; left hand, burn on palm; three inch cut on top of head.

"Herbert Hunt, the instructor, announced the treatment, as it was applied. In a few words, he stated that more persons were injured in 1932 than were killed and injured during the World War. He advised all firemen to take up First Aid and everyone to avoid accidents when possible."[cccxx]

Once America had entered the war, the War Effort had an immediate impact on how firematic conventions and tournaments were run. This time, it seems, the tradition of continuing merrily along while the rest of the world struggled did not hold true. The sport was affected on all levels. On a local level, for example, winners of tournaments in Monroe County were given blue ribbons instead of medals in order to ration steel, and at the Northern convention in Carthage in 1943 the decision was made to eliminate the use of motor vehicles during the parade and tournament:

"Despite the decision of firemen meeting in Carthage for the 47th annual convention of Northern New York Volunteer association to use no fire apparatus in races Saturday morning nor in the annual review of departments Saturday afternoon, the events, witnessed by a large crowd, were interesting and successful."[cccxxi]

The decision to suspend the use of motor vehicles continued for two more years, as trucks were not used at Northern tournaments again until their convention in 1946.

Out west, the effects of the war were felt as well, at their convention in Amherst:

"The blackout test on Tuesday evening came as a complete surprise and at a time when the celebrating of the visiting firemen was at its height. But the only thing that was dimmed out were the lights, and the fun continued.

"The blackout lasted for 24 minutes and was very successful; only a few violations were reported."[cccxxii]

On the state level, 1942 marked the 70th anniversary of FASNY, but rather than celebrate the officers instead did just the opposite. At their convention in Syracuse that year:

"In calling the 70th Annual Convention to order Mr. Britton A. Vannauker, Chairman of the Convention Committee of the Onondaga County Volunteer Firemen's Association, announced that this would be a 'streamlined convention.' Because of the war, the Executive Committee decided not to have the usual parade and tournament.

"This convention was originally scheduled to be in Rome, but due to the existing conditions – we were now in the Second World War – a last minute change was made. Several thousand people had moved there 'due to the Federal Government Air Depot program' and with gas and rubber rationing in effect the Local Committee felt that 'their efforts to put on the convention in the way they had planned would be definitely handicapped, and that the members would be disappointed with the results.' They asked that Rome again be considered as soon 'as the war is over and Peace has been again proclaimed in our land.' Six years later the convention was held there."[cccxxiii] (Under circumstances at which, as previously mentioned, would contribute to dramatic changes in the organization and running of tournaments in New York State.)

On Long Island, it appears that Suffolk County also ceased to hold tournaments during the War Years. Local newspapers that faithfully and consistently reported on the annual SCVFA tournament have no stories – or even mention - of them during the years 1942 through 1946. There is even an indication that tournaments were not just postponed or cancelled due to everyone being too busy with the war, but were actually banned until then, as this April 1946 article implies:

"The annual meeting and election of officers of the North Fork Volunteer Firemen's Association was held at Shelter Island last Wednesday night. With the ban lifted on the holding of firemen's tournaments, this phase of firemanic activities brings no little discussion. The matter of whether the North Forkers will stage one of these events, and where, is to be decided in the near future."[cccxxiv]

In 1946, it appears that the first tournament to resume taking place on Long Island was the Southern New York VFA tournament in Islip on June 27th, and indications are – based upon written accounts of invitations received by a Mattituck team to participate that year that didn't include it - that the Suffolk County drill didn't resume until the following year.

As for the state association, at the outset of the war they continued gamely on. There are no notes to indicate whether or not a tournament was held at FASNY's 1943 convention in Buffalo, but later that year, on November 27th, the State Tournament Rules Committee met in Hudson to hash out modifications of the rules used at state tournaments. In addition to technical changes in specific contests, the following was written:

"A lengthy discussion was held with reference to the present contest rules and it was the unanimous decision of the meeting that our present rules are cumbersome and should be streamlined. Mr. [Benjamin J.] Huskisson spoke on this subject, stating that reference to the number of lengths of hose and that they be so many feet in length, and that so many lengths be laid in each contest, should be dispensed with, as this information is described under Classification and Definitions, stating he had, several years ago, drawn up for Suffolk County a condensed set of rules, but as they were not adopted by the State Association, were not used in Suffolk County."[cccxxv]

It was as if, once again and all along, everybody was getting sick and tired of how complicated it was all becoming. The minutes go on to say:

"It was also brought up at this meeting that about six years ago Mr. W. K. Post of Bayport, had made up a set of rules which made the contest more along the lines of actual fire fighting methods, and not speed races. Several of the members present knew of these rules and told of their merits, and it was thought by the assembly present, serious thought should be given to these matters so that our tournaments could be made more interesting to the spectators, as well as making for better and more efficient firemen. It was recommended that copies of the Post Rules be obtained and that Mr. Huskisson work out a plan along the lines he mentioned for a simplified and condensed set of Rules, to be presented for future consideration."[cccxxvi]

The 72nd convention of FASNY took place as scheduled in Rochester in 1944, and in November of that year the Hudson-Mohawk Volunteer Firemen's Association approached the state association with a request to host the following year's convention in Albany. In May of 1945, however, FASNY President Albert J. Foley received word from Hudson-Mohawk that they were "releasing themselves of an unpopular movement at this time."

"The 'unpopular movement' was a directive from the Federal War Committee on Conventions denying the permit to hold the 1945 convention. The committee did state, however, that 'this action does not preclude you holding a meeting attended by not more than 50 persons from out of town, plus local attendance without numerical limitations.'

"President Foley explained that 'Our application was turned down because of the transportation problem involved, especially in this state. Most of the troops coming from Europe would be landed in New York and would go out from there by train, bus and air. They are necessary troop movements in the prosecution of this war. We as volunteer firemen need no other reason to refrain from holding our annual convention.'"[cccxxvii]

They did refrain, in a fashion. On August 21st – 22nd, 1945, a "convention" of sorts was held, but it was attended only by certain committees, including the Tournament Committee. Whatever was discussed – no doubt it included the dropping of the Atomic Bomb two weeks earlier - was eclipsed less than two weeks later, when Japan surrendered on September 2nd, and the war was ended.

As one might imagine, the end of World War II had a significant impact on all phases of life in the United States – in mostly positive ways. One description, found on the world wide web, states:

Ladder Practice, 1946

"After the war ended and the soldiers came home, America began to recover. It was a great time to celebrate; the economy was back on its feet, jobs were plentiful, and families were reunited. The 1940s ended with America, particularly Florida, moving into prosperous times."

Wikipedia, an online encyclopedia, echoes that sentiment:

"America became a superpower. Our economy had grown. Our society had remained completely intact (Compared to the thousands of German/British/Japanese/Soviet/French schools, hospitals, banks, and other institutions that had been destroyed) so that by 1945 we were bypassing other nations in every field of endeavor."

Or, as another author put it:

"When the war came to an end in 1945, many Americans wondered if the higher standard of living brought about by a wartime economy could be sustained in peacetime. Having experienced the Depression, they feared yet another crisis once millions of soldiers reentered the civilian workplace in an economy no longer stimulated by massive wartime production. Yet in spite of major dislocations caused by the reconversion to peacetime business and manufacturing, no new crisis occurred. Several factors cushioned the economy after the war. Demobilization occurred slowly, and military expenditures remained high. During the war Americans had saved billions of dollars, which they spent on new homes and newly available cars and appliances once the war ended. Wartime profits gave businesses money to invest in plants and equipment for civilian production. The United States was the only major industrial nation to emerge unscathed from World War II, and American loans to war-torn European nations gave these countries funds to purchase American-made goods. The Servicemen's Readjustment Act of 1944 gave veterans one year of unemployment compensation, financial assistance for job training and education, and low-interest loans to buy homes, farms, and businesses. This aid to veterans and their families, known as the GI Bill of Rights, helped nearly one-quarter of the population and further stimulated the economy."[cccxxviii]

Although these three quotes may seem to be off-topic in respect to an accounting of the history of a particular part of the fire service, they are included for a reason: they are presented to give the reader a better sense of the mindset of the populace – including firefighters – at the time the war had ended, as this will come to bear on subsequent events. In light of this, there was a down side – one could even call it a dark side – to the end of the war:

"Once they returned home veterans wanted above all to live 'the good life.' To these men who had lived through the Depression as well as the war, this way of life meant security above all else. Though white middle-class men faced few obstacles in the pursuit of educational and career goals, the jobs they entered brought little opportunity for self-fulfillment. The American workplace had changed during the war years, even for businessmen and professionals. Veterans found jobs in large, impersonal organizations that demanded conformity in return for job security and advancement. While the experience of service in the large, impersonal wartime armed forces had prepared them for this sort of postwar employment, veterans' peacetime work no longer provided them an arena for the exercise of 'manly' aggressiveness, authority, or individualism. Giving up autonomy for security and good benefits in the workplace, American men turned to their families for fulfillment and happiness."[cccxxix]

Keep in mind the phrase, "...veterans' peacetime work no longer provided them an arena for the exercise of 'manly' aggressiveness, authority, or individualism." It should come as no surprise that the fire service, as what is usually considered to be a quasi-paramilitary organization, is known for accepting – if not promoting – these qualities in their membership. Now, combine this opinion with the following quote, from the same book:

"A perceived rise in juvenile crime paralleled an actual increase in youthful autonomy in the war years; many aspects of adolescents' lives were free from adult scrutiny for the first time. As the demands of the war brought teenagers into the labor force, they also weakened the influence of family and community over their behavior...Charges of disorderly conduct increased almost two hundred percent in the war years, and charges for so-called moral offenses increased markedly as well. Observers explained that these young people were 'channeling' their youthful emotions 'into a burning feeling of patriotism' that had no immediate or legitimate outlet. Wartime young people were simply more agitated than the typical youth in peacetime. There were reports increasing restlessness, emotional instability, and hostility to adult authority figures among teenagers everywhere."[ccclxxx]

The result is a possible theory as to why firematic associations began to have more and more problems with unruliness and mayhem at their conventions. It may not be definitive, and even pure speculation, but it does offer an explanation as to why the gatherings were becoming harder and harder to keep under control.

A case in point was the Western convention in Batavia in 1945, where this story was featured on the front page of the local paper:

PAID FIREMEN GO INTO ACTION WHEN VISITORS RAISE BLAZES;
MAIN STREET TAKES A BEATING

"Those firemen came to town and 'went to town' with a bang.

"Peace and quiet reigned again, after their departure in the wee hours this morning with a few battered tempers in their wake. But their escapades were undertaken by a fun-loving, hearty band which has been doing the same thing (and more) at conventions for 45 years.

"Engine No. 1 was summoned to Main Street six times between 8 o'clock and 10:30 o'clock last night to extinguish rubbish fires set by celebrators. Police stood by hopelessly outnumbered while the men in vari-colored uniforms staged the usual convention shenanigans at Main and Jackson streets.

"They plunked their beer kegs down in the direct center of the street and imposed their hospitality upon passing motorists of all kinds, hustling glasses of beverage to drivers. Uncounted rolls of tissue went into use in wiping windshields and was handed out freely to passengers in cars.

"Only a kind fate prevented injury among those milling in the thoroughfare because some motorists didn't choose to be bothered and sped through the congested area without regard to life or limb.

"Dragging boxes and cartons out of the alleys, the first bonfire was set at Main and State streets. Flames towered probably 15 feet into the air before one of the 'spectators' pulled Box No. 22 at the intersection and brought Engine No. 1 to the scene. Firemen used a booster line to squelch the flames shortly after 8 o'clock and a few spectators got wet in the process.

"When it was all over, the assemblage of volunteers gave three cheers for the Batavia department. Apparently, they were curious to see the paid firemen in action.

"And so it continued for five more flare-ups. Next at Center Street, then Jackson, back to Center, again at State and one final one at Bank before they called it quits. Public Works Department used a 1-1/2 ton truck to clean up the debris.

"Suspicion was rife that those who set the fires then strolled down in front of Fire Headquarters to see Engine No. 1 take off for the scene. The Batavia firefighters knew well in advance that they could expect trouble, but had feared false alarms from distant areas – something which the visitors didn't do – rather than bonfires on Main Street.

"Most of the box alarms which came into headquarters – and five of the six calls were from boxes – were double ones. That is, two Main Street boxes were touched off at almost the same time. Alarms came into headquarters just two minutes apart in most cases.

"Some of the visitors bought a load of kindling material from a collector for $10 and proceeded to apply the torch.

"Sirens sounded throughout the day as squad cars whooped it up. One outfit from Wendelville, which, they explained is 12 miles north of Buffalo, set up their beer keg in the shade of a tree at the rear of the County Building and proceeded to invite all comers.

"They had other tricks, too. Like pouring Woolworth's best perfume over unsuspecting victims, taking members for rides in baby buggies, dragooning females into grills for drinks and blowing countless whistles.

"Well, what did you expect at a convention? Especially a convention of visiting firemen."[cccxxxi]

If one were to read between the lines of the above story, one might conclude that the overall sentiment was somewhere along the lines of, "Visiting firemen came to town and behaved badly. We tolerated it, since we owe a debt to our volunteers, but are at the same time we're not happy about it." The ambivalence of the last sentence - "Well, what did you expect at a convention? Especially a convention of visiting firemen" – summed it up perfectly. In hindsight, it is a sad commentary on the firefighter's conventions of the day.

Chapter 13: Rome and Gomorrah

Regardless of public consternation over firefighter's conventions, the years following the cessation of the Second World War were extremely important ones for the sport of firematic competition. Whether it was for reasons listed above, or other reasons, the sport took on a degree of popularity among firefighters – if not the general public - that had not been seen since its inception.

Each team has its own story about how things "took off" then, that is probably not all that dissimilar to West Sayville's in 1946:

"The Dutchmen did not see fit to compete in the Southern New York drill at Islip so soon after the war but many attended as spectators. After seeing the competition, the team members present caught racing fever. Shortly thereafter, West Sayville fielded a team and competed in the state drill at Riverhead. In the absence of a team uniform, the men raced in their Navy undress whites."[cccxxxii]

That Riverhead tournament, FASNY's 74th, was only one of eleven different state-sanctioned tournaments that year, according to State Statistician Henry Buck. The following year, 1947, that number jumped to sixteen different tournaments; things were on a roll.

The Riverhead tournament itself was one for the books, and fortunately there was a good amount of press coverage to describe the day. In terms of fire service attendance, it was a record-breaker:

"There were more than 2,000 firemen in uniform, hundreds of members of ladies' auxiliaries and members of bands and drum and bugle corps and about 150 pieces of apparatus in line. About 100 fire departments and companies participated.

"Three trained ducks were in the parade. Two of them 'marched' with the Floral Park Center firemen, and the other was with the Eastport Fire Department, known as the 'Flying Ducks.'"[cccxxxiii]

The extraordinarily large number of teams participating also had, understandably, an impact on the running of the contests themselves, which was further affected by inclement weather:

"It took about 12 hours to run off the 11 events, so numerous were the entries and so dismal was the weather..."[ccccxxxiv]

"The number of entries in the tournament contests was 436, or about 140 greater than for any tournament in the history of the State Association."[ccccxxxv]

"The contests were suspended for about 20 minutes during the 'cloud burst', and several of them were run off in the rain. The approximately 2,000 persons, including hundreds of women, remained in the grandstand and others along the course during the greater part of the 12 long hours. It is estimated that the contests were witnessed by more than 5,000 persons."[ccccxxxvi]

"...Pouring rain and a slippery track made the going difficult, holding down the times and resulting in numerous injuries, but one record was broken and competition was keen throughout."[ccccxxxvii]

One Eastport firefighter suffered a broken collarbone, and "...Three physicians were kept busy treating the sprains, bruises and abrasions sustained by more than 40 contestants."[ccccxxxviii]

There is also evidence that there was a bit of a Nassau-Suffolk rivalry going on that day, which manifested itself in a bit of "psychological warfare":

"The crack Hicksville F.D. racing team pulled a pretty piece of last-minute strategy on the eve of Thursday's tournament. Known officially as the 'Hicks,' they changed their name in a way slyly calculated to throw fear into the outfits from potato-growing Suffolk. The new name: 'Golden Nematodes.'

"The golden nematode, as every farmer knows, is a very low form of animal life which has been found near Hicksville, which has cut the yield on a large acreage. Scientists are now striving with state funds to eradicate the pest and prevent its spread."[ccccxxxix]

There is evidence too, however, that the teams from northern and western New York also brought with them some of the rowdy attitudes and shenanigans that they were becoming known for when they came to the island. There was even a delegation from Point Pleasant, who chartered their own special train to bring a crew of 350 to the show, along with a healthy "Bug House Parade" contingent. The Riverhead News had quite a bit to say about it, starting with how they wasted no time getting the party started:

"It was 'roll out the barrel' at the Riverhead Town Hall Monday as the Point Pleasant Pea Pickers utilized their racing cars to ferry kegs of beer between the railroad car which brought their equipment to town and their headquarters in the Odd Fellows lodge rooms."[ccccxl]

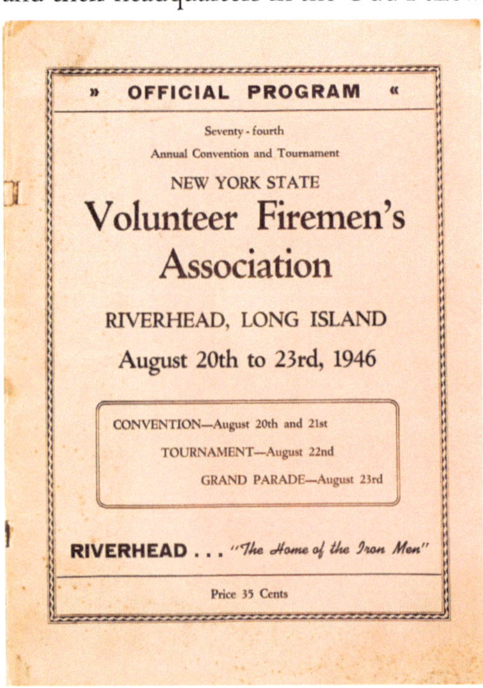

The citizens of Riverhead were apparently ready for this, though, and seemed even to condone it:

"When Supervisor Joseph V. Kelly of Riverhead presented the 'key to the city' at the opening session of the convention Tuesday morning, he did so, apparently, without reservation. Parking and other ordinances have been unofficially suspended to give the visiting firemen, here to have some fun as well as to transact some serious firemanic business, the greatest possible leeway, and local police officers obligingly look the other way should a group of carefree vamps start a dice game on the sidewalk or a bon fire in the street. The spirit of carnival reigns, especially in the evening, and it is quite obvious that the local citizenry is having just as good a time as are their guests.

"Tuesday afternoon some of the boys set up shop in front of Sears Roebuck and dispensed free beer to all comers. Orange crates served as a bar, and also as a convenient backstop when somebody started rolling the six-sided dominoes.

"Traffic was gloriously messed up, the local cops were brought to the point of nervous prostration, and everyone had a great time Wednesday night, when the red-flannel underwear corps from Point Pleasant staged their second 'bughouse parade.' Their sudden starts, stops and aboutfaces completely demoralized the traffic situation, but no one seemed to mind."[cccxli]

Their only concern, it seemed, was how much the whole thing was costing the teams:

"Suspicion grows that some of these out-of-town outfits either limit membership to millionaires or have billionaire 'angels.' We hear that one department is spending a cool $5,000 to bring its men to Riverhead, provide for their entertainment here and see them safely home. Another from up Westchester way is reported to have a bank balance of over $100,000."[cccxlii]

It should be noted here that this mention of money and concern about expenditures – in relation to drill teams and tournaments - is one of the very first to ever appear in print. Although for the most part it remained in the background, and reared its head only rarely, it was an issue that would remain a thorn in the side of the sport all the way up until the modern day.

Like the Riverhead tournament, rain also paid a visit to the Western convention and tournament in Williamsville in 1947, and also like Riverhead, it had little or no effect on the craziness of the festivities, which may even have ratcheted up a notch from the previous year:

"Rain hamper firemen? Not the Western New York Convention held in Williamsville this week. Everything from firemen's heavy raincoats to shorts were worn to protect them from the unseasonable weather.

"Amid the blasting of sirens, bells and yells, an incomplete Grotesque Parade was held Tuesday night. There were only a few floats but they had a wonderful time. Most of them were dispensing beer to anyone that happened to be near. Sonny Rist had a band on his truck, playing the good old German drinking songs – as if the boys needed them! While another float had a girl in a bathtub with a fellow sitting at the opposite end of the truck watching her from a 'john' seat. Although there was only one horse-drawn vehicle, it made enough noise and caused enough riots for a couple of them. The boys from Main and Transit way did themselves proud with their two beauties, weaving in and out of traffic and making U-turns in the middle of traffic, missing cars by inches.

"The most brilliant thing to come out of the evening was the visiting firemen's lighting fires in the middle of Main street. Although several were going at one time, only one remained at the early hours; that was at the corner of Main and Cayuga where boxes from nearby drugstores kept the pile blazing till the early hours of the morning. The Town Police and their new members did their utmost to keep order but it was an impossible task. After a few rounds at the numerous local pubs, the visitors tried to help out law enforcement officers by directing traffic everywhere except in the right direction. About 10 cars were re-directed by Cayuga. A couple of cars started to give them trouble, so they just held them up and bounced up and down on the bumpers.

"Firemen at Company headquarters had a gay time, especially back of Felix Bilger's station, where several tents were pitched for the occasion. The only real hope of the boys was not that the rain would stop but that their beer wouldn't give out.

"Pity the poor people that lived on Main street and tried to get some sleep – an impossibility. Sirens, singing, bands with only a couple of instruments, bells and yelling went on most of the night but with everyone having fun. The town gave them a wonderful reception, this convention being noisier than the last one, if memory serves right. Or is it that the proximity of this one pales the memory of others?"[cccxliii]

These "activities" were not going unnoticed by the top

The Hempstead "Yellow Hornets" compete on the Madison Avenue street course in Hempstead on September 25th, 1948

brass at FASNY, who, as the de facto overseeing body regarding conventions and tournaments, began to grow concerned – if not outright alarmed – at how things were showing signs of getting more and more out of hand. Whether or not the general public was worried about the excessive noise, pranks and mayhem that more and more accompanied the conventions each year was only part of their concern; of greater importance to them was how, as an overseeing body, they were ultimately responsible for anything bad that might happen. No one had harmed or sued anybody yet, but the possibility was definitely there that something could happen, and it was making them nervous.

So much so that at that year's State Convention in Catskill on August 21st, 1947, they decided to start doing something about it:

"In his opening remarks to the 75th Annual Convention, Mr. Roy C. Moon, the General Chairman of the Committee of Firemanic Activities of the Village of Catskill, stated that they were 'doing everything we can to see that that every one has a good time here this week.' He went on to explain that 'there is only one thing we ask in return, and that is that every one respect private property. Something that we cannot have is any malicious damage to property in Catskill. There is no reason for it and I am sure you all agree that you wouldn't want any damage done in your own town. We don't care how much noise you make – make all the noise you want. We are not going to worry if some one imbibes a little bit freely. That is all right. But what we can't have is destruction of property.'

"Mr. Moon, aware of the fact that during past conventions some of the younger and over zealous members had overstepped the bounds of propriety, warned the members and delegates that 'one rule which our local committee has been forced to adopt in an attempt to prevent destruction of property is this: If any outfit – not just an individual – but any outfit is found maliciously destroying property, that outfit will be barred from competition in the tournaments and in the parade. Now if your squad car is mixed up in it we feel that your outfit is involved, so make sure that you have responsible men in charge of your squad car.'"[cccxliv]

Then, they followed their words with action:

"At the Executive Committee meeting on the afternoon of the opening day of the convention the following resolution was adopted:

> RESOLVED, that the Executive Committee of the Firemen's Association of the State of New York gives its full and unequivocal support to the efforts of the Committee on Firemanic Activities of Catskill, New York, Inc., to prevent acts of vandalism during the 75th Annual Convention of the Association and depreciates the sounding of false alarms and the damage and destruction of property by firemen or others in attendance at the convention. And be it further,
>
> RESOLVED, that the Executive Committee approves and recommends the disqualification from the tournament and parade of any organization of volunteer firemen that may have been found to have been partners to acts of vandalism or sounding of false alarms of fire or other conduct unbecoming to volunteer firemen and the volunteer fire service."[cccxlv]

They weren't kidding around. Apparently the "Eagles" of Croghan didn't pay attention, as they were disqualified the next day for racing back and forth across the Rip Van Winkle Bridge, refusing to stop at tollgates or obey the attendants. As a result, they were barred from participation.

In a sense they were better off, as both the tournament and the parade that year were a mess. According to their history on the West Sayville Fire Department website, "Unfortunately, the water pressure at the drill was so bad that in order to trip the target, contestants had to jump on the hose to propel the water. The poor pressure caused several Nassau County teams to walk off the track. Later that night, the Nassau contingency started a huge bonfire in the middle of town to show their displeasure. To get even for the prank, the hosts extended the parade route to five miles, most of it through the high rolling hills of the area. That made the event the longest tournament parade ever and gave birth to the rule that limits parades to two miles."

Still, the number of tournaments sanctioned by the State Association continued to grow: 25 for 1948, followed by 32 in 1949. The years 1947 and 1948 also saw a huge jump in the number of newly formed teams as "competition fever" continued to grow in leaps and bounds. In fact, the last two decades had seen an ever-increasing boon in team numbers: according to recorded statistics, 18 new teams were formed in the 1920's, to be followed by a huge jump to 48 new teams in the 1930's, followed in turn by an incredible 86 new teams that were formed in the 1940's.

This development was good for the sport, but at the same time it may have actually contributed negatively to it, as many of the teams were no doubt made up of a number of "younger and over zealous members" (as quoted above) who wanted to do nothing more at a convention than run some races, drink beer and go on a tear.

It's a safe bet to assume that the Western convention in Warsaw in July of 1948 was just as rowdy as ever. Unfortunately – but quite humorously – this author discovered that the Wyoming County Times, in its coverage of that convention in its July 29th issue, copied a good portion – word for word – of the exact same article it wrote when reporting on the Western convention in Warsaw in 1925! Either someone in the editorial staff was incredibly lazy, or it was again a case of "the more things change, the more they stay the same."

West Sayville marches in the Parade at FASNY's 76th Convention in Rome

Which, in truth, was not the case. The mayhem at conventions got worse, and seems to have reached its zenith in August of that year at FASNY's 76th Convention and Tournament in Rome. The parade was reported to have had over 30,000 spectators, and things apparently got so out of control that even newspapers on Long Island, some 300 miles away, felt obliged to run stories about it. From the Riverhead News:

"The four-day firemanic rally, which terminated with a huge parade on Friday, produced the usual 'high-jinks' and the visiting vamps tried the patience of Rome authorities by starting bonfires in the streets, wetting down spectators, and turning in false alarms which all but drove the city's paid firemen to drink. The Riverheaders had their fun but steered clear of all disorders."[cccxlvi]

In its report, a relatively new Long Island newspaper to the time, *Newsday*, was not as kind:

L.I. VAMPS ROMP AS ROME BURNS

> "Volunteer firemen attending a state convention here went on a rampage last night and Rome literally burned. Fires were set at every main intersection in the city and instead of putting out the blazes, the vamps turned their hoses on spectators. Local authorities said that the majority of the delegates were from Long Island.

"STORMING THE BASTILLE"—A screen door, left, is smashed as a mob attempts to gain entry to the police station here from the adjoining alley. A volunteer fireman, taken into custody after he refused to obey an order to shut off his hose, was the object of the liberating throng. White 'flag' on the right is a man's shirt hanging from a stick. (Story inside).

Changes in Labor Law | Troops Raid

"After one vamp was arrested and a mob stormed Rome police headquarters, 50 additional MPs were rushed from Griffis Air Force Base to aid city police and undersheriffs in quelling the disturbance.

"Following a nine and a half hour exhibition yesterday on speed and efficiency in fire fighting contests in which Long Island garnered the five top spots, the vamps went looking for excitement. And while they were horsing around here, buildings at Utica Airport, 18 miles away, were burning to the ground.

"Every off-duty patrolman was recalled by Chief James C. Dunn. The Military Police town patrol ordered every soldier out of the city or back to the base.

"Tournaments are new to upstate New Yorkers. The majority of the larger cities have paid departments and those operated by volunteers have not developed the competitive events familiar to Long Island camps. The contests yesterday attracted only about 2.600 spectators, but last night's show brought thousands to watch at every street corner.

"The incident leading to the arrest of a vamp occurred at the intersection of Liberty and James Sts., the hub of the city. Capt. Patrick J. Smith ordered the unruly delegates to put out the fire there. While some of them continued to drag material to feed the flames, the hoses were played on a crowd of over 3,000 persons.

"The fireman, unidentified because he was later released, was taken to police headquarters. A wild mob stormed headquarters, ripped the screen door off and created such a commotion that police returned the man to the crowd. Then the riot call was sent out."[cccxlvii]

An even more detailed report of the ugliness could be found in the local paper, the Rome Daily Sentinel:

"The off-duty police, already burdened with a 12-hour day, were called out at midnight, as were 50 additional Military Police from GAFB after a mob of volunteer firemen and Rome citizens stormed police headquarters demanding the release of one fireman taken into custody by Patrick J. Smith.

"Wanton hoodlumism led to the chopping up of two lengths of Rome firehose as city firemen were prevented from putting out the James and Liberty St. bonfires.

"Public Works Commissioner Ira L. Elmer personally supervised employees called from their homes shortly after midnight to cope with the flotsam and jetsam of huge bonfires that were erected at nearly every downtown street corner.

"No bids as of yet have been extended for next year's convention.

"Rome paid firemen, attempting to extinguish the mammoth hour-long blaze at James and Liberty Sts. were repulsed by volunteers wielding hoses. Trucks belonging to volunteer companies were parked crosswise in the streets around the giant pyre.

"One GAFB airman, bowled over by the cascade from a N. James St. hydrant when a hose uncoupled, reportedly suffered a broken leg.

"All soldiers were ordered back to the base as the re-inforced MP's patrolled the city.

"Although a warning personally issued by Police Chief James C. Dunn, that he 'would arrest the next man who started a fire' dampened the spirits of the mob somewhat, water battles between firemen and Romans continued until well after 2 a.m.

"Chief Dunn took personal command of his force after he had been summoned from his home.

"By midnight, the setting of bonfires had assumed a standard procedure. Firemen rushed in with combustibles, ranging from tires to outhouses, started fires, and, as soon as a crowd had collected, began spraying the spectators. The fires were or were not extinguished as caprice dictated.

"After the hour-long fire at the corner of James and Liberty Sts. had tangled traffic, impeded city fire apparatus, and attracted a throng estimated at between 1,000 and 2,000, Capt. Smith ordered the firemen either to train their hoses on the blaze or leave.

"When his request was refused, he took one fireman into custody shortly before midnight.

"In a latter-day storming of the Bastille, volunteers, abetted by an equal number of Romans, jammed into the alley alongside police headquarters and demanded the release of the fireman.

"A screen door was wrenched from its moorings and smashed as MP's and Rome police prevented the vanguard of the crowd from entering the station.

"Capt. Smith released the fireman, after which the crowd, its spirits considerably lowered, milled about the James and Liberty intersection watching the extinguishing of the bonfire.

"Several snow fences, used in yesterday's tournament and which had been carefully rolled up and left in the street, were snatched from the fire, almost totally destroyed. An outhouse, brought in from an outlying farm just before the blaze was quenched, still reposed at James and Liberty Sts. early this morning.

"Earlier in the evening, Romans had excitement to burn, as citizens gathered at every street intersection gathered and re-gathered, only to be disbursed, soaked and laughing, by the volunteer's hoses.

"City police and firemen were nearly helpless in grappling with the situation, although a lumber truck was engaged at 8 o'clock last night to carry off potential bonfires before they were ignited.

"Volunteers turned their hoses on Rome firemen as they tried to extinguish the bonfires. They refused to uncouple hoses to allow the Rome Dept. near the blazes and later chopped two lengths of Rome hose.

"One insurance man reported that a frantic merchant called him shortly before closing time with a request for malicious mischief coverage.

"Several St. James St. grills locked their doors while hundreds jammed the streets to watch a fire at the corner of James and John Sts. shortly after 9:30. Deputy Chief Howard Bentley received an éclat of applause when he arrived to quell the blaze.

"Most durable of last night's fires, and the largest, blazed at N. James and Liberty Sts. shortly before 11 o'clock.

"Originally started under the traffic light, which it threatened to burn out, the fire was subdued for a short time and moved a slight distance north. There, it burned out, instead, several hundred dollars worth of asphalt pavement.

"At one time during the James and Liberty blaze, four hoses were being directed into the audience as the flames reached a height of nearly 50 feet.

"Several parades, complete with brass bands vied with bonfires as confusion-generators. Best organized of last night's processions was one put on by the Barnard Fire Co., Greece, which unaccountably marched from George St. to Black River Blvd. without mishap about 9 o'clock last night.

"Later, about 11:30, the Barnard parade re-appeared, after having been lost for several hours, on Liberty St.

"An irresistible force met an immovable object when the Barnard company marched head-on into the James and Liberty mob. At last minute command, the marchers turned, cut a corner, and headed down James St. picking up marchers like the Pied Piper of Hamelin.

"One city official remarked, as public works men shoveled up the smoldering rubbish after the blaze, "I wonder where they'll hold this thing when they've used up all the towns in the state. Certainly no city would ever want them back a second time.""[cccxlviii]

Fortunately, by the following day things had returned pretty much back to normal:

"The police blotter belies any rumors of riots, abductions, or other outbreaks of lawlessness about the city yesterday.

"There is no arrest recorded from 8 a.m. yesterday morning until noon today.

"There were several street corner fist fights last night, which policemen were called upon to break up, but no arrests were necessary.

"Although policemen were stationed at strategic areas around town and another force of policemen and deputy sheriffs stood by at headquarters, no large scale difficulties were experienced.

"Clubs were taken from several groups of firemen and Romans who were evidently spoiling for a fight, but by midnight soaring tempers on both sides had been cooled by the intermittent parading of individual band units and general fatigue.

"Police Chief James C. Dunn, indulging in understatement, called last night "quiet.""[cccxlix]

There is an incredible irony to this story as well. During the business session of the 1948 convention, for reasons that are incomprehensible and which seem to be in complete spite of the resolutions made in Catskill during the previous year's gathering, "...the convention turned down a proposed amendment to the state association's bylaws, advanced by Riverhead Fire Chief Joseph E. Walsh, president of and delegate from the Suffolk County association, which among other changes, would provide penalties for fire units guilty of rowdyism."[cccl]

Whether or not FASNY just wanted to "look the other way" where this was concerned, or even the reasons it may have wanted to, will never be known, but one might go so far as to speculate that the huge spurt of growth that the sport was experiencing may have had something to do with it: with so many fire departments and so many new teams suddenly getting involved, FASNY was experiencing, as a result, an unprecedented amount of growth of its organization as well...so why ruin a good thing by clamping down on the firefighters' enthusiasm? Nonetheless, they did not completely look the other way: as a result of the fiasco in Rome, FASNY cancelled the 1949 State Tournament entirely, and for the first time in the history of the sport – barring the years we were at war - no State Champions were crowned.

On Long Island, the sport's popularity within the fire service was at an all-time high. At Hicksville's 1948 Labor Day tournament, an amazing 85 teams competed. At the 1949 Southern New York tournament in Lindenhurst there were 60 teams entered, and the Town of Hempstead tournament that same year had 62.

As previously discussed, many of the reasons why this was happening had to do with the national economy and attitudes that were prevalent in post-WWII America. However, there were also more local, demographic events that took place that had an influence on the sport as a result during that time as well: the birth of the Suburbs.

In 1949, Abraham Levitt and his sons, William and Alfred, decided to convert a Long Island potato field into a prefabricated "carbon copy" community, and called it "Levittown." This type of living arrangement had never been done before. For the first time, "For $7,990 ($60 per month and no down payment) the consumer can purchase a four-room house with attic, outdoor barbecue, washing machine and 12-1/2 inch built-in television set."[cccli]

What this meant, in effect, was that just about anyone with a relatively decent income could now afford a pretty nice house outside of New York City – one of the main ingredients of what was then the American Dream. This "anyone" also included a quite large demographic within the fire service: New York City firefighters.

"Around the mid-fifties, middle class life became more than just a dream for New York's firefighters as they scanned the classified pages of the Long Island newspaper Newsday and discovered that they could afford to buy a three-bedroom Levittown house on a 40x100 lot for less than $10,000.

"Before World War II, firefighters and other city employees were required to live within the five boroughs. But the changes brought about by highway construction, suburban development, a severe postwar housing shortage in the city, and the home finance opportunities that were part of the GI Bill led the UFW to demand for a loosening of residency restrictions. Firefighters sought their piece of the new American dream of suburban home ownership. With an understandable lack of enthusiasm, starting in the 1950s the city allowed firefighters to live in Nassau and Suffolk Counties on Long Island and then, in the early 1960s, extended permitted residency to several upstate counties.

"Once the city eased its restrictions, however, the exodus from the urban boroughs of Manhattan, Brooklyn, and the Bronx was dramatic."[ccclii]

The Clayton "Calumets", 1949

What this translated to was a large, sudden increase in the number of city firemen who were now living on Long Island and commuting into the city for work. While off duty, however, many of these same firefighters – passionate about the job as they were - joined their local volunteer fire departments, and subsequently their drill teams, creating a large, new influx of talent into the sport.

The current enthusiasm for firematic competition was not limited to the modern-day manifestation of the sport either. Regarding old-fashioned musters, Dixon states, "After 20 years of hard times, muster activity was on a roll...A booming economy and boundless enthusiasm proved to be the right combination in getting the sport back on its feet."[ccliii]

At the beginning of the 1950's on Long Island, so great was the interest in the parades and tournaments that they were held just about anywhere, including some quite unusual places. The 55th convention of the Southern New York Volunteer Firemen's Association in June of 1950 was even held at the Belmont Park Clubhouse - a horseracing track - in Elmont, which if not unusual enough in itself was made even more so by the fact that the tournament was held on the Park's entrance ramp, on a track that was far from usual: "That drill was unique in that the hose contests were run on a curved road that was the entrance to the Belmont Racetrack. As a result, teams could not see the hydrant from the starting line. Even more unusual was the award presented to the Flying Dutchmen for winning one of the contests . . . a rocking chair."[ccliv] This unique situation didn't deter anyone; 3,500 firefighters representing 73 different fire departments marched that day in the parade, and 65 different teams competed in the drill.

There were some other unique occurrences at the time as well. The 14th Annual Brookhaven Town Tournament in Center Moriches on August 27th, 1949 featured "...the first parade appearance of the Gordon Heights Fire Department, the only all-Negro vamp unit in Suffolk"[ccclv] and at the 78th FASNY convention in Syracuse in 1950, a newspaper reported that, "...In the afternoon's special event, a women's bucket brigade, the Hinsdale auxiliary won the ladies' auxiliary state championship with a timing of 84.4 seconds Runner-up was the Mattydale auxiliary with 93.4 seconds. This event established a precedent in being the first time auxiliaries had taken part in the tournament."[ccclvi]

Unfortunately, the latter was overshadowed by events and issues that were rising more and more to prominence with each passing tournament. During the Three Man Ladder contest, "...Harry De Back, 20, of Point Pleasant, the third member of a ladder climbing team, lost his hold on the top rung and fell 25 feet to the asphalt pavement.

"Spectators screamed as he fell, and breathed an audible sigh of relief as he executed a complete back-flip in mid-air and landed on his feet. Red Cross attendants said De Back suffered a slightly sprained ankle."[ccclvii]

He seemed to have been extremely lucky, but a story in the same paper the following day explained why: "...De Back...formerly has been a tumbler. He thinks his training saved him from serious injury."

And, on a more serious, yet sadly familiar note, it seems as if the cancellation of the 1949 State Drill by FASNY as a result of the debacle in Rome the year before had little effect on the behavior of the firefighters in Syracuse in 1950:

"A crew of Long Island firemen were banished from Syracuse today and warned never to darken our fireboxes again.

"Judge Truman H. Preston in Police Court fined John F. Blomberg, 20, and William S. Klinke, 23, both of Central Islip, $50 each and suspended a 30-day sentence.

"The delegates to the State Firemen's Association, meeting here this week, admitted turning in a false alarm early yesterday.

"Judge Preston ordered them to leave the city by 3 p.m. and not to return for a year. By arrangement with convention heads, the entire Central Islip delegation is leaving also."[ccclviii]

Not surprisingly, this court action became – once again – a topic of conversation for the FASNY officials:

"A sobering note to the general humor and excitement surrounding this 78th annual convention of the New York State Volunteer Firemen's Association was sounded yesterday by retiring president Arthur M. Hefner of Rochester.

"In a sharp warning, rebuking water throwing antics and false alarm ringing by some of the delegates, Hefner expressed concern that the association might have to beg for a convention site if members did not put an immediate stop to such tactics.

"Admitting that 'I am talking to the wrong people, since the water-throwers and false alarm ringers don't come to convention meetings,' Hefner declared two delegates arrested for turning in a false alarm early yesterday morning would not be aided by the association. Teams to which they belong would be penalized at today's firemanic tournaments, and their departments may be asked to take action against them, Hefner said.

"The retiring president pointed out that only Saratoga Springs and three hotels in New York City have asked for the 1951 convention, and lamented, 'If no one asks us, I don't know where we'll go.'

"Despite the fact that two volunteer firemen from Central Islip, L.I. are jailed until Saturday for turning in a false alarm, rowdy delegates doused passersby with water thrown from the Yates Hotel windows last night, until police intervened."[ccclix]

Chapter 14: Assurance of Insurance

Hefner's concern about finding hosts for the Annual Conventions was well founded. In addition to FASNY's growing problem of misbehavior at the conventions, another headache was coming to the fore: that of firefighters – and spectators, for that matter – getting injured – or even killed - while attending the tournaments.

Injuries, or even deaths, of firefighters were not new to the sport. In fact, they had been occurring since the very beginning. The reader will recall the death of a Utica firefighter at the brakes of his "machine" in 1843, and that of Sag Harbor's Thomas Quail during ladder practice in 1895; the head injury to juvenile competitor John Pollack at Riverhead that same year, and how George Wheelock was run over by a hose cart at the Northern tournament of 1915. Considering the hundreds of tournaments – and thousands of individual contests – held over that period of time, however, one must agree that for a sport as physically strenuous and demanding as it was – even taking into consideration all the scrapes, bumps and bruises that went unreported – the number of deaths and injuries associated with the sport was relatively small.

But beginning in the late 1920's – early 1930's, that began to change, if not literally then at least potentially, as motor vehicles became an increasingly important part of competition on the track. As everyone knows, the faster you go the more dangerous things become, and hence the greater the potential for not just injury, but even tragedy, with the 1928 accident involving the Roosevelt "Brush Rabbits" driving off the course into the crowd while trying to avoid a policeman being one example.

These risks increased manifold after the conclusion of World War II, as the sport grew suddenly and exponentially as a result of the huge influx of new teams. The number of new participants wasn't the main problem, however; it was what they brought with them.

It is self evident that the contests at tournaments are all about speed, and achieving a given "task" in less time than the competition. Up until the late 1930's and the start of the war, any team advantage to finish a particular contest more quickly had less to do with the trucks as it did with the individual contestants themselves: someone who was a fast runner, or who was particularly adept or practiced at attaching a hose coupling to a hydrant would be one example. Not that there weren't little tricks or techniques - using a 2x4 on a clutch pedal to have the truck already in gear at the starting line before the men jumped on it and kicking it away once they were, or having one team member climb partially up the ladder while it was still on the truck traveling down the track are two examples – but for the most part the greatest advantages gained were skill-related.

After the war ended, however, this changed. Whether as a result of knowledge and skills learned while in the service of our country, or by simple scientific progress, there was a large amount of evolution and growth in the field of mechanics. Suddenly, it seemed, it was all about the trucks, and how you could make them go faster than the next team's. Armed with these new mechanical skills and knowledge, the trucks went faster and faster – and the sport became more and more dangerous as a result.

As a result FASNY, while obviously concerned for the health and safety of the firefighters while they were competing and the spectators who were watching them, was starting to become even more worried about its culpability – and even liability – should one of them get hurt. It was a worry that showed no signs of going away.

At the time, however, their main problem was just as Hefner stated: finding a willing host for their 1951 convention and tournament after all the unruliness that had been going on.

In this regard, they ended up taking the offer from New York City, and in spite of all the security issues, did their best to put on a positive attitude and get everyone pumped up for the event, as indicated by Tournament Committee spokesman Mario Pollini's report to the FASNY administration that year:

> "I will only take about two minutes of your time. About some three months ago, when we knew efinitely that we were going to have a Convention here in New York, trying to make a full fledged set-up, as your First Vice President stated yesterday, some of us decided to see what we could do about having a tournament with the convention.
>
> "We tried Nassau, we tried Suffolk, and we rode around New York, Central Park, Randall's Island – every place that we could possibly think of that could give us a possible course.
>
> "We had no luck at all until about six, seven weeks ago. We heard that there was a spot in Coney Island. We went and looked it over, and it was just the thing we wanted. But the red tape was something that we hadn't anticipated; and it wasn't until two weeks ago Monday, 3:00 o'clock in the afternoon, before we got the final okay, and, believe it or not, in the last two weeks we have put together a tournament which usually takes anywhere from three to four months to do.
>
> "The tournament tomorrow is going to be a red hot one. I am going to tell you that right now. It is going to be a very, very fast, very level course, probably one of the most beautiful set-ups for a tournament any one of us has ever seen.
>
> "We have forty of the best teams that have ever competed in any tournament. So records will fall if the water pressure stands up, as they have predicted.
>
> "It is going to be on an unnamed street. If anyone knows Coney Island at all, it is just east of the Cyclone. It is a little short street running through a park avenue between Surf Avenue and the Boardwalk. There is going to be seating capacity for 5,000; no charge for the seats. And first come, of course, first served. It is in the park area, and, of course, no charge can be put on the seats due to the fact that the City does not stand for that.
>
> "It is going to start at 11:00 o'clock sharp. It will probably run on until about 5:00 or 6:00 in the afternoon. There will be no intermission at all. There will be eight events; and with forty teams we are going to see plenty of action.
>
> "The arch will be put up late this afternoon, and if there are any of the men who can help, we would appreciate it. We have an arch that is being supplied by the Safeway Scaffold Company, and it is

The Bay Shore "Redskins" compete in the Motor Pump Contest at the 1951 FASNY tournament at Coney Island

> all in sections, and it will have to be installed. We will be there from 3:30 on. Anyone that can possibly help us, we would appreciate it very much. That is it for the tournament."[cccl x]

The very fact that Pollini states, "...We tried Nassau, we tried Suffolk, and we rode around New York, Central Park, Randall's Island – every place that we could possibly think of that could give us a possible course" gives one a pretty good indication of the Public Relations difficulties that FASNY was facing at the time. After all: Nassau – and even *Suffolk* – with all their many tournaments and course locations, not offering any of them to FASNY?

In hindsight, it may not have been such a bad thing after all, as having New York City as the convention host ended up being somewhat historical in nature: "Mr. John J. McConnell, Chairman of the Parade Committee, noted that 'It was the first time since 1886, on the occasion of the dedication of the Statue of Liberty, that volunteer firemen had paraded in Manhattan. An estimate of 10,000 persons witnessed and were immensely impressed with the spectacle of modern and efficient apparatus and equipment displayed."[ccclxi]

For some reason, too, there seemed to be a short break in the "action" of mischievousness at the time...at least out west. Regarding the Western convention in Williamsville in 1952, the *Amherst Bee* reported, "...It was one of the most successful as well as most colorful conventions the group ever held, according to the older members of the Association, and while the visiting firemen had a wonderful time, there was no destructive damage, and all events ran off smoothly."[ccclxii]

The Albertson Hook & Ladder, Engine and Hose Co. team competes at a drill in the early 1950's

But the relative serenity of the Western conventions was short-lived, and FASNY was in no way about to let the matter rest. At their Executive Committee meeting in Binghamton on August 17th, 1953, the FASNY Secretary reported the following:

> "At this time Attorney [Albert J.] Foley read the proposed settlement relative to tournaments sanctioned by this Association and advised the consideration of a bond or liability insurance policy to protect the Firemen's Association. The matter in question was fully discussed by the Executive Committee. Mr. Christy spoke in behalf of the Committee appointed to go into this matter. The Committee met with representatives of both Nassau and Suffolk Counties and the matter being considered by the Executive Committee was explained at that time. Our prime interest was protection, and it is recommended that the State Statistician does not grant sanctions unless insurance is furnished indemnifying the State Association. The amendment is as follows:
>
> "'He shall appoint a tournament statistician whose duty it shall be to represent this Association at all tournaments sanctioned by this Association. No sanction shall be granted until and unless the applicant shall provide a bond or insurance policy indemnifying the Firemen's Association of the State of New York in an amount of not less than $100,000.00 for personal injury that may be sustained by any one person, and $25,000.00 property damage.' This is an amendment to Article III, Section I.
>
> "Mr. Christy moved, seconded by Mr. Nold, that amendment above as read by Mr. Foley be adopted. Carried."[ccclxiii]

Soon after that, on November 14th of the same year, at a FASNY Executive Committee meeting in Albany, a report from the Temporary Commission on Fire Laws was read. Among the statements made was:

The West Brighton "River Rats" compete in Sea Breeze in the early 1950's

> "Tournaments, parades, conventions, and inspections, present various problems to the fire service, both from operational and fiscal standpoints. They undoubtedly will be discussed by the Commission in connection with other subjects, such as firemanic benefits."

Although they both involve liability, there were actually two issues involved here. What FASNY was trying to resolve was whether or not they – as an overseeing body - would have enough liability insurance coverage should either a spectator or firefighter be injured or killed while attending a convention or participating in a tournament. This may seem at first seem to be a fairly pedestrian matter, but in fact for FASNY it had much larger ramifications: if hit with a lawsuit of any decent size, they conceivably stood to lose much of what they had worked long and hard for over the years - most notably their crown jewel, the State Firemen's Home.

The Copenhagen "Cubs" compete in the Efficiency Contest at a 1950's drill

To be sure (and which has been illustrated previously in this text), firematic tournaments were not especially important to them in the overall scheme of things, and they were not about to let a relatively minor adjunct to their operations take down the whole mountain.

The second issue had to do with what the FASNY Temporary Commission on Fire Laws referred to as "firemanic benefits." From a firefighter's perspective, the question was being raised, "If I am injured or killed while competing at a tournament, am I covered under any form of insurance?"

This had equally huge, if not even larger, ramifications for the sport. Although it is a sad fact that issues such as these helped contribute to ours becoming more and more of a litigious society, it was also a sad fact – and indeed a reality – that if there was no insurance coverage for firefighters at the tournaments, it was a pretty sure bet that eventually none of the firemen would ever want to compete in the drills – or even attend them at all, for that matter. It would not have been too much of a stretch to say that in this sense the very future of firematic competition was at stake.

As if to flaunt the issue in FASNY's face, the trouble at conventions continued. At the 1954 Western convention in Tonawanda, the local paper was filled with all sorts of tales of antics by the firemen:

"Local fire departments reported six false alarms within a two hour period during the night, and six visiting firemen paid a visit to North Tonawanda City Court."[ccclxiv]

A 1950 western teams group photo, including the Spencerport "Red Onions", the Sea Breeze "Indians" and the Point Pleasant "Pea Pickers"

"A Niagara Falls Blvd. Resident, Edward Zimmerman, complained to North Side police at 8:27 p.m. yesterday that men aboard an Albion fire truck squirted water inside his car, soaking the clothing of his companion."[cclxv]

"Many companies beat the problem of keeping refreshments cold by renting coolers in downtown stores. The Live Stream Hose Company of Mt. Morris, however, displayed real ingenuity. It came to the convention in an emergency wagon, which had no emergency tools. Instead, compartments were filled with ice and converted into coolers."[cclxvi]

"Before police caught up with some offenders, 11 rubbish fires had been started on Young St. Earl Beiler, 58, of Fairport, today faced Tonawanda City Judge Nelson Staley for starting one near Young St.

"'I was just having fun, and I got caught,' Mr. Beiler explained. Knowing this was exactly the case, Judge Staley preached to him on the dangers of fire then gave him a suspended sentence. Five other volunteers received the same treatment."[cclxvii]

In all fairness to the firemen, however, there were also several anecdotes from the parade that were quite humorous:

"Twin Cities Auto Parts entered a battered fire truck of perhaps 1910 vintage and painted a sign on it 'Old Fire Trucks Never Die.' The truck, however, did die, right under the Tremont St. viaduct and held up the parade for at least 15 minutes."[cclxviii]

"The Albion Fire Department won the prize for having the best antique apparatus. It seems, however, the 19th century jumper does not belong to it at all. As the sign on the jumper said, it was 'Lost, Stolen or Strayed from the Holley Fire Department.'"[cclxix]

"A 12-year-old boy became angry when a Brighton Fire Department volunteer squirted water in his face with a water pistol. He challenged the fireman to a duel as he pulled out his own pistol and began to run after the truck carrying the fireman. He was met with a stream of water from a 1-1/2 inch hose. 'What happened?' he asked his buddy, who was bringing up the rear."[cclxx]

"Spectators between Miller and Wheatfield Sts. Claim they watched the parade in intermittent rain. Seems firemen decided to unload all squirt guns and hoses before the parade ended."[cclxxi]

"Thorold Canadian Corps was led by perhaps the prettiest and snappiest baton twirler in the parade. All eyes on the judges' stand watched her maneuver. When she had passed, a voice from the stand asked, 'What unit was that that went by?'"[cclxxii]

And, there was one particular item that was even heartwarming:

"Retiring [Western] association President Fred Miller insisted that he watch the parade from the judges' stand despite the fact that he has been ill for more than a year. As the parade neared the judges' stand, the heat and excitement proved too much for Mr. Miller, and he collapsed.

"Convention Chairman William S. Hawthorne reported later. 'The first thing he (Mr. Miller) said when he got home was, "Turn on the radio and let me listen to the parade."' 'It seems that all his life has been tied up in firemen and their activities.'"[cclxxiii]

The latter is not at all unusual. Firefighters then, as now, were passionate about their vocation, both at work and at play. It should not come as a surprise, then, that as FASNY remained stubborn about hosts providing insurance at tournaments, and issues regarding benefits for the firefighters who competed at them were being raised, the teams themselves had some opinions of their own on the matter.

The reader should first be reminded that the mid-1950's was a period of significant growth and change in the sport of firematic competition. New teams were being created every day it seemed, and the technology of motorsports – for that, in truth, was what the sport was becoming – was growing in leaps and bounds. Whether because of – or as a result of – all of this, the sport, on a whole, was being taken – and treated – more and more seriously with each passing year.

Nowhere was this more evident than on Long Island. For the very first time, permanent, "purpose-built" tracks were now being constructed for teams to practice and compete on. Whereas in the past, teams would practice on a quiet back street or deserted stretch of road, and tournament hosts would close a street for a day and set up a temporary arch (such as described in Pollini's report on the 1951 FASNY tournament in Coney Island, above), facilities were now being built that could be used on a permanent, year-round basis. Hempstead holds the honor of building the first true track of this nature, when their facility was officially dedicated on July 5th, 1954[8].

Hempstead's new track, at the time, was the envy of all. The arch alone – of solid concrete and steel construction – was reported to have cost $10,000 to build, which at that time was about the same as a new house in Rochester. This fact is known because word of this new arch that had been just built had spread so far and wide that members of the western area Sea Breeze "Indians" team, upon hearing about it, took a two-day road trip from Rochester all the way down to Long Island - just to see it with their own eyes. Once there, they

[8] Sag Harbor had a permanent, purpose-built arch as early as 1922, but it was constructed over a pre-existing horse race track, so it would not qualify as a full purpose-built firematic competition facility, *per se*. It has since been dismantled.

The West Seneca Hose Co. existed for a short period of time in the mid-1950's in western New York as an all-female exhibition-only fire company that would put on demonstrations of their speed and prowess at setting up and using firefighting equipment (hoses, etc.) at a "fire". They apparently were not members of any real fire department, and did no actual firefighting; their true function, therefore, isn't entirely clear

asked the Hempstead team, "This tournament stuff is going to keep on going, isn't it?" "Yes it is," they said, "it's going to be big, and it's going to bring young people into your fire department."

But it wasn't just fancy new tracks and arches that Long Island teams brought to the sport. On July 31st, 1952, the very first tournament to be held at night, under lights, took place at the North Fork Firemen's Association tournament in Riverhead. According to the Riverhead *News-Review*, "The use of the generator and flood light equipment in connection therewith was donated by the Riverhead Raceway."[cclxxiv] They went on to say, "...the first tournament held at night on Long Island was an unqualified success. Present and former officials of firemanic organizations of Nassau and Suffolk Counties who were among the approximately 1,500 persons who attended, predicted that most of the major tournaments in the future of the two counties might be held at night."[cclxxv] This same sentiment was expressed a year later, when the Suffolk County Volunteer Firemen's Association held their annual tournament at night for the first time on July 18th, 1953: "This was the first time that the association's contests had been held under floodlights and they were enjoyed by several thousand persons. The innovation was so successful that it is expected that the contests of the association in the future will be held at night."[cclxxvi] While this did not entirely come to pass – there were not that many tracks at the time that had the facilities to host a nighttime drill – the idea always remained popular with the various drill team associations, and has even experienced a resurgence of popularity of late. The annual Islip Town Tournament is very often held at night to this day.

Another feature of the North Fork convention in 1952 that was originated on Long Island, albeit almost 30 years earlier, was the Bug House Parade. As described by the News-Review that year:

"The program of activities will start on Thursday evening with the grotesque or bug house parade. From the number of entries already received by Supervisor Klipp, co-chairman with John J. Sherwood of the convention committee, this promises to be one of the most unique and humorous events ever staged in the village of Greenport. Fire companies, fraternal and civic organizations, groups of people or individuals are invited to participate in this fun-making event. Old-time costumes, humorous costumes, grotesque costumes or floats, in fact anything to provide a laugh and a real carnival spirit, will be acceptable. Suitable prizes will be awarded to the winners."[ccclxxvii]

Having grotesque parades held in conjunction with firematic conventions was nothing new; the "Darktown Brigade" contingent that won a prize in Lockport in 1910 is just one example. The tradition of the Bug House Parade specifically goes back as far as 1926, when Elmont firefighters created the Bug House Hose Company #1 as a fraternity for Nassau County firemen. Almost every "Bangtails" team member was also a member of the Bug House, as were members of 57 other Nassau County teams. Local Bug House chapters were soon formed in both Suffolk County and Western New York (the reader may recall a reference to the Point Pleasant "Pea Pickers" marching around in red flannel underwear in Riverhead in 1946), and their tradition still remains strong as they continue to bring fun and laughs to the Nassau County parade each year to this day.

But it wasn't just on the track that Long Island once again led the sport of firematic competition. In the early to mid 1950's, in the same manner as Albert Sammis did in the late 20's, representatives from Nassau and Suffolk counties continued to be the vanguards of the sport in the way it was administered and practiced; simply because their teams had more practice and experience – and, as a result, even interest - with it, compared to the other areas of the state. A testament to this can be found in the *Syracuse Herald-Journal*, in reference to the FASNY State Tournament that was held there in 1950: "It was noted that 51 of the 56 competing companies were from Long Island, where the tournaments are weekly affairs."[ccclxxviii] It is most likely that it is for this reason that FASNY allowed two Long Island representatives, Theodore Jaffie of Lindenhurst and Wilbur Young of Oyster Bay, to act as a Travel and Course Committee to the Association during this time, whose job it was to travel to wherever the next FASNY convention would be anywhere in the state and seek out suitable hosts and facilities where the tournament could be held.

Both firemen were highly respected members of their drill teams. Jaffie served as the first President of the Nassau and Suffolk County Tournament Team Captains' Association, and Young had already been by that time a *force majeure* in the sport for several years. Well-loved and respected by all the teams, he would go on to become Chairman of the Rules Committee after making a reputation for himself as a solid, level-headed and unbiased leader. "He had common sense," says Tom Champion, a member of the Sea Breeze "Indians" at the time. "You could bring in all these rule changes, and he would sit down and give it a few minutes' thought, and then question you – and you would get the *right* rule. That's the knack he had."

Their job, in effect, was to represent the interests of the drill teams insofar as the FASNY State Tournaments were concerned. By all accounts, this was not easy. FASNY's only real concern each year was to find a host for their convention that wouldn't mind having them and who had sufficient housing available; whether or not there was a suitable location nearby to hold a tournament was, to them, fairly inconsequential. For the teams who wished to compete in that tournament, however, those priorities were reversed: their ability to compete safely, and – ideally - quickly was much more important than any particular town that the tournament happened to be held in. As one might imagine this often became an issue, especially at convention locations outside of Long Island where cities and towns didn't hold the sport in as high esteem as it was on the island, and consequently didn't have the facilities to match. A perfect example of this was the State Tournament in Catskill in 1947 referred to earlier, where the water pressure was so bad that contestants had to jump up and down on the hoses to propel the water through them. As a result of this the Nassau teams were so annoyed that they ended up walking off the track, and later started a bonfire in the middle of a downtown street as a form of protest. The poor conditions at the drill in no way justifies the teams' behavior, but it does go to show just how frustrated the teams had been when certain, basic conditions for holding a tournament couldn't be met.

Chapter 15: A NYSVFDTCA is Born

As the years went by in the early 1950's, it became more and more apparent as time went on that the Tournament Teams Captains' Association and FASNY were each on different, diverging paths: what was essential for one was simply not as important for the other, and vice-versa. Perhaps seeing that there was not going to be any resolution regarding this anytime soon, the Team Captains' Association finally decided to take matters into their own hands. According to Eugene Bryson, a member of the Massapequa "Indians" team at the time, a meeting of all the team captains was called in Point Pleasant by "Pea Pickers" Captain Bob McGrail some time in 1953. During that meeting, at which a dozen or so team representatives from throughout New York State had gathered, McGrail brought up the idea of having the team captains form a new, separate association of their own, wherein they could have complete oversight and control of their tournaments…in light of FASNY's obvious lack of enthusiasm for doing so. All present agreed it would be a great idea, and so on March 2nd, 1954, at the FASNY Executive Committee meeting in Albany, it was

> "…reported that Nassau and Suffolk County Tournament Team Captains' Association have a plan which they propose for their organizations to take over the operation and administration of tournaments. Under this plan representatives from the various tournament team organizations and other interested groups would conduct tournaments, set up rules, survey tournament facilities, authorize and sanction tournaments, and in short, accept complete responsibility, thereby completely divorcing and relieving Firemen's Association of the State of New York. There was a great deal of discussion. The Executive Committee authorized the present sub-committee to pursue the matter farther and cooperate in every way. It was suggested that S. Walter Van Nostrand work with this committee."[cccLxxix]

The West Sayville "Flying Dutchmen" compete in the Motor Hook & Ladder #2 ("Dummy") Contest in 1953

Wilbur Young, some time in the mid-1950's

Van Nostrand did get the job, and apparently got to work right away. Five months later, on August 19th, at FASNY's 82nd Annual Convention at the Statler Hotel in Buffalo, he reported the following:

> "The first act of this Committee was to change the State Rules to tie in with the changes voted at the last convention with regard to protecting the State Association with Public Liability and Property Damage Insurance before State sanction is given. They also made other changes in the Drill Rules. The new rules were effective as of May 1st, 1954."[ccclxxx]

These changes in Drill Rules, with which Wilbur Young played a large part, were significant at the time simply because it was the first time in quite a long time that some attention had been paid to them. Those who were there at the time recall that prior to then anything that wasn't in the Rule Book was fair game and open to "interpretation", and cheating was rampant.

It took a year and four days from the Buffalo convention to work the details out for its Constitution and By-Laws, but on August 23rd, 1955, The New York State Volunteer Firemen's Drill Team Captain's Association was officially brought into being, "...to establish a closer relationship between the Volunteer Firemen's Competitive Drill Teams, to advance the training and efficiency of the individual firemen, and to improve the tactical, social and mutual benefits of the volunteers while providing universal competitive drill training throughout the state."[ccclxxxi] The first set of Official Drill Rules and Regulations was adopted not long after, at a meeting of the new association in Lake George on September 14th, 1955. Interestingly, Robert McGrail of Point Pleasant was named as its first President – but for a very logical reason: being from western

The Hempstead "Yellow Hornets", 1956

New York, he could remain neutral on matters – including rule changes – that Nassau and Suffolk might argue about. And, incidentally, it was Eugene Bryson who succeeded him in 1957.

Oddly, and in what could only be described as ironic, the very first drill that was held after the formation of the NYSVFDTCA (which hereafter will be referred to only as the "DTCA") was the Annual Labor Day Tournament at Hicksville on Long Island – and it was plagued with a huge problem related to conditions at the track:

"A leak in a fire hydrant forced postponement today of the twenty-seventh annual Labor Day Fire Tournament. Only two of the seven contests were held.

"The leak, which reduced water pressure to only a trickle, was not discovered until the start of the first hose contest. The two events held were motor hook and ladder contests in which no hoses were used. Officials called off the remainder of the tournament until next Saturday morning.

"However, the day was not a complete loss for the 50,000 spectators who filled two large stands and lined both sides of Levittown Parkway for several blocks. Besides the two ladder-climbing events, which supplied several thrills, they were treated to a three-hour parade."[ccclxxxii]

In fairness to the DTCA, however, they had not yet officially started overseeing the drills at that point. FASNY administered their 83rd tournament in Lake George in September of 1955, at which time S. Walter Van Nostrand reported, "This will be my final report to the Association." At their Executive Committee meeting in Albany on November 12th,

"Mr. Christy reported that the New York State Association of Drill Captains, which has been formed, are about ready to take the drills (tournaments) over and relieve FASNY from responsibility and supervision. It is their intention to incorporate. Attorney Foley was instructed to cooperate with the Drill Captains Association in their efforts to organize and incorporate."[ccclxxxiii]

The Southern New York tournament at Islip, 1956

The last tournament ever to be sanctioned and administered by FASNY was held during their 84th Annual Convention, on August 23rd, 1956, and took place at Floral Park on Long Island.

The convention itself, however, took place once again in New York City, and apparently the parade, which ran down Broadway from Bowling Green all the way down to City Hall, was quite a spectacle:

. "...The four day conclave was climaxed by a two hour parade. In the line of march were 164 pieces of apparatus and 2941 marchers led by 27 musical units. The parade was headed by the color guard, band, apparatus and uniformed personnel of the New York City

The Port Washington "Runts" practice during the 1950's

Fire Department. The spectacle was witnessed by over a million spectators. One of the units to draw unusual attention was the Onandaga Indian Reservation Volunteer Fire Company, whose members and wives were arrayed in full tribal regalia of brilliant color and dress."[ccclxxxiv]

FASNY may have been stretching the truth a bit with the phrase "a million spectators". According to the *New York Times*, "...About 125,000 persons lined the route. Office workers in the financial district floated their morning papers down in scraps."[ccclxxxv] There is no doubt, however, that it was something to have seen.

When the State DTCA was first formally organized in 1955, Nassau, Suffolk and Western New York were the first three "areas" to become part of the new Association. Possibly as a result of the area being that much more rural and thus a bit harder to organize, the Northern New York area did not become officially part of the State DTCA until sometime after September of 1956. Up until that time, "...Teams [were] being organized on a regional basis by George Briggs of Norfolk and Bill Reid of Massena, both of St. Lawrence County. The inclusion of the Northern New York Drill Teams in the State Association will raise the total number of member teams to one hundred and fifty."[ccclxxxvi]

While all of this was going on, it was still Business As Usual at all of the regional conventions and tournaments throughout the state. There was still some rowdiness at the 1955 Western Convention in Dansville, but it seemed to be a bit less serious than in previous years:

"Each fireman was provided with a map of Dansville as he registered, showing all the important spots he'd need to reach – except the saloons. They weren't marked on the maps – nor was such marking necessary."[ccclxxxvii]

"It is estimated that if all the empty beer cans left by the firemen were laid end-to-end they'd stretch all the way from the first hangover to the last. Or, if heaped up in one place they'd make a pile as big as some of the heads of visitors."[ccclxxxviii]

"Squirt guns became as rare as a cool breeze yesterday as supplies of even the most optimistic merchants were exhausted. But if you think that daunted the lads you just don't know your firemen. Those new squeezy plastic ketchup bottles were pressed into service – and even fly spray guns. Not all the guns were filled with water, however. Two firemen were lustily squirting the sprays in each other's mouths. The liquid was rum, a reportorial tasting revealed."[ccclxxxix]

But as proof of the fact that firefighters, although rowdy and somewhat careless at times, really do have hearts of gold and will always put the safety of their communities ahead of their own, there was this account as well:

"That grass fire didn't have a chance Tuesday afternoon.

"Summoned at 3 p.m. to the Joseph McIntyre property on North Clinton St., Dansville firemen quickly doused the fire and got credit for a putout.

"But the official scorer gave an assist to the conventioneers of the Western New York Firemen's Association.

"Armed with water pistols and driving an assortment of squad cars and emergency ambulances, the visiting firemen helped make short work of the flames.

"Only damage reported by Fire Chief Robert Cain was a jammed trigger on a water pistol. 'There wasn't any property damage,' he explained, 'because we had one fireman for each burning blade of grass.'"[cccxc]

It may seem at first to be just more and more of the same, but if one compares the tenor of the above anecdotes to that of those in years past, one can see that the tone of the stories was starting to change: from a journalistic perspective at least, all the firefighters' hijinks and revelry at the conventions was worth reporting, but at the same time the behavior had become so commonplace and predictable that it was no longer sensational, and more and more it was just tiresome, to the point of being nothing but a pain in the ass. The following year, during the first day of the 1956 Western convention in Batavia, the paper wrote:

"'Grin and bear it.' That was the advice for Batavians today in the wake of the restless first night of the 57th annual convention of the Western New York Volunteer Firemen's Assn. Fire sirens wailed until the early morning hours, being heard as late as 3 a.m. as fun loving volunteers scampered through city streets.

"Police had about a dozen complaints from residents who said their sleep was affected. City officials also reported calls from Batavians. Those familiar with firemen's conventions took it all in stride and got ready for a couple of more sleepless nights.

"Be a nice time to spend at the lake, wouldn't it?"[cccxci]

Then, the following day:

"The overnight antics of the 'visiting firemen' were just as noisy as the previous night despite pleas from police and city officials for them to quiet down. A few 'extras', such as water fights and a minor fire on East Main St. added to the activity Tuesday night.

"Police reported about a dozen calls from residents protesting about the blowing of sirens. The most serious casualty was a big front window at the F. W. Woolworth Co. store at Main and Jackson Sts. Which fell victim to an errant hose as volunteers engaged in a small but friendly 'water' fight during the early evening."[cccxcii]

Just as wearing a lampshade on your head at a party is hilarious at midnight but boring and stupid looking by 4 a.m., the overall desire by the firefighters to raise hell at the conventions began to lose more and more steam as time went on. Whether it was affected by public attitude or administrative protest, or if it was simply a case of the firemen "growing up," by the early 1960's descriptions of misbehavior had all but disappeared from local newspaper coverage. In fact, in 1962 those same Western firemen even managed to garner some praise:

"It has been an unusual three days for Warsaw in that, while we expected a lot of disturbance, there has been nothing serious. It can be frankly stated that the police had more trouble with our local yokels who were trying to barge in on the fun and have the firemen get the blame.

"We wish to congratulate the Warsaw Fire Department for this excellent convention, the many firemen from outside, who behaved exceptionally well, the fire police who donated their services and all others who contributed to its success. It can honestly be said that the behavior of this convention has been so much better than previous experience that the 'dread of a convention' has been removed.

"One woman who made several complaints was told off and told what taxes our volunteer firemen save through their services and she finally allowed that for that amount of money, 'I guess I can stand a little noise for three days.'

"Firemen, we salute you."[cccxciii]

The formation of a State Drill Team Captain's Association in 1955 was great news for both the teams and FASNY. In addition to the teams now being able to organize drills for themselves in a manner much more conducive to their own needs, it took the insurance liability headache out of FASNY's hands, effectively allowing them a huge sigh of relief.

But that was not the only good news to happen to the teams at the time. In reference to the *other* big insurance liability headache – that of insurance coverage should a firefighter be hurt or killed while on the track – a significant occurrence took place that in many respects allowed the sport of firematic competition to move forward unhindered from that particular concern, although it took a few years to finally come to fruition.

That event came courtesy of one Hans Weiss, who was a firefighter with the Franklin Square Fire Department and a member of the "Square Shooters" drill team. While competing at the Southern New York Firemen's Association tournament in Greenport on June 28th, 1952, Weiss took a spill and injured himself while stepping off the back of a truck during one of the contests.

What Weiss probably knew – or at least believed - at the time was that insofar as his volunteer work with the fire department was concerned, he would be entitled to receive compensation and benefits if he was to have been injured while he was engaged in any activities that were related to the fire department, such as fighting a fire, washing a truck at the fire station…or participating in a tournament. He was basically correct: for many years there had been laws on the books to that effect; most notably the Third section of Section 205 of New York State General Municipal Law, which states, in part:

> "Any such volunteer fireman who shall receive injuries while performing his duties as such, while in the fire house, while going to or returning from a fire by any means of travel, transportation, or conveyance whatever, or while working at the fire or answering a call or fire alarm or while officially engaged in testing or inspecting the apparatus, or equipment, or while engaged as a member of a fire department, or fire company, emergency relief squad or fire police squad, or fire patrol or while attending a fire school, or while instructing or being instructed in fire duties or while attending any drill or parade or inspection in which his company or department is engaged, or while engaged in emergency work not related to fire extinguishment or prevention or going to or returning there from when duly authorized to participate therein, so as to necessitate medical or other lawful remedial treatment or prevent him from following his usual vocation on account thereof, shall be reimbursed for such sums as are actually and necessarily paid for medical or other lawful remedial treatment, not exceeding five hundred dollars. He shall also be compensated for the time he was actually and necessarily prevented from following his vocation or for the time of his

disability on account of such injuries, at the rate of thirty-six dollars per week not to exceed one thousand eight hundred dollars."

After being injured at the drill, and with this in mind, Weiss filed a claim with the Workmen's Compensation Board to receive benefits. At first they honored his claim, but then, upon review, they changed their minds and denied him any benefits. After going back and forth with them for a year and a half, and then feeling like he got the short end of the stick, Weiss decided to fight their decision in court. As described by court documents for Weiss vs. Franklin Square and Munson Fire District:

"...petitioner made a claim for workmen's compensation and was granted an award by the referee. This award, however, was subsequently reversed by the Workmen's Compensation Board on the ground that the petitioner's participation in the tournament was voluntary and did not arise out of and in the course of his employment. The respondent resists the present claim on the ground that the determination by the Workmen's Compensation Board is *res adjudicata* and a complete defense to the instant claim."[cccxciv]

In English, what this meant was that the Workmen's Compensation Board (the "Respondent") had determined that since Weiss had decided to attend the tournament of his own free will, as opposed to by direct order from a fire department superior, he was not considered to be an "employee" of the fire department *per se* at the time, and thus not eligible for benefits. Moreover, they felt that as all their decisions were final ("*res adjudicata*"), Weiss had no case.

At issue in that case was whether or not Workmen's Compensation Law trumped General Municipal Law, or vice-versa: one said he was not eligible for benefits; the other said he was. In the end, on January 8th, 1954, the following determination was made:

"In view of the fact that the provisions of the General Municipal Law relating to compensation to firemen for injuries received by them are far more comprehensive than the Workmen's Compensation Law, in that the former authorizes compensation for injuries to volunteer firemen while attending any drill or parade, a determination by Workmen's Compensation Board denying compensation to volunteer firemen for injuries received as a participant in a firematic tournament following a parade, was not *res adjudicata* as to subsequent proceeding under the General Municipal Law."[cccxcv]

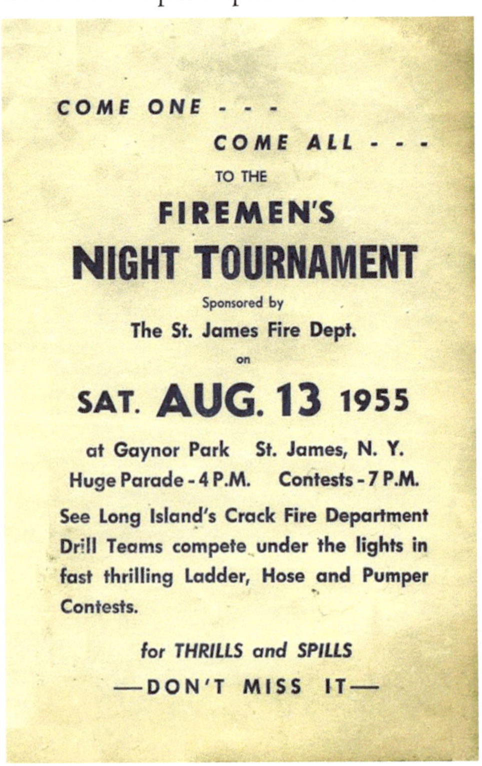

In other words, in the opinion of the Nassau County Court, the provisions of the General Municipal Law *did* trump Workmen's Comp Law – and Weiss was entitled to his benefits.

"Not so fast!" said the Munson Fire District and the Town of Hempstead, which in hindsight was probably self-insured, and therefore not too happy about having to pay out any dispensation. They appealed the County Court's decision, and the case was brought before the Appellate Division of the Nassau County Supreme Court. The decision in that case, which was handed down on June 7th, 1954, was not so good for Weiss:

"The Supreme Court, Appellate Division, held that the Workmen's Compensation Board's prior determination, that the injury, sustained by the petitioner in stepping from a fire truck at a parade and competitive drill, did not arise out of and in the course of his employment, was binding on the County Court."[cccxcvi]

For the time being, it seemed the firefighters were at a loss. However, the story doesn't end there, as the case was appealed once again – this time by Weiss, who took the case all the way to the New York State Court of Appeals. On July 8th, 1955, in a victory for the firefighters, the State Court overturned the lower court's decision. The case must have been significant enough that on September 9th New York State Attorney General Jacob K. Javits himself included a formal opinion about it. He said, in part:

"On appeal of the above [Nassau County Supreme Court] case, the Court of Appeals reversed the determination of the Appellate Division. The following appears in the opinion of the Court of Appeals, decided July 8, 1955:

"'There are essential differences between the two statutes. Section 205 of the General Municipal Law is a broader statute and not substantially identical with section 10 of Workmen's Compensation Law. The legislative intent that the statutes were to serve different purposes insofar as they protect volunteer firemen is evident from the enumeration in the General Municipal Law (Section 205, subd. 3) of activities other than those which might be designated solely as employment as a fireman. It specifically provides for compensation to volunteer firemen injured in the performance of his duties "while attending any drill or parade or inspection in which his company or department is engaged". Such activities are not referred to in the Workmen's Compensation Law.'"[cccxcvii]

In other words, the State Appellate Court agreed with the findings of the Nassau County Court: Weiss was entitled to his benefits. Happily, this decision went on to become the basis for the section of law that to this day covers firefighters if they ever get injured or killed on the track. Section 5, paragraph "i" of the Volunteer Firefighter's Benefit Law reads:

"While, within this country or in Canada and pursuant to orders or authorization, practicing for, or participating as a contestant or an official in, any competitive tournament, contest or public exhibition conducted for firefighters which is intended to promote the efficiency of the fire department, fire company or any unit thereof, including necessary travel directly connected therewith other than travel to and from such practice. The actual rendition of fire or other emergency service shall not be deemed 'practicing' within the meaning of this paragraph."[cccxcviii]

With apologies to the reader for the somewhat lengthy amount of legalese needed to explain it, the fact that this case was decided – and *when* it was decided – in favor of the firefighters had huge ramifications for the sport of firematic competition: now the firemen could feel less intimidated to compete, a factor that stood a good chance of getting in the way of the development of the sport.

And, that came none too soon. Tragically, on August 23rd, 1958, at the tournament held in conjunction with FASNY's 86th Convention in Amsterdam:

"A runaway Long Island fire truck roared into a crowd of spectators at the state firemen's tournament yesterday, killing an elderly man and injuring a dozen other persons.

"The truck, which was taking part in a hose-laying race, careened off the end of the tournament strip at about 40 MPH when its brakes failed. The driver shouted and waved to warn the crowd, and finally swerved into what appeared to be an opening in the ranks of spectators.

"The small, stripped down tournament truck crashed through a six-foot wire mesh fence squarely into the elderly spectator, who apparently froze on the spot while others scrambled for safety. The fence was thrown against several other persons including two young children. The children and two adults were hospitalized along with one member of the truck crew.

"The Rough Riders tournament team of the Oyster Bay Atlantic Steamer Co. – five time winner of the state championship in the last eight years – was operating the truck. The team had placed first in the first three events before the accident.

"Police said the truck's brakes failed half way down the 100-yard macadam race strip. The rest of the way, driver Wilbur Young, 41, of 42 Burtis Ave., Oyster Bay, shouted and maneuvered for an opening in the crowd. Fire Chief Osmund Smith of the Oyster Bay company said Young managed to shift down to second gear, slowing the truck a little.

14

RBC's exclusive story of what happens when hot rodders join the fire department.

DRAGGIN' FIRE WAGONS

By RICHARD I. ZIFF

Draggin' engines in the nineteenth century called for loud voices and strong backs (left). Today's draggin' firemen need perfect teamwork and, of course, a hot engine (above).

SPEED IS THE MOST important factor in fire fighting. Even in the days when clumsy, spindle-wheeled hand pumpers were dragged over cobbled-stone streets, fire fighters raced against time.

Since World War II, hot rodders have helped in this race to save precious seconds. Many Long Island, N.Y. home owners can thank the volunteer firemen-rodders and their straight pipe, hopped speed wagons, with red lights flashing and sirens screaming a path through town, for saving their lives and property. Hot rod know-how, that has cut the quarter-mile time to a slim 10 seconds, pays off at the fire house as it does on the drag strip.

Getting to the fire "fustest with the mostest" has been the driving spirit of all volunteer companies and trying to beat "the other guys" has kept generations of volunteers on their toes.

But, all too often, when the oldtime volunteer smoke-eaters of the 1800's got to a blaze, they fought other fire

Typical draggin' wagon belongs to the "Meadowlarks" of East Meadow, L.I. It has a Cad mill.

JANUARY 1957 15

Rod Builder & Customizer *magazine, January, 1957*

"He said Young's 'foot went to the floor' when he tried to brake. 'I heard him holler when he realized there was no brake,' Smith said. He said the spectators had been warned repeatedly not to stand at the end of the strip. Police said a hydraulic line to the left rear wheel broke. A coroners inquest was to be held today."[cccxcix]

Insurance claims notwithstanding, the accident rocked the entire sport and left Young and the rest of his team stunned: this was not something that that in any way they could have imagined might happen. Ultimately there was good that came out if it, for Young, in the fashion that he became so famous for, soon after fought for and won rule changes in the sport that required dual master brake cylinders on Class C trucks from that point forward – a fact that has not lost any significance to drivers who still experience the occasional failure of a brake cylinder to this day.

In hindsight, however, that incident was unfortunately an early indicator of just how the sport of firematic drill team competition would be characterized during the 1960's. Like the football days of old where the players wore no pads and just thin leather helmets, the competition was fierce and safety was not always the Number One Priority. As one former Oyster Bay team member put it, "it was pretty wild and wooly."

Speed, and making the trucks go faster, became everything. If the 1950's brought technology to the sport, the 1960's brought money to that technology, and there was nothing that teams wouldn't try.

Part of the "blame" – if there was any to be had – for this phenomenon had to do with a military tank test-driver for General Motors out on the west coast named Wally Parks, who in 1947 founded an organization called the Southern California Timing Association. At first he was only interested in seeing how fast racers could accelerate on the Bonneville Salt Flats, but by 1951 that sport had become so popular that

The West Sayville "Flying Dutchmen" compete in the 3-Man Ladder Contest at East Islip in 1958

Parks created the National Hot Rod Association – and the rest of *that* is history. Interestingly, Parks' intent seemed surprisingly similar to that of Albert Sammis in years' past: "…to 'create order from chaos' by instituting safety rules and performance standards that help[ed] legitimize the sport."[cd]

As the 1960's began, though, the sudden burst in motorsports enthusiasm made for some pretty rough and tumble times on the track. According to *Newsday*, at the State DTCA Drill in Riverhead in 1960:

"Twenty-three firemen were injured yesterday, including two who had to be hospitalized, during competition at the State Volunteer Firemen's Tournament. Fire officials said 21 of the 23 firemen were treated at the scene for minor injuries suffered during the contests. Two others went to Central Suffolk hospital."

And, again during the State Drill two years later in Colonie, local papers reported:

"Five firemen from a Garden City Park volunteer company were injured yesterday when their truck overturned and burned during practice for a contest at the state convention here.

"The injuries were not considered serious.

"Police in suburban Colonie said the drive shaft on the light hose truck had broken and pierced the macadam of a road near Latham, and the truck overturned in a ditch."

Although it was completely unrelated to anything to do with motorsports, at about the same time the New England States Veteran Firemen's League ironically suffered the worst "on track" accident in the history of their entire association. According to Dixon, on June 24th, 1961, "The most serious accident to ever occur while a handtub was pumping in competition took place at Salem, MA. Albert April of Dover, NH, Foreman of the Essex #2 of Essex, MA was killed instantly when the dome suddenly blew up, striking hum full force. Also injured were the Asst. Foreman and several pumpers. It was felt at the time that this tragedy would be a major setback for musters. However, as many times before, the mustering fraternity pulled up its bootstraps and continued on."[cdi]

As did the modern-day version of the sport, in spite of its setbacks on the track. In fact, its popularity even gave rise to a new competitive body, as on October 27th, 1961, the Central New York State Firematic Drill Team Association was born, to "encourage the establishment and continuance of drill teams around the Albany, N.Y. area." The formation of this new association made perfect sense: although there were four geographical Areas of the State Drill Team Association, none of the majority of the teams – and therefore tracks - of those four were situated close enough to Albany to make it practical for any of the local teams who were interested in competing to be able to commute back and forth easily. Teams from Elsmere, Delmar and Slingerlands in the Albany area, as well as those from Fuller Road, Latham and Boght near Colonie were joined by teams from Renssalear and Saratoga Counties.

SEVENTEENTH ANNUAL

Competitive Drill

OF THE

Fourth Battalion District

AT

FIREMEN'S FIELD

Rockville Centre

Sunday, July 10, 1960

at 1:00 o'clock

ROCKVILLE CENTRE HOST 1960

JOHN ANDERSON
Chairman

HENRY PETTERSON
Treasurer

ED. GRAEBE
Vice-Chairman

MALCOLM RICE
Secretary

EAST ROCKAWAY — ROCKVILLE CENTRE

LYNBROOK — MALVERNE — LAKEVIEW

Oddly, at the time they chose not to become part of the NYSDTCA. "When that association first formed in 1961," says Elsmere "Wanderers" Team Captain Rick Webster, "they actually had representatives from New York State come in and explain to them what they were doing, and tried to get them to become a regional part of that association. But, they decided to do it on their own." No matter; the Central Association survived without them, and indeed thrived on its own for many years afterwards.

The entire sport, in general, was also thriving during the early to mid 1960's, but in ways that it never had before. Although public attendance at the tournaments was not nearly as large as it had been in years past, the sport was gaining new recognition in

DRAGGIN' WAGONS

A fire engine at speed, only this time the fire is substituted by a time clock. The Stumpjumpers from North Merrick are a feared team to run against as their times are usually among the very best. Riding crew uses seat pads to alleviate small hurts. Some competitors feel they get better handling of ladder at end of run by riding low as shown during the 450-foot run.

Yell fire in New York State and you'll likely as not find a supercharged Chrysler powered '32 Ford fire truck at your front door

Text and photos by Jack Eaton

"OK Bud, where's the fire?" Such a phrase is familiar to apprehended speeders the nation over. But in New York State, the question can lead to all kinds of complications, especially if the vehicle happens to look like a cross between a hot rod and fire truck. You see, New York is the home stomping grounds of some of the hottest fire equipment this side of Hades. And, too, few spectacles compare with the thrilling fire tournament days when these strong-hearted stallions are unleashed in exciting combat.

It all started many years ago. Back around the turn of the century, nearly every community relied upon its stout-hearted male citizens to man the local horse-drawn fire wagon. Naturally, Dobbin required a weekly workout to unlimber the kinks and the hand-operated water pumps and related wagon accessories had to be cared for. Inevitably, the question arose between neighboring villages as to which had the best "equipment." No one knows for sure when the first fire tournament started or where. But, in New York at least, it's here to stay.

The events are called several things, including Competitive Drills, Drill Tournaments, and Fire Tournaments. Virtually everything is encompassed in the tournaments except Mrs. O'Leary's cow Contests are held statewide with regional, county, and state tournaments set up to determine champions in various categories. The average competition draws from 50-55 fire companies and the crowds vary in size from 3000 to 25,000. The most consistent company is Oyster Bay, Long Island. They have captured the state championship 13 times and the fire house is loaded to the rafters with more than 800 trophies won since 1949. Islip, L.I., has been state champ twice;

(Continued on page 49)

Hot Road *magazine, April, 1962*

fire breathing engine

WHAT APPEARS at first to be a cross between a street roadster, super/stocker and an altered is actually a *Class C (fire department)* racing fire engine built to compete at official drills and tournaments. By now you probably think the CARS Magazine editorial staff has flipped its collective lid and can't tell the difference between a fire truck and a drag racing machine. Well, don't give us up for lost yet. We've got some surprises coming. So let us explain.

The hot roadster featured here was built in accordance with the rules laid down by the New York State Volunteer Firemen's Drill Team Captains Association, Inc., for vehicles competing in official drills and tournaments. Their Official Drill Rules and Regulations Handbook makes the NHRA rules book look like it's suffering from malnutrition! Everything from inspection to protest information is jammed into this weighty rules book.

The idea behind this *friendly* competition is to improve each department's efficiency in the art of fire fighting. This form of competition dates back to 1900, when the departments were using horse-drawn rigs. In a typical statewide tournament, there are eight competitive events Three Man Ladder, Motor Hook and Ladder Class B, Motor H&L Class C, Motor Hose Class B, Motor Hose Class C, Motor Pump, Efficiency, and Bucket Brigade. Most events make use of specially designed Class B&C rigs, but a few events were designed to test the human element only.

When the flag is waved the crew-laden, souped-up trucks peel out towards either a fire hydrant for the hose events or a permanent or makeshift arch for the ladder event. Just as drag racers do, the Class B&C pilots race against highly accurate timing clocks. However, in this sport the fastest accelerating truck can lose to a stocker because of a fumbling crew member It's not at all unusual to see a blown, injected Chrysler Hemi-powered roadster clocking slower overall times than a stock-powered rig. Class B rigs run 100-foot

Wilbur Young warms up the hot Stage III Rammy-powered engine while the rest of the crew takes a break. The Oyster Bay, L.I., Rough Riders have taken more New York State championships than any other team, retiring three separate (three wins required per retirement) championship trophies.

By MARTYN L. SCHORR

CARS • 49

Cars *magazine, December, 1964. Note Wilbur Young in the driver's seat*

The Glenwood "Greenhorns" compete in the Motor Pump Contest at the New York State Drill in Hempstead in 1964

other ways, through print and broadcast media. The April, 1962 issue of *Hot Rod* Magazine contained a full-length story about the teams and their unique "apparatus", as did the December, 1963 issue of *Argosy*, the December, 1964 issue of *Cars* Magazine, and the July, 1965 issue of *Super Stock & FX*. In 1965, the State DTCA drill in Utica was featured for the very first time on national television during a segment for ABC's Wide World of Sports, giving the New York State teams exposure on a national level that they had never achieved before. "...Crews from ABC television networks Wide World of Sports show ran out of film as they taped events for future showing," stated a local paper. "Five hours worth of time was taped."

But in spite of a smaller public attendance at the drills, the sport was still beloved by its fans. An interesting depiction of this, and what it was like to attend a drill back then, was described in a posting on a website forum maintained and frequented by the drill team community:

"It is good to see of the 53 teams entered at this S.N.Y. drill in 1962 that only 12 dept's have dropped out of competition. Another interesting drill that year was the Far Rockaway Inter-Community Drill (read that old fashioned). 51 teams from Nassau entered the drill and 41 actually ran that day. Hempstead Drill- 48 entered 40 ran. Nassau County Drill -42 teams entered -33 ran. They must have put a LLLOOONNNGGG day in, back then, with a parade in the morning that you paraded your "B" truck in every week. I doubt it discouraged the fans back in 1962. They didn't go home to air conditioning. They only had 6 or 7 channels on a black and white TV. Probably only had 1 TV at best with no VCR or DVD or video games to plug in to it. You couldn't even go to the Mall for Labor Day sales. Malls weren't invented yet. The good old days. You went to a tourny and drank your Schaefer/ Rheingold/ Ballentine beer, sweated it out and you were happy to be there. And that's the way it was in 1962."[cdii]

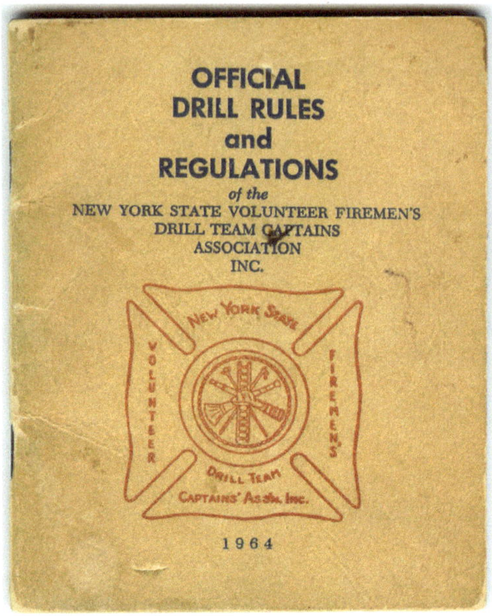

But as the 1960's reached mid-decade, in addition to the crowds at the tournaments, the number of teams participating at them also began to decline. There were several reasons for this. The first had to do not so much with conditions on the track, or within the various competitive associations, as it did with what was going on in the world at the time: the Vietnam War.

Like previous wars, the Vietnam War had a significant impact on the sport. Although U.S. Military Advisors had been involved since the early 1950's, the early 1960's saw a sudden escalation in America's involvement, and in 1965 the instigation of compulsive military service (also known as The Draft) started decimating the drill teams.

"The first six years of the 1960s were bleak for the Flying Dutchmen. A dwindling interest in racing resulted in fewer members on the team," states the West Sayville Fire Department website. "By 1966, the department had trouble putting a team on the track and did not actually compete until the East Islip Labor Day drill."

Others fared even worse. In 1965, the Plainview "Plainsmen" were on top of the world with what many considered to be the best team in the sport, but by 1966 they had all but disappeared. Port Washington, another long-time heavy hitter, lost not one but two teams – the "Runts" and the "Rangers" – probably as a result of the war.

If it wasn't the war, though, it *was* conditions on the track. As the NHRA and drag racing, and motorsports in general, reached new heights of technical sophistication, more and more teams were faced with the reality that if they were to remain at the top of the points ladder – or simply even competitive, for that matter – they would have to at least keep up with, or at best be at the forefront of, motorsports technology in order to be faster than the next guy. But as each team added new tricks, workarounds and "interpretations" of the current rules to maintain an edge, old rules were in turn changed by the DTCA and new ones were added to keep the playing field as level as possible. Whereas before a team might be able to purchase a stock model vehicle right off the showroom floor and take it right to the track and have it compete as a Class B or Class C competition vehicle, new rules now made that impossible, as modifications would have to be made before you would be allowed to compete. As a result, the costs to build and maintain the trucks started to grow significantly, and more and more it was becoming a situation where a team could either afford to field a state-of-the-art truck or they couldn't. Those that couldn't, no doubt understandably discouraged, often chose to just give up and call it quits.

Also during this time, as the level of competition in the sport increased and the teams that were able to "hang in" started concentrating more and more on improving their performance on the track, the growing split between them and FASNY over their respective directions and purpose began to grow even wider. Even though the teams had made life easier for FASNY by taking control over the tournaments with the formation of the NYSDTCA in 1955, the two associations still chafed against each other. In addition to still having the bulk of its attention taken up with matters of firematic administration within the volunteer fire service, FASNY, it seems, as part and parcel of same, was also still worried about the behavior of firefighters as they were seen by the public in general. At their 92nd Annual Convention held (once again) in New York City in 1964, a Mr. Donald Herbstman, of the New York Fire Insurance Rating Organization, spoke to the assembly about the necessity for FD's to screen FD applicants so as to keep firebugs out of their ranks, and about the necessity of "maintaining a good public image." In typical FASNY fashion, however, not all of the firemen attending seemed to care:

"As the public image was discussed inside the Olympia Ball Room [of the Manhattan Hotel], a Mineola firemen's whistle shrilled in the marbled lobby amid shouts of 'Everybody out of the pool.' One volunteer explained that this seemed a good thing to shout.

"'They've been pretty quiet. You should have seen Niagara Falls last year,' said Arthur Crute of Rocky Point, L.I. 'Of course, they did get a hose out of a window last night. They wet the sidewalk.'"[cdiii]

In contrast, at the same time, the DTCA was independently handling the sanctioning and running of all the drill team tournaments in New York State…mostly. The sole exception was the New York State Drill, which although it was organized and administered by the DTCA, was still "associated" at the time with FASNY, and for which FASNY still kept statistical records. In a sign of a growing trend, however, that too changed the following year at FASNY's convention in Utica in 1965:

"An amendment to the Constitution and Bylaws of FASNY was approved at the 1965 convention eliminating that section which provided 'that the President shall appoint a Tournament Statistician.' By this action the Association disassociated itself from all drills. Further action by the Executive Committee turned over all work in connection with drills to the new association, the New York State Volunteer Firemen's Drill Team Captain's Association, Inc."[cdiv]

If this action wasn't a divorce, it at least signaled that the two parties were becoming more and more estranged with each passing year. Indeed, their needs were becoming increasingly different: those firefighters who were mainly interested in attending a FASNY convention to meet other firefighters and attend seminars were most keen on going to a resort of some sort far away from home, while

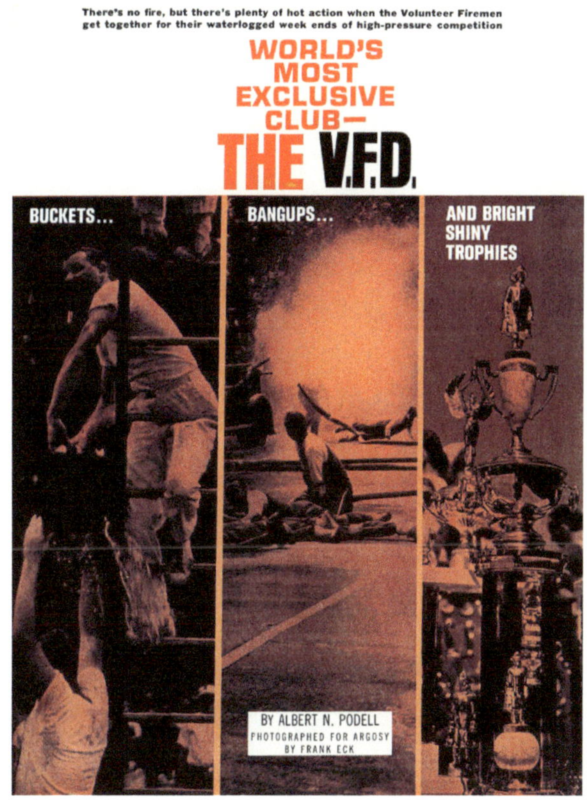
There's no fire, but there's plenty of hot action when the Volunteer Firemen get together for their waterlogged week ends of high-pressure competition

WORLD'S MOST EXCLUSIVE CLUB—THE V.F.D.

BUCKETS…

BANGUPS…

AND BRIGHT SHINY TROPHIES

BY ALBERT N. PODELL
PHOTOGRAPHED FOR ARGOSY BY FRANK ECK

Argosy *magazine, December, 1963*

NEWS COLORFOTOS BY WILLIAM KLEIN

Elmont Bangtails hit the target with their line in motor hose event of N.Y. State drill at Hempstead. But their time wasn't fast enough to win

'WATER!' IS THE CRY

Getting it on target fast is vamps' aim at firemen's drill

WHEN VOLUNTEER FIREMEN assemble for a drill—that unique carnival of souped-up, stripped-down tournament rigs, screeching tires, racing men, skinned elbows, teetering ladders, flying buckets, jerking hoselines and doused spectators—it's a wonderful outing for vamp and fire buff alike. One of the biggest drills was recently held at Hempstead, L.I., in connection with the 92d annual convention of the Firemen's Association of the State of New York. Sixty drill teams flaunting colorful nicknames like Pea Pickers, Rinky Dinks and Tail Draggers competed in eight events for scores of trophies and plaques donated by firemanic organizations and local businessmen. Though the competition was something fierce and the men seemingly reckless in their enthusiasm, hours of skilled practice went into the drill for weeks beforehand. And after it was over and the prizes were doled out—Oyster Bay's Rough Riders topped the field (TR would have loved that!)—the laddies got together for some rousing fellowship.

The New York Sunday News*, 1964*

The arch collapses at FASNY's 95th Convention at Albany

those who were going to the convention mainly to compete in the State Drill were more interested in staying on Long Island where the tracks were better and they wouldn't have to transport their teams and equipment as far. As one Historian of the sport put it, "the people who 'conventioned' and the people who 'drilled' became two separate groups." The reader will note that as far as the history of locations for FASNY conventions and tournaments went up to that point, there were relatively few that took place on Long Island. It appeared that as far as the tournaments were concerned FASNY really just didn't care all that much. As if to make its point – and in doing so add insult to injury - at the same convention in 1965 where FASNY voted to disassociate it self from the DTCA, the drill teams arrived the day before the tournament to find that no arch had been built or even arranged for. Understandably, this stuck in the craw of the DTCA, who as far as the State Drill was concerned felt they were getting the short end of the stick.

But there was also another aspect to the State Conventions that was starting to really bug the DTCA as well, one that would have even larger ramifications: the issue of housing.

For many years, FASNY had scheduled their conventions to take place over the course of the better part of a week. Attendees would typically arrive on a Sunday, with the business session then taking place over the next three days, Monday through Wednesday. The tournament would take place on the following day,

Thursday, and then the convention would end with a bang with a parade on Friday. To house the firemen, FASNY created a Housing Committee to arrange blocks of rooms in local establishments for the visiting firemen.

This arrangement worked fine for anyone wishing to attend the convention from start to finish. One would simply book the room, arrive on Sunday and enjoy their stay. For firemen wishing to make the trip only to attend or compete in the State Drill however, this became increasingly problematic: those trying to find local lodging available from Wednesday night on were forced to book the rooms through FASNY regardless if they were attending the convention or not, and once they tried to they more often than not found "no room at the inn" since the rooms were already occupied and/or were already blocked out by the Housing Committee for use by the conventioneers. In addition to this, the DTCA alleged that FASNY was padding the room rates – a dollar or so at first, then more as time went on – so that the teams were, in a sense, "blackmailed" into paying higher lodging rates to attend drills that were associated to FASNY's convention by name only. FASNY consistently denied this, but nonetheless the lines in the sand were drawn and sides were taken.

The problem was somewhat rectified in 1967, although it was done so in a rather inauspicious manner. On the morning of August 24th, at FASNY's 95th convention in Albany, the first team of the day to compete was on the starting line and set to begin their run when the temporary metal scaffold at the other end of the track, erected as an arch for the day's events, collapsed suddenly and without warning. Judges atop the stand scrambled as best they could for safety – some even leaped onto nearby trees – but not all landed safely. As FASNY reported it, "…The 20 foot high judges' stand collapsed injuring 24 men. The two men in serious condition following the accident were Walter Worthington of Riverhead and Fred Kendrick of Albany. They were admitted to the Albany Medical Center, and have since been released and are in satisfactory condition."[cdv]

Understandably, the competition was cancelled for the day, and officials elected to re-schedule it to take place in Lindenhurst on Long Island during the following month. In doing so, however, they scheduled the drill for September 23rd - a Saturday - instead of the traditional Thursday. And, a review of the records since that time shows that from that point forward State Drills were, with few exceptions, always held on a Saturday, with the parade on Sunday, the following day; a practice that continues to this day.

Although there is no evidence to indicate that the move to hold the rescheduled drill on a Saturday was in any way related to the collapse of the arch, the biggest advantage of this change was that since firefighters who wished to compete at the State Drill would now be arriving on a Friday as opposed to a Wednesday, finding rooms for them became much easier, and that fact made traveling to a State Drill that much more tenable.

Chapter 16: "Race" Relations

Between the Vietnam War, the assassination of John F. Kennedy, "counter-culture" and social revolution, the 1960's were a tumultuous time both off – and on – the track. As the decade turned and the early 1970's came into being, volunteer firefighters, and their respective drill teams, once again found they were starting to have problems with their public image. Conservative by nature, the fire service was traditionally slow to change and adapt to new mores that society was adopting as it grew and changed around it. As a result, once again, firematic tournaments were not immune to becoming stages upon which to act out issues of the day. On June 18th, 1966, at the Southern New York tournament in Merrick:

"A demonstration against alleged discrimination in volunteer fire departments disrupted a tournament of 70 suburban fire companies here today.

"Lincoln Lynch, associate national director of the Congress of Racial Equality, and Harold Trent, a Long Island CORE member who took part in the James Meredith March in Mississippi, staged the demonstration.

"At one point, the two leapt a trench in front of a grandstand and ran into the path of a fire truck, which swerved and barely missed them. The police escorted the demonstrators back to the grandstand but brought no charges.

"'We resent this last bastion of the bigot north of the Mason-Dixon line,' Mr. Lynch said later. He said the fire department's membership policy was 'openly segregationist.'

"He and Mr. Trent staged their protest at the tournament of the annual convention of the Southern New York Firemen's Association. About 1,000 firemen from Nassau, Suffolk, Westchester and Rockland Counties took part.

"More than 40 members of Long Island civil right groups, including women and children, set up picket lines outside."[cdvi]

The West Sayville "Flying Dutchmen" compete in the C-Ladder Contest at the State Drill in Watertown in 1970

This was not the only accusation. In 1969, when New York City Fire Department Commissioner Robert Lowery issued a Department Order forbidding F.D.N.Y. firefighters from serving simultaneously on a volunteer fire department, some questioned whether Lowery – who was the first black man to ever serve as a Fire Commissioner in a major city – had passed the order as a way to punish the volunteer fire departments for not being more open to racial integration. A Manhassett-Lakeville firefighter, Irving Cohen, was asked basically this exact same question in a newspaper interview in August of that year:

"[Cohen] was asked if he considers Lowery's action retaliatory, based on charges that Long Island departments discriminate against and exclude members of minorities.

"'I can't judge the Commissioner's motive,' he said, 'but if he had anything like that in mind, it would be grossly unfair. We don't turn down any man who offers to risk his life. We've had just one Negro applicant, and he was accepted.'

"He said the department would welcome applicants from Spinney Hill, the predominantly Negro section of Manhassett, and suggested that leaders in that community urge its members to participate."[cdvii]

The West Sayville "Junior Flying Dutchmen" at the Islip Town Drill at the Oakdale/Sayville Shopping Plaza in July of 1961

Whether or not the fire service was "segregationist" in the 1960's is debatable, but it cannot be argued that they didn't work to uphold the traditions of firematic competition by promoting the sport to younger generations. On July 12th, 1963 the very first all-Juniors tournament to take place in Suffolk County was held in Bayport, and six years later, on August 30th, 1969, the first ever Juniors State Drill was held on a track next to the high school in Bay Shore, with the youngsters from West Sayville forever holding the honor of being the first Junior State Champions. Junior racing was no novelty; it was – and still is – a vital, respected part of the firematic team community, which had as its core – like the senior firefighters – the overall health, safety and well-being of their communities ultimately as a goal, for which competition team skills provided a platform from which to build children of good character. "These young men are the future volunteers who will protect our community from the disasters of fire," wrote a Long Island paper regarding one team, the St. James "Junior Wildcats". "'It is noteworthy,' a department member said, 'that the men of the St. James Fire Dept. realize their responsibility to the community not only in fire protection and emergency health care but also being influential in the growth and development of today's youth – through such programs as the Junior Wildcats.'"[cdviii]

Aside from simply promoting the Juniors program, what is interesting about the above quote is the spin attached to it. This same type of self-promotion can also be seen in other articles published at the time, so it seems every effort was being made to let the public know that the firemen, who felt that they were being judged too quickly and unfairly, were actually conscientious, caring individuals who put the lives of their

1st Suffolk County
All Junior Firemen's
Parade and Drill

SPONSORED BY

BAYPORT
FIRE DEPARTMENT

"Home of the Red Devils"

FRIDAY JULY 12th, 1963
BAYPORT FIRE DEPARTMENT
BAYPORT, LONG ISLAND

Price 10 Cents

neighbors above their own – voluntarily, no less. The same newspaper article that raised the question of racism in the volunteer fire service had actually first raised the question if the volunteers were even fit to protect their communities at all, now that paid F.D.N.Y. firefighters could no longer participate. Firefighter Cohen was presented with this question as well, and the article's author relates his reply:

"'Our volunteers come from all walks of life,' he said. "We have lawyers, politicians, businessmen, steamfitters, accountants – you name it. Most of us work in Manhattan, but this is our home community, and we feel an obligation to protect it."[cdix]

Thankfully, the newspapers of the day didn't always side with those criticizing the fire service and its drill teams, especially when the subjects of the controversies bordered on the silly. In an Editorial published in September of 1973, *Newsday* wrote:

"While most Long Islanders were spending Labor Day in shady spots with cool drinks, members of fire department drill teams were competing in tournaments. Apparently the heat had little effect on the amount of effort expended, because teams from the West Sayville and Carle Place Fire Departments set new state records in individual events.

"The healthy spirit of competition and sportsmanship that surrounds these tournaments speaks well of the men and the organizations that participate. We wish this same spirit of give and take extended to all the firemen's activities. Unfortunately, we note that a silly quarrel over hair length is still creating turmoil in the Massapequa Volunteer Fire Department. Two members of the department spoke critically of a decision to suspend six of their fellow firemen for having hair longer than the department's regulations permit. Recently the two were called before an advisory board that heard charges against them.

"We realize that fire companies have to maintain certain levels of discipline, but teamwork is also vital to a fire company. And for a team to work well, fellowship and discipline are equally important. We hope the leaders of the Massapequa Fire Department will now put as much effort into restoring fellowship as they have exerted in maintaining discipline."[cdx]

And, two years later, after Long Island teams took home all the prizes at the State Drill in Niagara Falls, there was even praise:

"Effective firefighting demands a combination of planning, skill, teamwork and dedication. And that's why the apparent hijinks of volunteer drill teams are really a vital element in fire department training. Competition on the firemen's field helps build precisely these qualities, along with the more evident muscle, sweat and thirst.

"All Long Islanders should be pleased to know that in the statewide fire tournament at Niagara Falls last weekend, drill teams from Long Island swept all the titles. West Sayville won its first state championship, followed by Central Islip and the Hempstead Flukes. Drills and championships should never be the end-all of a local fire district. But they're an excellent beginning, and we congratulate Long Island's new winners."

If the 1960's were to be considered an era of social unrest and change, the late 1960's and early 1970's, as far as the sport of

firematic competition was concerned, were an era that could be considered in some ways an extension of that era of change (although it was much more positive in nature) and the sport in fact moved quite a ways forward in terms of degree of competition as a result.

The biggest changes that took place had to do with experimentation with the design and setup of the trucks. Following again the lead of the NHRA, the late 1960's saw many varied and unusual approaches to both the construction of the rigs – especially the Class C trucks, which were seen sporting dual front or rear wheel setups, or strange and unusual body designs made from stock bodies that were either chopped in unique ways, or even facing the wrong way entirely – and tinkering with the engines and power trains that included different types of fuel delivery all the way up to and including superchargers and blowers. This was by no means a new idea, since the concept of "hot rodding" had been around since the 1950's, but things were now reaching a level heretofore unseen in the sport. Horsepower was a god, and hole-shots – being faster than the next team off the starting line – and overall speed down the track were king, and teams began spending more and more money on the equipment and technology to make them happen.

Some of the teams, at least. The advent of the use of faster and faster trucks by those teams that could afford them began to cause a rift in the sport, between those teams and those who were not as able to pony up to build a state-of-the-automotive-art vehicle. Times on the track were becoming noticeably faster, but it was only to the advantage of certain teams...and the other teams began to cry "foul."

By several accounts, this situation came to a head in 1966, at the State Drill that was held in Oyster Bay. The "Alerts" of Great Neck had for a couple of years been competing with a C-truck whose supercharged engine was aspirated with a blower; at the time it was one of only two trucks in the entire sport to be fitted this way. During the Class C Ladder contest, the Alerts took to the track and completed a run that many who were watching considered to be mediocre at best, as their ladder climber had some trouble making it up the rungs to the top smoothly. It came as complete surprise to all, then, when the judges announced that a new State Record had been set – and by an incredible margin of 12 hundredths of a second, no less!

It became apparent that since Great Neck's climber hadn't been especially quick up the ladder, the gain in time had to have been made up in the truck's speed down the course. Not that there was anything wrong with that *per se*, but it was a glaring indication of what kind of advantage could be obtained if any given team had the means...and most, apparently, did not.

Fortunately, as had happened in the past, the general agreement was that if the sport was to continue to move forward and remain competitive, some things would have to change. They soon did, as during the winter of 1969-1970 the DTCA enacted rule changes that, among other things, did away with allowing blowers and other exotic kinds of aspiration, and required all C-trucks to be normally aspirated with a carburetor of American manufacture by the 1972 competition season.

The effect of these rule changes on the sport was profound and dramatic. Since all the teams were now somewhat regulated as to the setup of their motors, not only did it make things more affordable for the "have-nots", it leveled the playing field for all the teams as a result. Since achieving raw horsepower to win events was taken off the table, teams were now forced to work with changing only driving techniques, handling characteristics and brake and chassis setups in order to win. Consequently, the 1970's are remembered fondly by many as a decade of overall competitive parity, where there were a number of very good teams – and any one of whom could win on any particular day. If one team had to be chosen as being better than the rest, it would have to have been the "Hoboes" from Central Islip, who during the course of the ten year span took home three

Some examples of the radical designs created for Class C trucks during the late 1960's through the late 1980's. From top left: the Lindenhurst "Snails" "Reverse 'Vette", ca. 1968; the Snails' "Wedge", ca. 1973; the Dexter "Ramblers", ca. late 1970's; the Bohemia "Suicide Squad", ca. early 1980's; the Snails ca. early 1980's; the Inwood "Mud Ducks" ca. late 1980's

State Championships, for which Jim Lampley of ABC's Wide World of Sports referred to them as "the Yankees of the sport."

Lampley wasn't entirely naïve about the sport, either, as by the end of the 1980's Wide World of Sports had featured coverage of the State Drills no less than six times, in 1965, 1968, 1977, 1979, 1980 and 1984. In addition, the sport was honored again eleven years later as clips from those earlier drills were included in an ESPN special entitled "30 Years of Wide World of Sports" that was broadcast as part of Super Bowl Sunday on January 27th, 1991. The sport was also one of the first local Long Island sports featured during the mid-70's on Channel 6 of a brand-new little company called "Cablevision".

Television wasn't the only extra exposure the sport was getting in the 1970's, however. On September 22nd, 1973, a first in the history of the sport took place when a motorized drill took place outside of New York State, in Warminster, Pennsylvania, as part of the Hartsville, PA Fire Department's 50th Anniversary celebration. In an article that bore the headline, "Ah, Those Daring…NY Firemen" the *Bucks Courier Times* wrote:

"Gray skies yesterday afternoon didn't prevent the Hartsville Fire Co. from culminating its 50th anniversary festivities by staging a firemen's 'Olympic' competition.

"More than 400 firemen, representing 19 companies from Long Island, participated in the event at the Johnsville Air Development Center.

"This is the first time such a competition has been held in Pennsylvania. On one of the runways at the center approximately 5,000 spectators saw an eight-event competition which featured displays of skill, speed and daring.

"Ed Pfeiffer, Hartsville president, said the company decided to host such a competition, which took three years of planning, because 'we wanted to celebrate our anniversary like no one ever has before and because we also wanted to stimulate local interest in firefighting.'"

The experience was apparently unique not only for the Pennsylvania firemen, but for the teams from New York as well: the "track" they were using was actually an airport runway that had been shut down for the occasion, and although it made for a suitable racing surface, it was situated parallel to another runway that had not been shut down and was still active. Although the firemen from Long Island were used to riding the trucks down the track at speed, none had ever experienced doing so as A-4 fighter jets took off and landed right next to them!

It would not be the only time drill teams from New York State visited Pennsylvania during that decade: friendly relations between the firemen of both states gave rise to additional tournaments that were held in Allentown in 1975 and Levittown in 1976. Although these drills were not Sanctioned by the NYSDTCA in the official sense of the word, many of the teams enjoyed attending them because they were novel, and as they were both held in September they ended up being a second, unofficial State Drill – but without all the pressure. Unfortunately no more tournaments have been held in Pennsylvania since, but happily teams from New York took out-of-state excursions four more times since then to exhibit and promote the sport. The first trek was to Miami in 1980, when the "Forty Thieves" of Patchogue went to visit the Florida State Firefighter's Association Annual Convention and give an exhibition of their skills at their "Firematics Day", a visit they did again in 1985, this time with the company of the East Farmingdale "Hares" and the Wyandanch "Indians". Two years later the East Islip "Guzzlers" and the Bay Shore "Redskins" traveled north to Springfield, Vermont, to put on an exhibition of their skills as part of the town's anniversary, and in 1997 Las Vegas was the destination when six Long Island teams – the Bay Shore "Redskins", the Roslyn "Highlanders", the Bayport "Red Devils", the Selden "Slo-Pokes", the Oyster Bay "Teddy's Boys" and the Patchogue "Forty Thieves" - participated in two days of exhibition contests at the World Firefighter's Games[9] that were being

[9] This could be a whole other book. More information regarding these games can be found on the World Wide Web at http://www.worldfirefightersgames.com/

The Roslyn "Highlanders" put on an exhibition of the C-Hose Contest in Las Vegas in 1997

held at Cashman Field there that year, to the delight (and no doubt amazement) of firefighters and their families from the western part of the country and other parts of the world. Although most enjoyed making the trip, most who went also agree it was difficult, if not downright grueling. Aside from the grind of having to haul all the trucks across the country and back, other things were also not easy: after cutting one of a set of two split-rim tires while unloading their B-truck from its trailer, the Selden "Slow Pokes" found that split-rims were illegal in Nevada, and as a result no service station would repair them. Because of this, they were forced to compete without any inside tires on the rear.

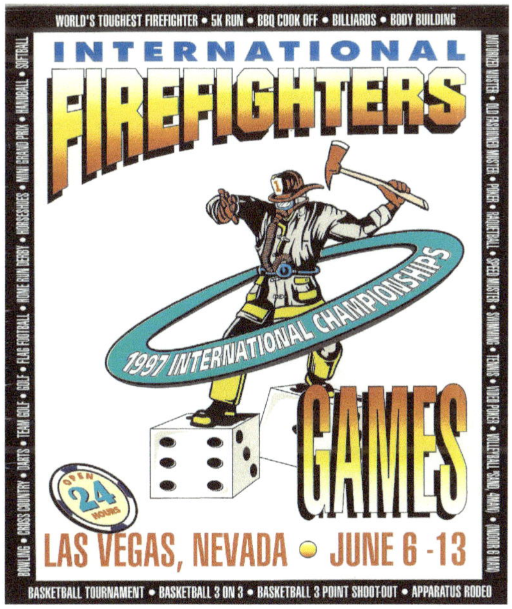

The 1970's also saw some additions and innovations to the sport of firematic competition that were both interesting and fun. At the Western Convention of 1971 in Bowmansville, firefighters held, for the first time, a Fire Police Competition. In this event, individual members of Fire Police companies were placed at the center of a busy local intersection and were given five minutes to control that intersection in as smooth a manner as possible, as determined by judges standing by. Contestants were each awarded points as he or she directed traffic, gave right of way to emergency vehicles, helped pedestrians cross the street, etc. and the top three participants with the highest number of total points were awarded prizes. Although this event was not part of –

The Sea Breeze "Indians" at Spencerport, 1979

and indeed had no relation to – any of the contests sanctioned by the State DTCA, it became a mainstay of the Western convention, and is still a part of them to this day. Another small, but significant change that also took place that decade had to do with the state association's official title: in 1975 the word "Parade" was added to it. After some growing dissatisfaction over the fact that parades at the time were still rather disorganized and lacked an established set of Parade Rules – especially in comparison to how far drills had come by that time - a meeting in Oyster Bay was called by the State DTCA. At that meeting, it was voted that all aspects of parades, including hosting, judging, enforcing of rules, etc., would be overseen by the state association from that point forward, thus formalizing and modernizing them up to a point more on par with their accompanying drills. In doing so, the state association's official now became the New York State Volunteer Firefighter's Parade and Drill Team Captain's Association.

On a more frivolous note, the mid-70's also gave rise to some more informal drill team tournaments that were held - of all places - on the beaches of Long Island. In 1974 and 1975, Davis Park, a small community located on Fire Island (a thin spit of land that runs parallel to the south shore of Long Island) first hosted these "old fashioned" non-motorized beach drills, which by at least one firemen's account were "a bunch of drunken fun." The Davis Park location proved to be problematic, however; in addition to all of the team members and their families having to arrange and pay for ferry service to get them there[10], an elevated boardwalk that ran parallel to the "track" often became so crowded during the tournaments that the safety of the public became a concern. Consequently the beach drills, which at the time were annual events, were moved to Sunland On The Beach at Blue Point, back on the "mainland."

An Efficiency Contest at Main-Transit, 1978

Owing that it was a more or less casual thing to begin with, the rules at these events were changed to reflect the conditions. Instead of racing hose, regular 1-1/2" double-jacket service hose was provided by each team, along with any old nozzles of any sort they could find lying around the firehouse. Instead of using modern-style ladder and hose carts equipped with large bicycle-style wheels, the host team provided three carts that were custom-built with large, balloon-type tires made for use on the sand. As each team finished their run they would simply bring the cart back to the starting line for the next team to use. Ladder An contests were easier in some respects but more difficult in others, as there was no set "patch" to spike the end of the ladder into as there was on a paved track surface – anywhere

[10] A significant portion of Fire Island has no roads, and many of the communities are only accessible by ferry.

The West Islip "Raiders" compete at a beach tournament at Sunland on the Beach at Blue Point in 1976

in the sand will do – but team members had to keep in mind that the ladder would usually end up sinking six inches or so once it was placed, so that had an effect on just where to place in comparison again to a paved surface. When it came to hose and bucket contests, things would get interesting: as the afternoon wore on and the tide moved in, the Dummy Hydrant (the hydrant used for competition, that was usually supplied by a hose coming from a real, permanent hydrant) would start to become covered with water, and the last contest of the day, the Buckets Contest, was always an exercise in frivolity as no water trough (as used at paved tracks) was needed: a team's bucket carriers simply went splashing into the sea to fill their buckets, to the merriment of all.

Back on land, as the 70's moved into the 80's, the sport of firematic competition still remained healthy – at least as gauged by the number of fans in the stands at the annual State Drill – but there were some items of note that took place during that period.

Most significantly, and unfortunately, the start of the decade saw the demise of the Central Firematic Drill Team Association. The CFDTA saw a heyday of membership and activity during the late 1960's and 1970's, when local tournaments were attended by a dozen teams or more, but by 1982 only two teams – the Elsmere "Wanderers" and the "Nighthawks" from Boght – were the only ones left that were still actively involved, and it seemed pointless to continue an association where there was no one else to compete against. Interestingly, though, the Central Association did not suffer its fate as a result of the same money (or lack thereof) issues that caused many Long Island teams to fold. Instead, and although there is no one real reason, indications are that it was simply an issue of not enough time and interest on the part of the firefighters: between now having to work second jobs to make ends meet, and ever-increasing demands from OSHA and the NFPA for additional and more thorough training and certifications in firefighting in order to simply remain an active member of a fire department, team members probably felt that the ends less and less justified the means. The officers of the Central Association did what they could to keep the association alive, including the re-drafting of a large number of rules (such as how points were allocated) with the hopes that it would re-stimulate interest, but in the end it wasn't enough, and everyone decided to just give it up.

When the CFDTA dissolved, Boght and Elsmere were left pretty much without anywhere to compete, so at the time they both opted to join the State DTCA. Boght continued on for a few more years, but then finally gave it up. Elsmere's team ended up joining the state association as well:

"The State Tournaments always followed FASNY, so you always heard about it [LI Tournaments] through FASNY at the time," says Rick Webster, Team Captain of the Elsmere "Wanderers". "A lot of the guys, myself included, used to go to the State Championships – to watch them; never to participate. We always thought that was the neatest thing, to go watch the Island trucks run. So, when the local association pretty much died out we said, 'you know, we can do that' so we converted our old truck into a B-Truck and went racing."

In doing so, however, they ended up living up to their "Wanderers" team moniker. When they joined the NYSVFP&DTCA in 1982, they were "assigned" to compete in the Nassau division, due to the fact that they were situated on the west side of the Hudson River[11]. After a couple of years, however, since they had only competed in Suffolk County events, they petitioned the state body to join the Suffolk County DTCA. The State DTCA's response was, essentially, "we don't care where you compete," so Elsmere then joined Suffolk, whom they were a part of until 1994 when they, too, stopped competing due to a lack of interest.

There was another reason, however, that Elsmere chose to compete on Long Island: they didn't have much of a choice. Outside of the Albany area, the closest DTCA Area to them was the Northern association. But Northern - as a clever reader may have noticed – had never included any Class-B contests at their tournaments in the entire history of their Association; instead running Class-C contests only. Some have attributed this to be for economic reasons, as in the poorer, more rural areas of the state, "you can feed one

[11] The reasoning behind this still remains unclear. After several attempts to define when or how specific boundaries for the four Areas of the NYSVFDTCA were determined, the author could find no records, or agreement between knowledgable parties, regarding this.

Car & Driver *magazine, May, 1980*

horse more than you can two." But as Elsmere only owned a Class-B truck to compete with, so rather than make the long distance trek to western New York, it was off to Long Island they went.

The situation changed for the better though on August 16th, 1994, when the Northern association finally included its first B-truck contest at a tournament. Never mind that it wasn't actually *in* Northern's area – it was the Deerfield "Hose Busters", who competed in the Class-B Ladder and Class-B Hose contests at the State Drill in Main-Transit – but nevertheless it remains the very first time a B-truck from that area had actually competed in a drill. With no one else in their area equipped with a B-truck to run against at the time, Deerfield continued to compete only in Class B contests at the State Drill until 1999, when teams from West Carthage and Black River joined the fray after building B-trucks of their own. As a result, the first ever B-trucks to run at a northern drill competed at the Northern Invitational Drill on June 19th, 1999.

These contests were at first run only as exhibition events, but starting at the Northern New York drill on June 7th, 2003 they were officially added as points-producing contests. Since Elsmere had already been B-equipped for some time, and based on the addition of Class B contests up north, they decided to re-form their team and join the Northern New York association, rather than drive all the way down to Long Island for each drill. In doing so, the Elsmere "Wanderers" were finally able to quit "wandering."

There is an interesting and humorous side note to this "Northern B-trucks" story, as told by Kevin Hayes, one of the historians of the sport:

"When they started running the B Trucks, some teams could not afford to install a real operational pump. To meet the standard of the rule that all B trucks must have a pump installed in the truck, some teams

Newsday's Magazine for Long Island • May 31, 1981
LI
THRILLS AND SPILLS
What Goes On at a Firemen's Tourney

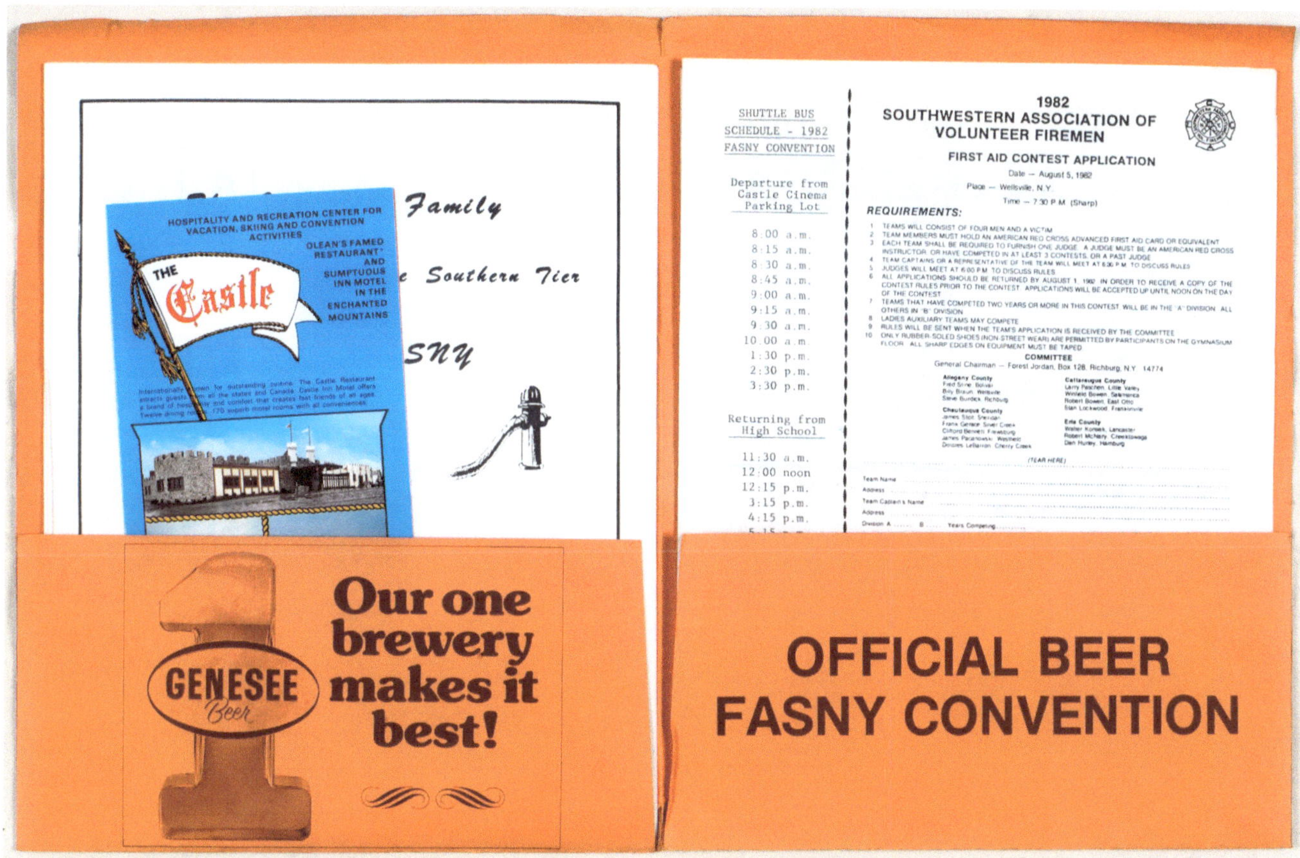

The conventioneer's packet from the FASNY Convention in Olean in 1982

installed a washing machine pump in the pump compartment. The rulebook does not say that the pump has to be operational - it only has to be there! That is one reason that the pump contest was not run with these B Trucks yet. Other teams had a "real" pump, but they were not hooked up to the drive train. The rulebook only stipulates that your truck must have a pump. The rule book said nothing about requiring PTO's or transfer cases."[cdxi]

In spite of the bad news regarding the dissolution of the Central New York association, the 1980's can also be remembered in a positive way as being a decade of increasingly close and intense competition for the sport. While advances continued to be made in the technology of motor sports in general – e.g. the NHRA, NASCAR, and so forth – the high cost of that technology for the most part made any kind of implementation or application of that technology all but impossible, so teams were continued to be forced to make do with whatever tricks or techniques they could come up with on their own base on the budget restraints of the team or the amount of time the individual members could put into it. That is not to say that money wasn't being spent on the teams and their equipment – the thorny issue of "haves" vs. "have-nots" certainly hadn't gone away – but nonetheless, as a result, the level of both quality

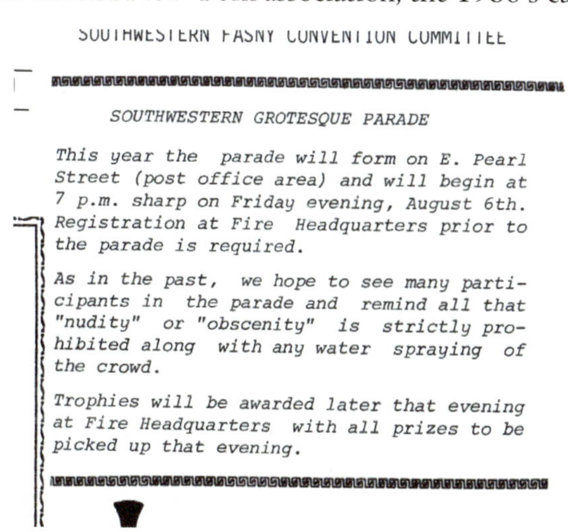

SOUTHWESTERN FASNY CONVENTION COMMITTEE

SOUTHWESTERN GROTESQUE PARADE

This year the parade will form on E. Pearl Street (post office area) and will begin at 7 p.m. sharp on Friday evening, August 6th. Registration at Fire Headquarters prior to the parade is required.

As in the past, we hope to see many participants in the parade and remind all that "nudity" or "obscenity" is strictly prohibited along with any water spraying of the crowd.

Trophies will be awarded later that evening at Fire Headquarters with all prizes to be picked up that evening.

The Islip "Wolves" compete in the B-Hose Contest at Patchogue's 7th Annual Invitational Drill on June 26th, 1982

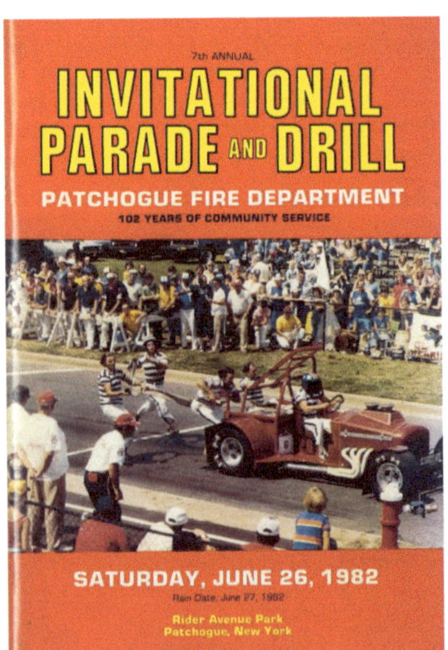

and equality between the teams – and in turn the level of competition - grew to a point in the 80's never before seen in the history of the sport.

One illustration of this was seen at the Islip Town tournament in Central Islip on August 8th, 1986. As *Newsday* described it:

"The August 8 competition ended in a three-way tie for the title, the first time that occurred in modern racing history. The score going into buckets saw Central Islip one point ahead of the Flying Dutchmen and two ahead of the East Islip Guzzlers. The Dutchmen were the last team to run the contest. East Islip had already posted a second place time of 24.62 seconds and Central Islip was third with a time of 27.44. East Islip's time seemed out of reach, but, incredibly, West Sayville ran a 27.36, narrowly edging out the Hoboes by a mere eight one-hundredths of a second. As a result, the Flying Dutchmen, Hoboes, and Guzzlers shared first place honors. All three teams drove their B trucks down the course side-by-side to pick up their awards, as the stands, still packed with fans, erupted in thunderous applause."[cdxii]

The Great Neck "Alerts" compete in the C-Hose Contest with their legendary Chevy Vega at Hempstead some time in the early 1980's

This increasing level of competition had one unfortunate side effect, however: the 1980's also saw some unusually serious accidents, both at the drills and during practice for them. One such incident took place at the 1986 Labor Day drill at Islip Terrace, when the Hagerman "Gamblers" C-truck lost control and went off the track, causing several injuries:

"Several people were injured when a fire truck went out of control and struck them during a tournament in front of the Islip Terrace Fire [Dept.] yesterday afternoon. Suffolk County police reported that the injuries occurred at 2:30 p.m. when a truck belonging to the Hagerman Fire Co. left an asphalt racing track and struck several people who where watching the tournament in front of the fire house on Beaver Dam Road. Approximately 10 people were injured, said Sgt. Bill Durham.

"John Falci, chief of the Islip Terrace Fire [Dept.],

The accident at the 1986 Labor Day Drill at Islip Terrace

said some of the injured were aboard the fire truck when the accident happened. 'The truck went astray; we don't know for what reason,' Falci said.

"A witness to the accident, Theresa Imbo, 23, of Islip Terrace, said the accident occurred at about 2:30 p.m. during the performance of a timed C-hose drill, a test of speed and proficiency. The truck went out of control after it dropped off two men at a hydrant, and before it was supposed to drop off two additional men who were to connect a nozzle to the hose, Imbo said.

"The tournament, in which fire companies from 18 local fire districts were participating, was immediately called off, said Falci. Hagerman Fire Chief Tom Brenza called the incident 'an unfortunate accident.'

"Falci said six or seven of the injured were taken to Southside Hospital in Bay Shore. A nursing supervisor at the hospital, Valerie York, said one person required surgery for repair of a fractured leg. None of the others was seriously injured; they were treated and released. The names of the injured were not disclosed"[cdxiii]

Fortunately no one lost their life as a result of this mishap, but tragically such was not the case with other incidents that took place during this time period. In 1979, Roslyn drill team member Ed Dykowski was killed after falling off a transport trailer, and in the decade that followed other firefighters died participating in the sport. During this particularly bleak period, four firefighters lost their lives within the space of five years, a statistic completely unheard of within the sport to that day or even since.

It began in June of 1986, when Oceanside firefighter Steven Wade fell while practicing for a ladder contest, falling 18 feet and hitting his head on the pavement. Then, during the first week of the 1988 season, Copenhagen "Cubs" team member Donald Henry was riding on the back of a C-truck during a "Duster

Run"[12] when the driver – his father – lost control of the vehicle and ran into a ditch, throwing Henry onto the corner of a house. Barely a month later, on July 6th, 1988, 18-yr-old Patchogue firefighter Robert Cushing was killed during a drill team practice "…after he fell beneath the wheels of a truck as his two older brothers, members of the same five-man team, watched in horror."[cdxiv] And then, only three years later on May 31st, 1991, Hempstead "Flukes" member Scott Laverty was gravely injured after falling off his team's truck and hitting his head on the curb while practicing for the Motor Hose Replacement contest…a contest that they normally did not run. After suffering a brain injury, he clung to life in the hospital until June 6th, when he finally succumbed.

While each of these deaths was a tragedy on its own, Cushing's death was as bitterly ironic as it was terrible: in a decade that was marked by how closely teams throughout the sport were matched, it was Patchogue's "Forty Thieves" that were the star exception to that rule, by winning the State Championship an incredible five times over the course of a ten-year period.

But without diminishing the gravity of Cushing's death, or the deaths of any of the other firefighters during that period, it was the crash at Islip Terrace two years earlier that actually had a larger impact on the future of the sport.

---

12 A "Duster Run" is where a truck makes a pass down the track in the opposite direction from which it would normally travel (i.e., beginning at the arch and driving away from it) so as to clear off any "dust" that may be on the track prior to beginning its timed pass in the other direction. This practice is most often seen in the northern and western parts of the state, where village streets are used to compete on in lieu of a purpose-built track.

Chapter 17: Breaking Up is (Not So) Hard To Do

The reader will recall the ongoing tensions that were taking place between FASNY and the NYSDTCA, as far as the State Drill was concerned. Ever since the formation of the State DTCA, and indeed even before then, representatives of the drill teams would travel to wherever FASNY was planning to hold their annual convention and work with them to ensure that there was an adequate – and safe – location at which to hold the State Drill. The reader will also recall that this wasn't always easy – in the minds of the team members some of FASNY's choices were nothing less than "horrible" – and this fact was one reason that the NYSDTCA was actually brought into being.

So it comes as no surprise that by 1989 many of the teams – especially those from Long Island - were feeling more concerned and cautious than ever about safety issues on the track. Those concerns came to a head at the State Drill that was held in conjunction with FASNY's 117th Annual Convention in Watertown on August 26th of that year.

Prior to the tournament, officials from the State DTCA had inspected the track at Watertown and declared it fit for competition, but between that time and the day of the tournament things had changed: the day before the drill a minor league baseball game had been held at a ball field that was adjacent to the track, and spectators at that game had used the track as an exit roadway after watching the game. Normally this would not have been an issue, but it had rained that day, and by the following day, the day of the tournament, the track surface was covered with dust and caked mud left behind from vehicles exiting the day before. A Local Committee did their best to sweep off the track surface prior to the start of the drill, but it was apparently not enough to satisfy many of the teams from Long Island, who in addition to that were upset over the use of snow fences at the sides of the track instead of guardrails – no doubt with the Islip Terrace incident fresh in their minds. After completing only three contests, they decided it was too unsafe and refused to participate any further.

PETER R. BARBER ■ WATERTOWN DAILY TIMES

New York State firematicsofficials gather at the center of the track at the Alex T. Duffy Fairgrounds during a four-hour delay in the competition Saturday afternoon. Teams from Long Island pulled out the the state competition after complaining that the track was unsafe.

Much to the annoyance of the teams from northern and western New York, who felt like they had just received a slap in the face: this course was one that they routinely practiced and competed on, and one which had been given a "seal of approval" by the State DTCA, so why should they be held to an apparent "higher standard" dictated by the Long Island firemen, who in their eyes probably felt they were faster and better than teams from anywhere else in the state? Sensing that the dirt on the track made for some good dirt in a story, the local press played up the conflict:

"The cancellation of the 117th annual firefighters' drill at Watertown this weekend created such a rift between teams from Long Island and those upstate that tournament officials predicted yesterday a statewide competition may never be held again.

"'This will break the state right in two,' said Clark Dano of Clayton, co-chairman of the tournament. The cancellation was the first in 40 years.

"Organizers blamed teams from Long Island – who account for 90 percent of all participants – for causing the competition to be scrapped after the early rounds. Some Long Island teams had refused to continue Saturday after complaining that the track had become too slick from fresh asphalt, dust and water, and there were no guardrails to protect about 5,000 spectators.

"'I couldn't sleep at night knowing I could hurt someone if my truck slammed into the grandstand,' said a Nassau County firefighter who, like most others interviewed yesterday, declined to give his name for fear of retribution by tournament officials.

Hagstrom
Suffolk County Atlas
Easy-read Large Scale
SUFFOLK COUNTY, NEW YORK:
• FULLY STREET-INDEXED
• LONG ISLAND EXPRESSWAY
• NORTHERN AND SOUTHERN STATE PARKWAYS
• SCHOOLS • HOSPITALS • PARKS
• STATE PARKS FACILITIES
• GOLF COURSES AND COUNTRY CLUBS
• CEMETERIES • AIRPORTS • ZIP CODES
• GOVERNMENT GUIDE • POPULATION STATISTICS
Committed to excellence since 1916
$11.95

The West Sayville "Flying Dutchmen" compete at the 1990 Labor Day Invitational Drill at Hagerman

"'We are not trying to be troublemakers,' said another. 'It just doesn't make any sense why they wouldn't put up any guardrails.'

"After the third event, West Sayville firefighters refused to participate, complaining of hazardous conditions. This created a domino effect, according to tournament officials, and many of the 48 teams also dropped out. Fifteen teams finished, including teams from Patchogue, St. James, Selden and Farmingville. But the tournament was declared unofficial Saturday because so many teams had refused to participate.

"Tournament officials said the complaints were unwarranted and should have been raised during several meetings held on Friday, a day before the competition began. 'We go through this with Long Island firemen all the time. They just don't want to come up here. They want to have the tournament on Long Island every other year.'"[cdxv]

As the article stated, it would be only the second time in the history of the sport (the cancellation of the State Drill in 1949 being the first) that a State Champion was not crowned. In official parlance (and in what will ultimately appear as historical record), "...the sanction was lifted due to safety concerns on the course."

In the end, however, neither "side" harbored any hard feelings, and by the following year's State Drill teams from all parts of the state had "made up" and competed with one another happily.

But nonetheless a precedent *had* been set, and from that point forward safety – for both the competitors and the spectators – would always remain a very important issue. So much so that after the Watertown drill the NYSDTCA decided to separate itself completely from ASNY as far as the State Tournaments were concerned. Although the State DTCA had been running the State Drills autonomously from FASNY ever since 1954, it was still FASNY that had been choosing the location of the State Tournament, based upon the location of their convention and their ability to find a suitable (and willing) host for the drill, at which point the NYSDTCA would inspect FASNY's choice and either approve or disapprove of it. The State Drill in Watertown in 1989 was therefore the last drill that was held *in conjunction with* FASNY's convention. The 1990 State Drill, held at Main-Transit, was planned, organized and ran, soup-to-nuts, by the New York State Drill Team Captain's Association.

As were the State Drills every year from that point on, although one might be led to think that that wasn't the case after comparing the locations of FASNY's conventions to where the DTCA State Drills were held, as throughout the early part of the 1990's the location of the drill was, with only one exception, always in the same Area as the convention. This was not coincidence, but nor was there any specific reason for it. According to one senior member of the DTCA, the tournament locations followed the convention locations simply as a matter of tradition.

This changed in 1993, when, after some discussion, the Executive Board voted to begin rotating the location of the State Drill between the four Areas of the state. Again, there was no specific reason for this other than equitability; by doing so each of the four Areas of the NYSDTCA would have a chance to host the drill every four years, which gave each Area a chance to adequately prepare for the event, as well as offer teams from throughout the state the chance to travel to the other three Areas on an equal basis, thus eliminating any potential argument that the Long Island teams would be "hogging" the majority of the State Drills. This rotation system has been in use, without trouble or controversy, ever since.

Were it that easy for the drills themselves! Another unfortunate by-product of the ever-increasing level of competition in the 80's and 90's was an associated increase in the level of rules governing the sport…and the number of protests on the track regarding them.

Although it is definitely considered a taboo subject, cheating amongst the teams is something that has taken place since the first two handtubs squared off against one another in Town Square on a sunny day at the beginning of the nineteenth century, and indeed the reader will recall several serious conflicts recounted throughout this book that were a direct result of it – or at least the accusations of same. It should come as no surprise, then, that as time went on and competition grew tighter and the difference in winning times between first, second and third place was starting to be measured in hundredths of a second, some teams would try to slip in some nefarious practices in order to gain an edge. To their credit, some of these tricks were extremely ingenious and clever – for example soaking a hose in ice water so it contracts and thus becomes shorter prior to a run, after which it expands back to normal length after warm water runs through it so it is the correct length afterwards - but as a result the NYSDTCA, and in turn the officials overseeing the running of the drills, were required to become proportionally more circumspect by adding and changing rules and becoming just that much more vigilant on the track.

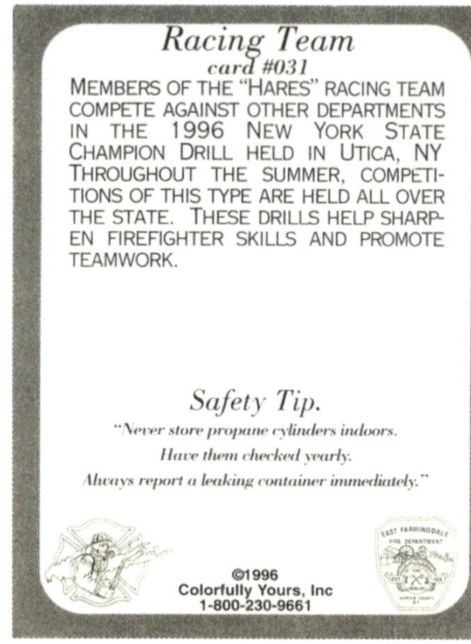

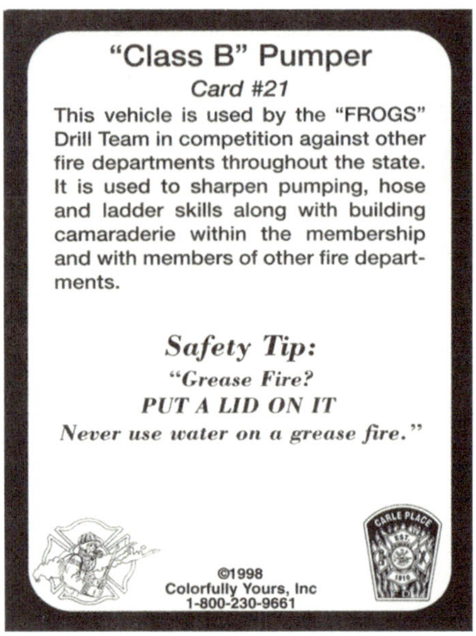

Firematic Drill Team baseball cards from the 1990's

An example of this took place at the South Shore Drill at Lindenhurst in 1989. When the Patchogue "Forty Thieves" ran a State Record time in C-Hose, officials opted to test the inside diameter of the hose that they used during the run to see if it was within specifications, with the premise being that if the inside diameter of the hose was smaller than the rules allowed water would pass through it more quickly and thus reach the target sooner. Using a unique tool developed only a couple of years earlier that will slide through a regulation-sized hose but not one that is narrower, officials ruled they could not slide the tool through Patchogue's hose, and the record run was disallowed.

Patchogue protested the officials' decision, and unfortunately the disagreement escalated to the point where it turned into a litigation that went all the way to the Suffolk County Supreme Court. Patchogue was eventually vindicated, but the incident served as an indicator of just how serious and captious things had become on the track, and in the firematic competition community as a whole.

The Islip "Wolves" celebrate after winning the 2003 New York State Championship Drill at Ridge

Although many argued that it was necessary for the sport to create more and more rules to keep the competition fair, just as many others argued that doing so was sentencing the sport to a slow, painful death. As track officials struggled with keeping timing equipment calibrated and accurate, measuring the time of the water from the hydrant to a nozzle using a Reference Standard hose, testing hose inside diameters and fielding seemingly endless protests from teams who felt they were slighted as a result of same, keeping the length of a tournament on any given day to a reasonable amount of time was starting to prove to be more and more difficult. This in turn, said many, was causing spectators to lose interest in attending the drills. Indeed, by the turn of the 21st century, many of the events – including the State Drill – were only shells of years before, at least as far as the number of spectators was concerned: drills that used to routinely see attendance of many hundreds of people were by that time struggling to fill the stands, much to the dismay of the teams…who in turn were becoming increasingly exasperated by the long days, innumerable new rules and seeming futility of it all, so that their participation at tournaments started to fall off as well. This phenomenon started to become evident at least as early as 1994, when as part of a story about a summer's day at Hempstead's annual drill *Newsday* wrote:

"On the field's thatch of sun-battered yellow grass, firefighters barbequed chicken wings, sang along to their car radios and readied their equipment for the events ahead.

"Near one such encampment a length of hose lay coiled like a snake under a hulking flatbed used for trailering the trucks; not far away a pair of blue chinos stretched out on the ground, as if their owner had melted away.

"That last image could almost be a metaphor for the slow decline of the Long Island fire drill, which in its last peak, in the late '70's, effortlessly drew young members eager to follow in their fathers' rubber-booted footsteps. Though today that dynastic imperative still helps fill the ranks, the firehouses' blue-collar base has been diverted by spiraling taxes and shaky economics.

"'We have two guys who hold down two full-time jobs in addition to being on our racing team,' says Patrick Cooke, a member of the Hempstead Flukes (which, along with the department's other team, the Yellow Hornets, hosted last weekend's drill) and first vice president of the Nassau County Volunteer Firemen's Parade and Drill Team Captain's Association. 'But it's hard to find volunteers when people are working more than one job and can't make the long hours needed to practice and compete in these events.'"[cdxvi]

The frustration of the teams only grew as time went on. Almost a decade later, it was exemplified in July of 2003, when the following appeared on "digger74", an internet blogging site used by teams members and those associated with the sport:

Author **Topic: Is FD Racing dying? (Read 2,710 times)**

Danger
Grievance Committee
member is offline

Is FD Racing dying?
« Thread Started on Jul 4, 2003, 12:50am » quote

Is it just me, or is the number of teams racing in the suffolk cty/nassau cty/dansville drills lower than usual? Will there be a lot of post entries? Or is motorized starting to go the way of old fashioned, a slow death by loss of interest?

Joined: Jul 2001
Gender: Male
Posts: 979
Karma. 0
[Exalt Smite]

Link to Post Back to Top Logged

PS This is what part of the alphabet would look like if Q and R were eliminated.

roadrashSJFD
Full Member
member is offline

Joined: Sept 2002
Gender: Male
Posts. 29
Location. ithaca college
Karma. 0
[Exalt Smite]

Re: Is FD Racing dying?
« Reply #1 on Jul 4, 2003, 11:42am »

quote

I think there is a definite lack of interest, which is a shame because the actual sport this year looks like it will be exciting. While there are a few clear favorites, no team is really dominating any tournament or any event and there are a bunch of teams that are starting to get competitive. The downfall in my opinion is all the protests and hose tests. I know that we have to run this sport fairly and go by the book, but before we seriously come up with a way to speed along tournaments, we have to realize that we need interest in this sport to keep it running or else the red book will be useless.

-pj

Link to Post Back to Top Logged

GO BOMBERS

sticktime
HARDCORE MEMBER !!!
member is offline

Joined: Jul 2001
Gender: Male
Posts. 296
Karma. 4
[Exalt Smite]

Re: Is FD Racing dying?
« Reply #3 on Jul 5, 2003, 11:42am »

quote

Well, in the last 3 Suffolk County Drills we've had

2002- 20 teams
2001- 19 teams
2000- 20 teams

This year there are 15 teams on the program. But that includes Coram and Bayport. Will they show? Who knows. But teams not included are Copaigue, Lindenhurst, N Babylon and Riverhead. So it looks as if we'll end up with around 20 again.

Is Racing Dying? No, it's not dying. Suffering may be more appropriate. The way I see it, Racing will begin to die out when the gracious hosts we now have, refuse to host drills any longer because the day has simply become too long for people to sit through. Less people means less money When a host sees the profits decrease, why would they put in all that hard work of putting on a drill, when the bottom line is going to be less and less? And in some cases it's not going to be up to a racing team if they will continue to host a drill, it'll be up to their department or district.

It's been said on this message board a million times before, but has never been put to action by anyone yet. The day of a drill needs to be shorter The downtime that occurs when there a dispute over a rule or when a decsion needs to be made is ridiculous. What's the solution? That is a good question. There isn't one thing that can be changed for the days to magically become shorter But it has to start somewhere. There is no rational reason a drill with 22 teams cannot end in 6 to 7 hours. But as we all know, most times that is not the case.

That's why come September when it's time for rule changes more thought should be put into the far reaching ramifications of a rule change, beyond the contest (or contests) it is being applied for That would be a good place to start.

Link to Post Back to Top Logged

Heard some people talking just the other day....and they said you were gonna put me on a shelf...

F-Man
Grievance Committee
★★★★★
member is offline

Joined: Jul 2001
Gender· Male ♂
Posts. 819
Location. New York
Karma. 0
[Exalt   Smite]

Re: Is FD Racing dying?
« Reply #18 on Jul 6, 2003, 8.52pm »

quote

You're right, these tournaments are dragging on more and more and half of it's to blame on more and more rules.
Anyone notice the new rule books?
We had to get these because of all the new rules over the past few years made it too thick for the old book covers to work.   Every year we have to make new rules to try and solve a problem that happened the previous year
We need to come up w/ one simple rule You're caught cheating, you're gone! No way around it, no lawsuits no nothing, You're gone!
Maybe that will make some people second guess how they want to compete.
The biggest problem is that we all have so much $$$ invested in this great sport, that we tend to think that we have to do whatever it takes to win
Sometimes I wish we could just revert back to 1960 when the rulebook was just 50 pages (Parade rules included)

Link to Post . Back to Top   Logged

Please stand clear of the doors. Por favor mantengase alejardo de las puertas.

Then, the following posting appeared farther down in the thread:

digtime
Digger Probie
☆
member is offline

Joined: Jul 2003
Posts: 1
Karma. 0
[Exalt   Smite]

Re: Is FD Racing dying?
« Reply #22 on Jul 7, 2003, 2 14pm »

quote

I remember when I was growing up being overwhelmed by the amount of people in the stands and general enthusiasm that accompanied drills but unfortunately in today's politically correct atmosphere this sport is not an acceptable means of recreation.
The enrollment numbers of vollie fire departments across LI are way down hence the number of teams and fans are down. The only way to increase interest is to go outside the FD's and publicize but then the catch 22 is if you go outside and publicize, increased interest is followed by increased scrutiny as seen last year at the Charity Drill when the Newsday photographer chased after the Farmingville C-Truck under the arch and into the street.
In attempting to publicize and attract outside audiences you are taking a great risk of cutting our own legs out from under us. We should focus on keeping the fans we already have and bringing back the fans we have lost over the years. I agree with the fact that there are too many drills and the ones we have run too long I personally love racing but I also love my summer's and its hard to get people to come out- week in week out for drills every Saturday for ten hours. For many it is a weekend killer

Link to Post   Back to Top   Logged

The worst, however, was yet to come, as in addition to all the haggling on the track and low attendance at drills, the sport was starting to slide into the center of a public relations nightmare. Fortunately, this did not catch the teams unawares, as there was, perhaps, a scent in the air or some handwriting on the wall that alerted them to the impending storm, as that last posting by "digtime" implies: "...unfortunately in today's politically correct atmosphere this sport is not an acceptable means of recreation." The reference to the *Newsday* photographer "chasing after the Farmingville C-truck under the arch and out into the street" is also an indication that they knew something was up; they just had no idea when or how hard it would hit them.

Chapter 18: The Controversy Lives On

What was brewing was a series of newspaper articles by *Newsday* wherein they, in their eyes, undertook an in-depth investigation of the volunteer fire service on Long Island in an attempt – as far as anyone can tell – to dissect it and lay bare its history and inner workings, and by doing so educate and inform the public regarding everything – ostensibly both good and bad – about it. The project was both extensive and long-term, as some firefighters even recall being interviewed for it well prior to the attack on the World Trade Center in 2001.

When the eight-part series of articles, entitled "Fire Alarm", was finally published on November 13th, 2005, members of the Long Island fire service – and not just those associated with the drill teams – were incredulous. A large majority of them agreed that while some deference was made to some of the real problems the modern-day volunteer fire service was facing ("...amid more demands on fire departments, including increased medical calls, the need for dedicated volunteers is ever present...") the majority of the stories were negative and detrimental in tone, focusing instead on anything and everything they could find wrong with it. Indeed, the tag line under the headline read, "Costs are spiraling. Response times lag. Volunteers are scarce. Can LI's troubled fire services be fixed?"

What was especially damaging to the drill team community was Part 5 of the series that appeared on November 17th entitled "Getting Your Money's Worth", wherein reporter Elizabeth Moore went after the budget practices of the local fire districts, who use tax revenue from local residents to cover their operating costs. While she can be seen on one hand as simply presenting a series of facts from which the public could draw their own conclusions, many felt that the *way* she presented those facts was nowhere near "fair and unbiased":

"There is another type of spending that fire officials call a training expense: tournament drill teams. Firefighters in about 45 Long Island districts compete by leaping off the back of speeding, specially modified racing vehicles to climb ladders or hook up hoses.

"Only a small part of each department's membership can be on a racing team, and district spending on them has been a frequent target of auditors. 'It makes them committed. It makes them in good shape,' said Islip Commissioner Benjamin Galletto, 72, who was on three state championship teams as a young man. 'They train well. They get to know various types of hoses and ladders. I know some people don't see it, but it's a benefit. We've never had a complaint from anybody, and they pack them in at the tournaments.'

"In its quest for its first state championship since 1965, Islip fire officials spent more than $250,000 on its Wolves racing team from 2001 to 2003, flouting a warning from their own auditor that the spending seemed 'inordinate.'

"In 2000, it paid $37,500 for a custom-built dragster called a 'C' hose tender, records show. The next year it spent $72,128 on the drill team. In 2002, it spent $139,500 for a cream-white tractor-trailer to transport the racing vehicles to tracks around Long Island. The district says the tractor also pulls their fire-safety trailer.

"In 2003, Islip paid $37,051 for engine work and racing tires and spent $15,329 on an electric-blue mobile locker room emblazoned with the Wolves logo (last year it was designated for general department use). It also bought $11,000 worth of Wolves T-shirts, jerseys, garment bags and Junior Wolves uniforms.

"The Wolves captured the state title in August of that year at the New York State Motorized Drill in Ridge.

"'I guess we got our money's worth,' Galletto said.

"Records show the district spent thousands more in taxpayer money celebrating the victory. Team members received championship jerseys, shirts and posters and a parade down Islip's Main Street to a party behind the firehouse where town and county officials delivered congratulations from a mobile stage while volunteers handed out free hot dogs, hamburgers, beer and soda."[cdxvii]

To their credit, once the entire series of articles had appeared in print *Newsday* created a forum on their website that allowed readers to respond with any comments they had about the series. To their disappointment, most likely, reaction to the articles was swift, intense and largely negative. As *Newsday* itself put it:

"Newsday's 'Fire Alarm' series prompted more than 1,100 e-mails, letters and telephone calls. The majority were critical of the eight-day series, with many characterizing it as an attack on volunteer firefighters. But a significant number praised the project as a comprehensive report on the costs and the quality of service provided by Long Island fire agencies."[cdxviii]

While the latter may have been true, anyone taking the time to review those responses could see that compared to letters praising the series, those that were critical of it were much more vehement in tone, as in this example:

"If part of the mission of Newsday is to report in the best interests of the public, can you explain to me how your witch hunt on volunteer firefighters is in the best interest of the public? These men and women volunteer their time. They hold regular jobs and are then called out at all hours of the day or night, regardless of weather, to face most people's greatest fear -- fire. They are pulled away from holidays, celebrations, quiet dinners and kids' sporting events to protect us and our property. They give up countless hours to train and risk life for zero pay. Has Newsday considered the reason people aren't volunteering is because they don't feel the appreciation?"[cdxix]

But as welcome as the letters of support were, the series had still done a significant amount of damage to both the public image of the Long Island firefighters and the morale of the members of its two drill team associations. They began to feel that they were now under intense public scrutiny, and that their respective fire districts would now be under a microscope when it came to anything having to do with firematic drill teams and the allocation of any kind of funding for them – regardless of whether they actually funded them or not. Many teams, especially the smaller ones, were self-funded and received no monies at all from their fire districts,

N PLAYOFFS newsday.com

Go to the Web for live updates on today's Yanks game and other extras.

PLUS FULL-COLOR TOM GLAVINE POSTER IN THE FOLD A48-49

Mets Win Again, Take 2-0 Lead SPORTS

NEWSDAY PHOTO / J. CONRAD WILLIAMS JR.

Yankees Fall To Tigers, 4-3 SPORTS

Newsday

★★★★ SPORTS FINAL

LONG ISLAND

NEWSDAY.COM | FRIDAY, OCT. 6, 2006 | LONG ISLAND EDITION | 50¢

ONLY IN NEWSDAY

Angry Commissioner On Fire Budget:

Money To Burn

Farmingville's $2.9M spending plan includes $55,000 for new race car A3

FILE PHOTO BY BRIDGET O'BRIEN

The Farmingville Bandits Drill Team competes at an event in 2005.

digger74 General Members Only Today's Newsday- 10/6/06

Page 1 of 4 » Jump to page 1 Go search reply send topic to friend print bookmark

Author **Topic: Today's Newsday- 10/6/06 (Read 2,096 times)**

digger127
Full Member
member is offline

Today's Newsday- 10/6/06
« Thread Started on Oct 6, 2006, 7:40am » quote

This kind of front page publicity is not good for this sport. They are hunting for us and will use every possible avenue to bring us down. Be careful out there!!

Joined: Sept 2005
Gender: Male
Posts: 29
Karma. 0
[Exalt Smite]

Link to Post Back to Top Logged

but even that didn't matter; in the heartland of the very area of New York State that created modern-day firematic competition, the actual death of the sport, spoken of before only in passing and with mild concern, suddenly now seemed like a very real possibility. Those involved in the sport became cautious about everything they did, and suspect of anything that might contribute to someone gaining the wrong impression about them; even going so far as to change the web address of the NYSDTCA from www.fdracing.com to www.nysdrillteams.com, with the idea being that the word "racing" in the URL was too politically incorrect. In short, following the publication of the story, the Long Island teams "circled the wagons", as it were, and hunkered down to weather the storm.

It was a wise decision. In spite of a largely negative reaction to their "Fire Alarm" series, *Newsday* – a newspaper that the reader may recall had several times in the past offered high praise about the local drill teams and their positive effect on the fire service – kept up a full-court press by printing another article less than one year later that criticized a particular fire district's allocation of funds involving their drill team:

"The Farmingville fire district has passed a 2.9-million dollar budget that sets aside hundreds of thousands of dollars for renovating its stations' recreation rooms – and a new race car.

"Approved last month by three of the five district fire commissioners, the budget has been criticized as excessive and misprioritized by the lone commissioner who voted against it.

"'As a taxpayer, I'm outraged and as fire commissioner, I'm baffled,' said Stephen Curtin, adding that the residents will see their fire district taxes increase by 26 percent, or approximately $75. 'Our priority should be on saving lives and property.'

"Other Farmingville fire commissioners said spending had remained unchanged for years despite rising costs. They said the allocations are judicious and will enhance public safety.

'We know the fire service better than anyone,' said Commissioner Norman Neill. 'We basically know what is and what is not needed.'

"Other Farmingville commissioners also defended their upcoming spending. 'Anyone can make a mountain out of a molehill,' said [Commissioner Edward Jr.] Stewart, noting one of the team's two cars – both are needed to compete – is damaged and more than 20 years old."[cdxx]

On the drill team blog, all team members could do after this most recent attack was vent their frustration over what they felt was an attempt at a smear campaign by *Newsday,* and their lack of any viable way to get the public to understand their side of things:

iclimbfordoc
Full Member
member is offline

Here we go Dan

Joined: Aug 2006
Gender: Male
Posts. 15
Location. Port
Karma. 0
[Exalt Smite]

Re: Today's Newsday- 10/6/06
« Reply #13 on Oct 6, 2006, 3:28pm »

quote

loo371 you have some valid points about how it looks from an outsiders view but you state "what are you going to do to stop it" The question is if you support your drill team like you say you do then you can help stop it as well as the rest of the association. I belong to the same team as RRrafflesales and we did raise our own money to buy the c-truck that we have as well as the trailer and pick up. Every drill team does not get everything that they want and need, some drill teams get very little support from their departments. Some of the drill teams have to submit a budget to thier department for appoval and that can be denied.
We as members of a very special brotherhood of volunteer firefighters should let the outsiders know that we are here to protect the very community that we live and work in and when most of joined the volunteer fire department our sole intention was not only because of the drill team but help our coumminty, and that drill teams have been around the fire department for a very long time and the sport has evolved into something that I don't think even Wilber envisioned, but the sport is to resemble firematic functions ie. the bucket brigade. The drill team is also a recruitment for a younger member who might be looking to join as well as retention for that slighty older member who may be thinking of leaving the fire service. Which most departments are having the problem with both recruiting and retention. The only answer that I can come up with that maybe the drill team association or even Southern NY Firemans Association write an article to Newsrag with all of the positives of being a member of the very Special Brotherhood the Volunteer Firefighter

As depressed as they may have been at the time, the teams did have one ace in their hand that coincidentally and quite happily became available in 2005, at just about the same time the "Fire Alarm" series hit the stands: a full-length, professionally produced documentary about the sport that presented it in both an honest and positive light. Called Race For Glory: The Story of Drill Team Racing, the film, which had been two years in the making and was produced by Triangle Park Entertainment of New York City, featured narration by actor William Baldwin, and in addition to highlighting the ups and downs of the Lindenhurst "Snails" during the course of an entire summer competition season, included rare archival footage of drills from 1927 and 1950, interviews with many different team members about what it's like to be part of a team, and exciting and different camera angles of contests (such as from the back step of a C-truck) that most people had never seen.

In spite of a bit of grumbling by some of the teams about how long it took for the production company to deliver the finished product, the film, by most accounts, went a long way towards the DTCA's goal of enabling the public to have a better understanding of what it is that they actually did, and the history and reasons behind it.

Unfortunately, however, the "ripple effect" of the newspaper articles had already started to spread. If it was *Newsday's* intention all along to be a "whistle-blower", in their eyes they probably succeeded. Most likely as a result of those stories, the New York State Comptroller's Office eventually decided to conduct a financial audit of six different fire districts - and not just on Long Island, but throughout the entire state. As their final report, issued in April of 2007, stated:

"The objective of our audit was to determine whether fire district officials at the six districts we audited used public resources entrusted to their care as economically and appropriately as possible for the period January 2005 through August 2006 (for administrative, travel and other expenditures), and for the period January 1998 through August 2006 (for competition/drill team expenditures). Our audit addressed the following related questions:

- Have fire district officials established appropriate internal controls over travel and other administrative expenses?
- Do fire districts have policies and procedures for installation/inspection dinners?
- Are fire district expenditures for competition/drill teams actual and necessary to accomplish the districts' missions?"[cdxxi]

To its credit, the comptroller's office was very careful to remain as neutral and objective as possible, and not draw or imply any conclusions regarding its findings. Regarding the first and second questions, the report stated:

"General Municipal Law and Town Law allow for fire districts to send officers and employees to conferences and conventions and to provide refreshments at official functions. Town Law also authorizes districts to hold annual inspection/installation dinners. Beyond this statutory authorization, there is little detailed guidance provided to fire districts on exactly what expenses are appropriate and allowable. In addition, very little guidance is provided by law with regard to competition equipment.

"Districts spend varying amounts of money (often significant) sending officers and employees to conferences and conventions; holding annual inspection/installation dinners; providing refreshments at official functions; and providing funding for competition equipment. Due to the extent of these expenditures, and the potential for abuse that exists in these areas, it is incumbent upon the Board of Fire Commissioners of each district to establish appropriate policies and procedures and an effective system of internal controls. This will provide some assurance that associated funds will be expended only for proper municipal purposes and that such funds will not be expended in amounts greater than necessary to achieve the goals of the organization.

"This audit's results and recommendations are resources for Fire District officials to use in effectively managing operations and in meeting the expectations of their constituents."[cdxxii]

If the third question sounds familiar, it shouldn't come as a surprise. The reader will recall that this was just another variation of a question that had been asked, in one form or another, dozens of times over the course of the last 200 years, ever since the first two volunteer fire companies in the New World raced each other to put out a fire: is competition between firemen – and the purchase and maintenance of equipment for same - a malfeasance, or is it a healthy and legitimate way for firefighters to sharpen their skills while at the same time promoting recruitment in the fire service?

Again, the State Comptroller's Office stepped carefully, sticking mainly to the facts:

"There is no express authority for the purchase of equipment exclusively related to competition teams for fire districts. We found that five of the six districts expended taxpayer dollars on racing/drill team activities. The amount spent by districts varied greatly, ranging from $0 to about $115,000 during the scope period. Further, none of the five districts have passed resolutions indicating the benefits of competitive racing or adopted formal policies to govern related expenditures. Of note, two of the six districts have discontinued or suspended competitive racing activities at this time."[cdxxiii]

"Fire districts may expend funds only for purposes that are expressly authorized by or are implied by statute. There is no express authority for the acquisition of equipment and vehicles for use exclusively in competitive racing events. Town Law Section 176(11) authorizes the Board of Fire Commissioners to 'provide for public drills, parades, funerals, inspections and reviews of the fire district fire department, or any company or unit thereof.' Further, Section 176(13) authorizes the Board of Fire Commissioners to 'purchase apparatus and equipment for the extinguishment and the prevention of fires'. In addition, Volunteer Firefighter's Benefit Law Section 5(e) states firefighters are covered for worker's compensation and other benefits during these drills.

"New York State Comptroller's Opinion 87-79 addresses the appropriateness of purchasing competitive racing equipment. The opinion states: 'There is no express authority for the acquisition of equipment and vehicles for utilization by the fire department in competitive racing events, nor is there express authority for fire department personnel to participate in such events at fire district expense.'

"While it may be permissible to provide funding for competitive racing events and related costs in instances where the Board of Fire Commissioners has determined that the team activities contribute to the efficiency of the fire department in the performance of its official duties, none of the districts audited could document that such expenditures represent a prudent use of public funds.

"All six fire districts have drill teams and own equipment that is used exclusively for competitive racing/drill activities. None of the districts have ever used these vehicles to fight fires; in some cases these pieces of equipment are not even 'street-legal.' Therefore, the districts generally have to acquire a transport vehicle and these vehicles can be expensive. For example, the Manhasset/Lakeville Fire District spent approximately $80,000 on their tractor-trailer to transport their competition equipment.[13] Team membership related to the competition teams ranged from 10 to 30 volunteers, corresponding to 8 to 20 percent of the total volunteer membership.

"Out of the six fire districts audited, only four had significant expenditures related to the racing teams. Barnard did not utilize any district funds on their drill team and Spencerport only spent about $3,800 on racing-related activities. The other four districts on Long Island spent about $322,000 from January 2005 to August 2006. West Islip kept records as far back as 1998, Syosset from 1999, and West Sayville and Manhasset from 2000. Expenditures during these prior periods totaled approximately $683,000.

"Three of the six districts indicated that the racing team and associated activities served to attract volunteers and had a training purpose. Two of the districts indicated that the racing activities served as morale boosters and improved camaraderie among the members. However, at these five districts, nothing was presented to indicate that the Board, by resolution,

[13] Taken literally, the Comptroller's statement, "None of the districts have ever used these vehicles to fight fires" is not 100% accurate. In 1981 the East Meadow Fire Department used the tractor of the "Meadowlarks" tractor-trailer transport rig to tow a box semi-trailer away from a commercial building that was burning. West Islip had a similar occurrence in 2004.

documented how the team activities contributed to the efficiency of the district in the performance of its official duties. In addition, the Board of Fire Commissioners at all six fire districts, did not adopt a formal policy to clearly establish the circumstances under which related expenditures were to be paid, the types of expenditures that would be paid, and the extent to which such expenditures would be paid."[cdxxiv]

In closing, the New York State Comptroller's Office had this recommendation:

"The Boards should document the cost benefit of maintaining a competition racing team. In the event the Board determines that the benefits of such activities exceed the costs, the Board should adopt a formal policy with respect to competition equipment. Such a policy should indicate the circumstances under which such expenditures will be paid, the types of expenditures that will be paid, and the extent to which such expenditures will be paid. All disbursements should then be made in accordance with the policy."[cdxxv]

And, it concluded the document with this note:

"STATE COMPTROLLER'S NOTE: The Comptroller's Office has not taken an adverse opinion to the existence competition/drill teams provided for by fire districts. Comptroller's Opinion 87-79 states that: *a fire district may not expend moneys to fund a fire department racing/drill team for purposes of competitive tournaments unless the board of fire commissioners determines that the team activities contribute to the efficiency of the fire department in the performance of its duties and responsibilities.* The conclusion presented by this Office is that the Districts have not determined the benefit of their competition team and/or adopted resolutions of policy that speaks to benefits of this team."[cdxxvi]

When the initial study was completed, a copy of that report was sent to each Board of Fire Commissioners of the six fire districts that had been audited, and they were asked to provide comments on the findings. Included in those comments were these statements:

"The Board believes that the operation of the drill team provides a unique opportunity for members of the Fire Department to increase their training skills in firefighting techniques and operations. The competitions and training hone the physical skills of the volunteer firemen. Ultimately, these skills are utilized in the every day operations of the Fire Department."

"Competitive drills are events that often attract other members of the community and expose them to the work of their fire department. Pride in the local fire department often translates into pride in the community making the community a better place to live. If these teams are to be deemed unnecessary for a local community, it should be a local choice. If these teams continue to remain important vibrant parts of the fire department, the local board should have the discretion to continue to support them through the purchase of vehicles and equipment."

And, more significantly:

"The District has not done a cost-benefit study which would be the basis for a written policy. This study will be completed in the near future after all other aspects of the audit have been rectified."[cdxxvii]

The Coram "Hellraisers" compete in the Cart Ladder Contest at the 2008 Keith Fairben Memorial Old Fashioned Drill at New Hyde Park

While the opinions and sentiments between the six fire districts audited regarding the effectiveness and overall conclusions of the study may have differed, the net result was the same: everyone associated with the sport agreed that after having been held under a microscope, to both public and official scrutiny, they had passed (ahem) "muster", and would live to compete another day. True, some changes would need to be made, both in policy and attitude, but the Comptroller's opinion – combined with the large showing of support from the public in response to the *Newsday* articles – heartened the NYSDTCA and its teams, and left them feeling like things would finally be OK…for now.

In truth, however, the sport was never really in any danger of actually *dying*. What was seriously wounded (aside from some pride) was the progression of the *motorized* segment of the sport. Lost in the shuffle of accusations and finger-pointing was the fact that Old Fashioned contests and drills of the sport had always continued on, out of the limelight, through thick and thin, ever since – and in spite of – the introduction of the automobile. For many smaller towns and villages across the state that were either unable or unwilling to support a motorized team, fielding an Old Fashioned team was a viable, affordable alternative, and one which many felt still provided just as much fun, excitement and competition as the motorized teams did, while at the same time "sticking to tradition".

Like everything else, the popularity of these Old Fashioned tournaments rose and fell over time. In spite of the sport becoming fully motorized in 1923, interest in the Old Fashioned contests, such as Cart Ladder, Cart Hose, Efficiency Replacement, etc. never died out, and Old Fashioned tournaments – *sans* any

West Islip "Raiders" Hydrant Man Steve Ferrante goes airborne while competing in the B-Hose Contest at the 2007 New York State Drill at Ridge

kind of motor – continued to be held, separately, ever since that time. Although they would never reach the level of popularity as the motorized segment of the sport did, the early 1960's and early 1970's saw renewed interest in these types of contests, and in 1976 the creation of an Old Fashioned State Drill, apart from the motorized version, allowed for Old Fashioned State Champions to be crowned each year. That year it was held in Lindenhurst, with the "Frogs" of Carle Place holding the honor of winning that very first title.

Happily, the old fashioned musters of the New England States Veteran Firemen's League have also stood the test of time, and continue to this day. Their version of the sport went through several ups and downs as well: the 1960's saw a seemingly contradictory gain in popularity in the musters but fewer participants and the 70's saw a jump in the average number of musters held each year (16, which was attributed to the influence of America's Bicentennial celebration), but the 80's were just the opposite. According to Dixon, "New England League Musters in Rhode Island, Connecticut and Massachusetts drew less than ten handtubs each. This was one of the lowest points in the 97-year history of the League."[cdxxviii] Not long after, however, things turned around for them: "On the brink of extinction just twelve months earlier, the New England League experienced a life-saving reversal in 1990, which started its Centennial Celebration."[cdxxix] Today, an average of four musters are still held each year in villages in Connecticut, New Hampshire, New York, Massachusetts and Maine, where in 2008 the old handtub "Governor Bradstreet" took First Place honors in the Class A division after "playing" a stream of 195' 3-3/4". The victory was made even sweeter for the fact that the muster was held in Bath – the location of the very first recorded New England muster, on July 4th, 1849.

It was a different story with the State Firemen's Association. Ever since the NYSVFP&DTCA began rotating the location of their State Drill in the late 1990's, FASNY never again offered any kind of tournaments at their conventions – at least none that had to do with any kind of firematics. Starting in 1998 they began to offer alternative forms of competition, such as "Winterfests" at Lake Placid, where firefighters could compete with one another individually or in five-man teams in contests such as tobogganing, bobsledding, tug-of-war or hose-relay races, where teams would be required to ski a slalom or downhill course while all five members simultaneously held on to a 50-foot length of fire hose. During the summer months, FASNY offered Driver Challenges to their membership, which were not that much different in principle than Emergency Vehicle Operations Courses that all fire apparatus drivers in the country are required to have taken in order to be legally qualified to drive a fire truck. The State Association continues to hold these "Firefighter Games" to this day.

But while not associated with their conventions, there was, in fact, a period of time beginning in the 1980's when FASNY once again became involved with firematic competition.

It began in 1984, when brothers Mark and Jeff French of the Mellenville Fire Co. began thinking of ways to put on a fundraiser for their fire company. Being that they were located only a few miles east of Hudson, the site of FASNY's New York State Firemen's Home, they approached the Home's Board of Trustees with the idea of organizing a Firemen's Field Day, whose proceeds would benefit not only Mellenville but the Firemen's Home as well. The Board was enthusiastic in its response, and soon after the brothers were able to recruit retired FDNY firefighter and noted author Dennis Smith to act as their first Honorary Chairman for event.

The first New York State Firefighter's Competition and Field Day was held in July of the following year, and it was an immediate success. Touted as a "family affair", volunteer firefighters from not only all parts of New York but other states as well descended upon the grounds of the Firemen's Home to socialize and take part in four different "old fashioned" firematic contests, including a Ladder Climb, a Midnight Alarm (Dressing) contest, and two types of Dry Hose contests.

Everyone, including FASNY, was thrilled over how well things had turned out, so with their blessing the French brothers continued to hold the Field Day at the Firemen's Home on an annual basis. The event reached its epitome four years later on July 11th, 1992, when it was held in conjunction with the 100th Anniversary of the Firemen's Home. It was a day everyone would remember: over 100 different fire companies from 26 different states made the trip to Hudson, including firefighters from Canada and Brazil. There was a massive parade that included scores of antique and modern apparatus (100 and 200 pieces, respectively), and attendees were even able participate in breaking a Guiness Book world record for the longest Bucket Brigade in history, after 2,271 people passed 50 buckets of water along a 11,471-foot course.

That series of musters represented a form of "modern heydey" for FASNY and the Firemen's Home, with over 100,000 attendees visiting the home over the course of eight events, but it was not to last. Much like it had with the NYSVFP&DTCA, FASNY's relationship with Mellenville became bogged down with disputes over particulars, and no one could seem to agree. Simple things such choosing a date for the following year's event became a headache. As a result, the 100th Anniversary muster was the last that Mellenville was associated with after the French brothers decided to call it quits as far as organizing the Field Day was concerned.

The muster continued to be held for two more years after that, with the City of Hudson taking over the responsibility of organizing it, but by that time the writing was on the wall: attendance both years had dropped dramatically each successive

The 2008 Southern New York Drill at Ridge

year, so after the 1994 event was finished Hudson gave it up as well.

In 2007, the French brothers were contacted again by the Trustees of the Firemen's Home, who asked if they would be willing to once again help organize a Field Day there. While they were not initially opposed to the idea, talks between them and FASNY again got mired down when it came to choosing a date that worked for both Mellenville and the Firemen's Home, and as a result the whole thing once more fell through. It remains to be seen if the Firemen's Field Days in Hudson are a thing of the past, or will perhaps be held again one day in the future.

Regarding motorized competition, the events of the last five years seem to have ultimately been beneficial to the drill team community, who, if not a bit humbled, have taken all that recent experiences have had to offer and applied it in positive ways. As this book is completed there is a renewed sense of optimism in the sport. While it may never again reach the level of popularity it enjoyed at the turn of the 20th century, and it may take a few years for the number of teams competing at the State Drill to once again reach the levels seen during the 70's and 80's, the 2008 State Tournament at Main-Transit was, by all accounts, the best that can be remembered in many years.

It has been said many times before, but it still bears repeating: "the more things change, the more they stay the same." In 1849, when those two gentlemen of New York City's "high society" said:

"And do you really mean to tell me, Mortimer, that these men get no pay for dragging those engines through the streets? You must be joking."

"No, Traxton, they do not receive a farthing for their labor, but they are low fellows and call it sport."

...it was apparently no different than the attitude of the public towards volunteer firefighters in the 21st century, who still question the fabric and morals of the firemen, and the legitimacy of drill team competition as a "sport". FASNY struggled over the worthiness of holding tournaments at their conventions for 84 years without ever reaching a definitive conclusion; it is therefore no wonder that the modern drill team associations face that same skepticism to this day. However, as controversial as it may be, the sport of firematic competition will, if history is any indication, continue to survive, adapt and grow – as it has for close to 200 years.

APPENDIX I: Firematic Competition Timeline

1726	First fire brigade on Long Island (Hempstead Town, June 17th)
1824	First reference to firefighters marching in a parade (New York City, July 4th)
1826	First parade of firefighters on Long Island (Independence Day)
1835	First account of formally arranged competition between companies in New York City (Co. No.13 and Co. No.11)
1849	First recorded organized group muster (Bath, Maine, July 4th)
1855	Contest between hand-pumped engine and steam engine (New York City, February 9th)
1860	First mention of Horizontal (vs. Vertical) Muster (Springfield, Massachusetts)
1865	Competition between companies is banned in New York City as it becomes a paid fire department
1872	Firemen's Association of the State of New York (FASNY) formed (Auburn, N.Y., September 6th)
1872	1st Annual FASNY Convention (Auburn, N.Y., October 1st)
1876	First written reference to a Cart Hose Contest (4th FASNY Convention, Elmira, August 16th)
1876	First Juvenile Contest (4th FASNY Convention, Elmira, August 16th)
1876	First use of team uniforms at a tournament (4th FASNY Convention, Elmira, August 16th)
1881	First delegate from Long Island to attend a FASNY convention (Saratoga)

1882	Babylon marches at Patchogue during Decoration Day festivities (Patchogue, May 29th)
1882	First inter-county gathering of Long Island firefighters to march in a parade (Babylon, August 30th)
1883	Tournament feature first introduced on Long Island (Babylon, September 21st)
1884	First "official" tournament at a FASNY convention (Utica, August 21st)
1884	First reference to any Long Island team attending a tournament upstate (Utica, August 21st)
1886	Suffolk County Volunteer Firemen's Association (SCVFA) formed (Patchogue, August 25th)
1886	First SCVFA Tournament (Patchogue, September 9th)
1887	First "wet hose" contest (2nd Annual SCVFA Tournament, Riverhead, September 15th)
1889	FASNY crowns the Forbes Hose Company of Canastota, N.Y. as the first "State Champions" (17th Annual Convention, Plattsburgh)
1889	Last recorded Perpendicular Muster held (Providence, Rhode Island, September 5th)
1890	American Indians from the Sioux tribe compete in a tournament (Pierre, South Dakota, June 18th & 19th)
1890	New England States Veteran Firemen's League formed (Boston, November 20th)
1892	Hook & Ladder Contest #2 (now Bucket Brigade contest) first introduced (SCVFA Tournament in Greenport)
1893	First FASNY tournament held on Long Island (Coney Island)
1894	First recorded use of grandstands at a tournament (SCVFA 9th Annual Tournament, Southampton)
1894	Special Hose Contest (now Efficiency Contest) first introduced (SCVFA 9th Annual Tournament, Southampton)

1895	Jefferson County Firemen's Association formed (would become the Firefighter's Association of Northern New York)(Adams, May 29th)
1895	First use of a Hose Cart instead of a Hose Reel (SCVFA 11th Annual Tournament, Sayville)
1895	First Jefferson County Firemen's Association Annual Tournament (Adams, September 18th & 19th)
1897	First Town of Hempstead Tournament held (Freeport, June 16th)
1898	First All-Juniors Tournament (Rockaway Beach, March 21st)
1899	Nassau County is created after Greater City of New York is established the previous year
1899	Western New York Volunteer Firemen's Association formed (Lockport, November 12th)
1900	1st Annual Western New York Volunteer Firemen's Association Tournament (Lockport, June 13th)
1900	Firefighters from Kansas City compete in International Firemen's Tournament (Paris, France, August 11th & 12th)
1900	First use of pressure-regulated water supply at a tournament (15th Annual SCVFA Tournament, Lindenhurst, September 12th)
1900	First time 2nd and 3rd Place prizes are awarded at an SCVFA tournament (Lindenhurst, September 12th)
1903	Nassau County Firemen's Association (NCFA) formed
1903	First Hose Contest for Juniors held (18th Annual SCVFA Tournament, Patchogue, September 9th)
1909	First mention of an automobile in a parade (St. Lawrence and Jefferson County Firemen's Association Annual Convention, Potsdam)
1909	Chemical Contest first introduced (24th SCVFA Tournament, Southampton)
1910	Last appearance of Steamer apparatus at a tournament (25th Annual SCVFA Tournament, Sayville)

1910	Nassau County Firemen's Association joins FASNY
1911	Suffolk County Volunteer Firemen's Association joins FASNY (June 13th)
1913	First instance of motorized apparatus in a competition (28th SCVFA Annual Parade and Tournament, Bay Shore, August 27th)
1913	Women Suffragists march at 28th SCVFA Parade (Bay Shore, August 27th)
1915	First motorized contest (Motor Contest) introduced (30th SCVFA Tournament, Bayport, September 1st)
1921	Motor Hook and Ladder #2 Contest (now Dummy Contest) first introduced (33rd SCVFA Tournament, Riverhead, August 31st)
1924	Long Island Railroad ceases running Special Tournament Excursion trains
1924	First all-motorized tournament (36th SCVFA Tournament, Bay Shore, August 27th)
1926	SCVFA and NCFA adopt list of 21 Standardized Contests (Patchogue, May 12th)
1928	First Bi-County Association Motor Hose Race adopted into schedule by Jefferson-St. Lawrence Bi-County Firemen's Association
1929	First instance of State Champions being crowned for individual contests (57th FASNY Convention, Saranac Lake, August 22nd)
1929	First inter-county Long Island Championship Tournament held (Mineola, September 17th)
1933	<u>The Suffolk County Volunteer Firemen's Blue Book</u> is published by J. Fred Flugrath
1936	Western N.Y. team members march with WPA protesters in picket lines (Rochester)
1939	67th Annual FASNY Convention and Tournament held in Flushing, Queens, in conjunction with 1939 World's Fair (August 17th)
1939	First instance of a team from outside New York State (Bernardsville, N.J.) competing at a N.Y. State Tournament (Flushing, August 17th)

1940	First instance of Three Man Ladder Contest at a SCVFA Tournament
1940	Suffolk County Running Captain's Association formed (February 2nd)
1941	Introduction of First Aid Contest at 69th FASNY Convention (Geneva, August 21st)
1942	Tournaments in Suffolk County banned/suspended due to WW II (through 1946)
1945	No FASNY convention held due to WW II
1946	First post-war tournament held on Long Island (Southern New York Tournament, Islip, June 27th)
1948	Rioting takes place at FASNY's 76th Convention and Tournament in Rome (August 19th)
1949	No FASNY Tournament held; first instance of no State Champion being crowned for reasons other than war
1950	First instance of a Ladies Auxiliary competing at a State Tournament (Bucket Brigade, 78th Annual FASNY Convention, Syracuse)
1952	First scheduled nighttime tournament held under lights (North Fork Firemen's Association Tournament, Riverhead, July 31st)
1953	FASNY requires all tournament hosts to show proof of insurance before sanctioning a tournament (Binghamton, August 17th)
1953	Team reps from throughout New York State meet in Point Pleasant to discuss the formation of a State Drill Team Captains Association
1954	FASNY hears proposal from drill teams to form their own association to organize and run tournaments (Albany, March 2nd)
1954	First completely purpose-built track for tournaments inaugurated (Hempstead, July 5th)
1955	New York State Court of Appeals finds in favor allowing Workman's Compensation claims for injuries sustained at tournaments (July 8th)
1955	New York State Volunteer Firemen's Drill Team Captain's Association (NYSVFDTCA) founded (August 23rd)

1956	Last tournament sanctioned by FASNY (84th Annual Convention, Floral Park, August 23rd)
1956	Northern New York becomes fourth Area of the NYSVFDTCA (October)
1958	Spectator killed after C-truck crashes at 86th FASNY Convention in Amsterdam
1959	Western New York Drill Team Captains Association formed (formerly Monroe County)
1961	Central New York State Firematic Drill Team Association founded (October 27th)
1963	First All-Juniors Tournament held in Suffolk County (Bayport, July 12th)
1965	NYSVFDTCA State Drill featured on ABC's Wide World of Sports (again in 1968, 1977, 1979, 1980 and 1984)
1965	FASNY discontinues appointment of Tournament Statistician and disassociates itself from all drills (93rd Convention, Utica)
1966	Protests against racism interrupt Southern New York tournament in Merrick (June 18th)
1967	Arch/Judges Stand collapses, injuring 24 (95th FASNY Convention, Albany, August 24th)
1969	First Juniors State Tournament held (Bay Shore, August 30th)
1971	First Fire Police Competition held (Western New York Convention, Bowmansville)
1973	First NYSVFDTCA tournament held outside New York State (Warminster, Pennsylvania, September 22nd)
1974	First tournament held on a beach (Davis Park)
1975	Parade organization and rules come under the auspices of the State DTCA. The word "Parade" is added to NYSVFDTCA title
1976	First Old Fashioned State Tournament (Lindenhurst)

1979	Roslyn firefighter Ed Rykowski dies after falling off a transport trailer
1980	Patchogue's "Forty Thieves" give exhibition at Florida State Firefighter's Association Annual Firematics Day (again in 1985)
1982	Central Firematic Drill Team Association disbands
1986	Oceanside firefighter Steven Wade is killed during ladder practice
1986	First three-way tie for First Place at a tournament (Central Islip, East Islip, West Sayville) (Islip Town Tournament, Central Islip, August 8th)
1986	Hagerman "Gamblers" C-Truck goes off course at tournament, injuring several spectators (Labor Day Drill, Islip Terrace)
1987	East Islip "Guzzlers" and Bay Shore "Redskins" give exhibition drill in Springfield, Vermont as part of town's anniversary celebration
1988	Copenhagen "Cubs" team member Donald Henry is killed after a C-truck he is riding on crashes during practice (June)
1988	Patchogue "Forty Thieves" team member Robert Cushing is killed during practice (Patchogue, July 6th)
1989	Teams walk off track in protest of poor track conditions; no State Champion crowned (NYSVFP&DTCA State Drill, Watertown, August 26th)
1990	First NYSVFP&DTCA State Drill held not in conjunction with FASNY (Main-Transit, August 18th)
1991	Hempstead "Flukes" team member Scott Laverty passes away after being gravely injured while falling from a truck during practice six days earlier (June 6th)
1991	Television coverage of NYSVFP&DTCA state drills included in an ESPN special, "30 Years of Wide World of Sports" (January 27th)
1993	NYSVFP&DTCA Executive Board votes to rotate the location of the State Drill between the four Areas
1994	First instance of a Class-B truck from the Northern Area competing in a tournament (NYSVFP&DTCA State Drill, Main-Transit, August 16th)
1997	Six teams travel to Las Vegas to give exhibition tournament at World Firefighter's Games

2005 *Newsday* publishes Part Five of an eight-part series entitled "Fire Alarm" that denigrates Long Island fire service (November 17th)

2006 Documentary film, Race For Glory: The Story of Drill Team Racing, is released

2006 *Newsday* publishes story criticizing Farmingville's drill team expenditures

2007 New York State Comptroller's Office publishes Administrative Expenses of Fire Districts, recommending changes in how fire districts oversee and administer finances related to drill teams (April)

APPENDIX II: NYSVFP&DTCA State Champions (Motorized)

1946	East Meadow "Meadowlarks" at Riverhead
1947	Islip "Wolves" at Catskill
1948	Riverhead "Ironmen" at Rome
1949	(No Tournament)
1950	Atlantic Steamer "Rough Riders" at Syracuse
1951	West Hempstead "Westerners" at Coney Island
1952	Atlantic Steamer "Rough Riders" at East Rochester
1953	Atlantic Steamer "Rough Riders" at Amsterdam
1954	Atlantic Steamer "Rough Riders" at Buffalo
1955	Atlantic Steamer "Rough Riders" at Lake George
1956	North Bellmore "Rinky Dinks" at Floral Park
1957	St. James "Wildcats" at Elmira
1958	East Meadow "Meadowlarks" at Amsterdam
1959	Islip "Wolves" at Watertown
1960	North Bellmore "Rinky Dinks" at Riverhead
1961	Islip "Wolves" at Rochester (Penfield)
1962	Elmont "Bangtails" at Albany
1963	Central Islip "Hoboes" at Niagara Falls
1964	Atlantic Steamer "Rough Riders" at Hempstead
1965	Islip "Wolves" at Utica
1966	St. James "Wildcats" at Oyster Bay

1967	Atlantic Steamer "Rough Riders" at Lindenhurst
1968	North Bellmore "Rinky Dinks" at Syracuse
1969	Port Washington "Road Runners" at Buffalo
1970	Central Islip "Hoboes" at Watertown
1971	St. James "Wildcats" at Niagara Falls
1972	St. James "Wildcats" at Syracuse
1973	Port Washington "Road Runners" at Rochester
1974	Oyster Bay "Teddy's Boys" at Lindenhurst
1975	West Sayville "Flying Dutchmen" at Niagara Falls
1976	Central Islip "Hoboes" at Watertown
1977	West Sayville "Flying Dutchmen" at Syracuse
1978	Central Islip "Hoboes" at Hempstead
1979	Patchogue "Forty Thieves" at Rochester
1980	Patchogue "Forty Thieves" at Hempstead
1981	North Lindenhurst "Piston Knockers" at Albany
1982	Patchogue "Forty Thieves" at Olean
1983	Patchogue "Forty Thieves" at Watertown
1984	Hempstead "Yellow Hornets" at Central Islip
1985	Hempstead "Yellow Hornets" at Hempstead
1986	Patchogue "Forty Thieves" at Syracuse
1987	Lindenhurst "Snails" at Rochester
1988	Patchogue "Forty Thieves" at Deerfield
1989	(No State Champion crowned)
1990	Central Islip "Hoboes" at Main-Transit

1991	Patchogue "Forty Thieves" at Hagerman
1992	Bay Shore "Redskins" at Main-Transit
1993	Central Islip "Hoboes" at Hempstead
1994	West Sayville "Flying Dutchmen" at Main-Transit
1995	North Lindenhurst "Piston Knockers" at Clayton
1996	North Lindenhurst "Piston Knockers" at Deerfield
1997	North Lindenhurst "Piston Knockers" at Hempstead
1998	North Lindenhurst "Piston Knockers" at Clayton
1999	West Sayville "Flying Dutchmen" at Central Islip
2000	West Sayville "Flying Dutchmen" at Main-Transit
2001	West Sayville "Flying Dutchmen" at Merrick
2002	Hempstead "Yellow Hornets" at Deerfield
2003	Islip "Wolves" at Ridge
2004	West Sayville "Flying Dutchmen" at Main-Transit
2005	West Sayville "Flying Dutchmen" at Hempstead
2006	Hempstead "Yellow Hornets" at Deerfield/Hempstead
2007	West Sayville "Flying Dutchmen" at Ridge
2008	Islip "Wolves" at Main-Transit

APPENDIX III: NYSVFP&DTCA State Champions (Old Fashioned)

1976	Carle Place "Frogs" at Lindenhurst
1977	(No Tournament Held)
1978	(No Tournament Held)
1979	(No Tournament Held)
1980	(No Tournament Held)
1981	(No Tournament Held)
1982	(No Tournament Held)
1983	(No Tournament Held)
1984	Melville "Time Bandits" at Oceanside
1985	Hempstead "Yellow Hornets" at Oceanside
1986	(No Tournament Held)
1987	Hempstead "Flukes" at Hicksville
1988	Freeport Truck 1 "Dead End Kids" at Hicksville
1989	Freeport Truck 1 "Dead End Kids" at Central Islip
1990	Malverne "Chipmunks" at Central Islip
1991	Freeport Truck 1 "Dead End Kids" at Lindenhurst
1992	Malverne "Chipmunks" at Rockville Centre
1993	Freeport Truck 1 "Dead End Kids" at Copiague
1994	Freeport Truck 1 "Dead End Kids" at Setauket
1995	Freeport Engine 1 "Wide Awakes" at Baldwin
1996	Freeport Engine 1 "Wide Awakes" at Bellmore

1997	North Bellmore "Rinky Dinks" at Bellmore
1998	Freeport Truck I "Dead End Kids" at Valley Stream
1999	Freeport Truck I "Dead End Kids" at Wantagh
2000	Rockville Centre "Woodlanders" at Freeport
2001	Rockville Centre "Woodlanders" at Lynbrook
2002	Rockville Centre "Woodlanders" at Baldwin
2003	Rockville Centre "Woodlanders" at Rockville Centre
2004	Rockville Centre "Woodlanders" at Lynbrook
2005	Rockville Centre "Woodlanders" at Rockville Centre
2006	Freeport Engine I "Wide Awakes" at Babylon
2007	Rockville Centre "Woodlanders" at Hagerman
2008	Rockville Centre "Woodlanders" at Merrick

APPENDIX IV: NYSVFP&DTCA State Champions (Juniors)

NOTE: No teams from outside of Long Island had been allowed to compete for points until the 2008 season, at which time the State DTCA enacted a rule change allowing them to do so.

1969	West Sayville "Flying Dutchmen"
1970	Hewlett "Smokey Stovers"
1971	Patchogue "Forty Thieves"
1972	Lindenhurst "Snails"
1973	Lindenhurst "Snails"
1974	West Sayville "Flying Dutchmen"
1975	St. James "Wildcats"
1976	St. James "Wildcats"
1977	Lindenhurst "Snails"
1978	Lindenhurst "Snails"
1979	East Islip "Guzzlers"
1980	Central Islip "Hoboes"
1981	St. James "Wildcats"
1982	Bay Shore "Redskins"
1983	Bay Shore "Redskins"
1984	Bay Shore "Redskins"/St. James "Wildcats" (Co-Champions)
1985	Lindenhurst "Snails"
1986	Bay Shore "Redskins"
1987	Bay Shore "Redskins"
1988	Bay Shore "Redskins"

1989	Central Islip "Hoboes"
1990	East Islip "Guzzlers"
1991	Hagerman "Gamblers"
1992	Hagerman "Gamblers"
1993	Bay Shore "Redskins"
1994	Bay Shore "Redskins"
1995	Hagerman "Gamblers"
1996	Hagerman "Gamblers"
1997	Bay Shore "Redskins"
1998	Bay Shore "Redskins"
1999	Central Islip "Hoboes"
2000	Central Islip "Hoboes"
2001	Central Islip "Hoboes"
2002	Bay Shore "Redskins"
2003	Bay Shore "Redskins
2004	Bay Shore "Redskins"
2005	Central Islip "Hoboes"
2006	Bay Shore "Redskins"
2007	East Islip "Guzzlers"
2008	Central Islip "Hoboes"

APPENDIX V: State Convention & Tournament Locations

1st	1872	Auburn
	1873	(No Convention Held)
2nd	1874	Oswego
3rd	1875	Poughkeepsie
4th	1876	Elmira
5th	1877	Lockport
6th	1878	Ithaca
7th	1879	Canandaigua
8th	1880	Auburn
9th	1881	Saratoga
10th	1882	Rochester
11th	1883	Kingston
12th	1884	Utica
13th	1885	Syracuse
14th	1886	Buffalo
15th	1887	Troy
16th	1888	Cortland
17th	1889	Plattsburgh
18th	1890	Watertown
19th	1891	Herkimer
20th	1892	Niagara Falls

21st	1893	Coney Island	
22nd	1894	Oswego	
23rd	1895	Troy	
24th	1896	Lockport	
25th	1897	Schenectady	
26th	1898	Binghamton	
27th	1899	Yonkers	
28th	1900	Syracuse	
29th	1901	Buffalo	(as part of Pan American Exposition)
30th	1902	Hudson	
31st	1903	Gloversville	
32nd	1904	Dunkirk	
33rd	1905	Glens Falls	
34th	1906	Jamestown	
35th	1907	Elmira	
36th	1908	Oswego	
37th	1909	Poughkeepsie	
38th	1910	Watertown	
39th	1911	Rochester	
40th	1912	Newburgh	
41st	1913	Utica	
42nd	1914	Geneva	
43rd	1915	Oneida	
44th	1916	Rockaway Beach	

45th	1917	Flushing	
46th	1918	Lockport	
47th	1919	Syracuse	
48th	1920	Glens Falls	
49th	1921	Binghamton	
50th	1922	Freeport	
51st	1923	Hudson	
52nd	1924	Fredonia	
53rd	1925	Mamaroneck	
54th	1926	Alexandria Bay	
55th	1927	Niagara Falls	
56th	1928	Ithaca	
57th	1929	Saranac Lake	
58th	1930	Nyack	
59th	1931	Troy	
60th	1932	Lake Placid	
61st	1933	Poughkeepsie	
62nd	1934	Geneva	
63rd	1935	Greenport	
64th	1936	Tonawanda	
65th	1937	Potsdam	
66th	1938	Saranac Lake	
67th	1939	Flushing	(in conjunction with 1939 Worlds Fair)
68th	1940	Dunkirk	Tournament: at Fairgrounds in Flushing

69th	1941	Geneva	
70th	1942	Syracuse	
71st	1943	Buffalo	
72nd	1944	Rochester	
73rd	1945	Albany	(No tournament due to World War II)
74th	1946	Riverhead	
75th	1947	Catskill	
76th	1948	Rome	
77th	1949	Albany	(No tournament due to 1948 riot)
78th	1950	Syracuse	
79th	1951	New York	Tournament: Coney Island
80th	1952	East Rochester	
81st	1953	Binghamton	
82nd	1954	Buffalo	
83rd	1955	Lake George	
84th	1956	New York City	Tournament: Floral Park
85th	1957	Elmira	(First NYSVPDTCA-run tournament)
86th	1958	Syracuse	Tournament: Amsterdam
87th	1959	Watertown	
88th	1960	Riverhead	
89th	1961	Rochester	Tournament: Penfield
90th	1962	Albany	
91st	1963	Niagara Falls	
92nd	1964	New York City	Tournament: Hempstead

93rd	1965	Utica	
94th	1966	Rochester	Tournament: Oyster Bay
95th	1967	Albany	Tournament: Lindenhurst
96th	1968	Syracuse	
97th	1969	Buffalo	
98th	1970	Watertown	
99th	1971	Niagara Falls	
100th	1972	Auburn	Tournament: Syracuse
101st	1973	Rochester	Tournament: Williamson
102nd	1974	New York City	Tournament: Lindenhurst
103rd	1975	Niagara Falls	
104th	1976	Watertown	
105th	1977	Syracuse	
106th	1978	Hempstead	
107th	1979	Rochester	Tournament: Williamson
108th	1980	New York City	Tournament: Hempstead
109th	1981	Albany	
110th	1982	Olean	
111th	1983	Watertown	
112th	1984	Cheektowaga	Tournament: Central Islip
113th	1985	Uniondale	Tournament: Hempstead
114th	1986	Syracuse	
115th	1987	Rochester	Tournament: Leicester
116th	1988	Utica	

117th	1989	Watertown	(Tournament cancelled due to dangerous track conditions)
118th	1990	Niagara Falls	DTCA Tournament: Main-Transit
119th	1991	Syracuse	DTCA Tournament: Hagerman
120th	1992	Cheektowaga	DTCA Tournament: Main-Transit
121st	1993	Melville	DTCA Tournament: Hempstead
122nd	1994	Rochester	DTCA Tournament: Main-Transit
123rd	1995	Watertown	DTCA Tournament: Clayton
124th	1996	Utica	DTCA Tournament: Deerfield
125th	1997	White Plains	DTCA Tournament: Hempstead
126th	1998	Niagara Falls	DTCA Tournament: Clayton
127th	1999	Hauppaugue	DTCA Tournament: Central Islip
128th	2000	Syracuse	DTCA Tournament: Main-Transit
129th	2001	Rochester	DTCA Tournament: Merrick
130th	2002	Buffalo	DTCA Tournament: Deerfield
131st	2003	Alexandria Bay	DTCA Tournament: Ridge
132nd	2004	Melville	DTCA Tournament: Main-Transit
133rd	2005	Lake Placid	DTCA Tournament: Hempstead
134th	2006	Liverpool	DTCA Tournament: Deerfield/Hempstead
135th	2007	Cheektowaga	DTCA Tournament: Ridge
136th	2008	Rochester	DTCA Tournament: Main-Transit

APPENDIX VI: SCVFA Tournament Locations

1st	1886	Patchogue
2nd	1887	Riverhead
3rd	1888	Sag Harbor
4th	1889	Sayville
5th	1890	Islip
6th	1891	Patchogue
7th	1892	Greenport
8th	1893	Babylon
9th	1894	Southampton
10th	1895	Sayville
11th	1896	Amityville
12th	1897	Riverhead
13th	1898	Huntington
14th	1899	Bay Shore
15th	1900	Lindenhurst
16th	1901	Port Jefferson
17th	1902	Riverhead
18th	1903	Patchogue
19th	1904	Sag Harbor
20th	1905	Babylon
21st	1906	Greenport

22nd	1907	Huntington
23rd	1908	Islip
24th	1909	Southampton
25th	1910	Sayville
26th	1911	Amityville
27th	1912	Sag Harbor
28th	1913	Bay Shore
29th	1914	Riverhead
30th	1915	Bayport
31st	1916	Greenport
	1917	(No Tournament Held)
	1918	(No Tournament Held)
	1919	(No Tournament Held)
32nd	1920	Lindenhurst
33rd	1921	Riverhead
34th	1922	Sag Harbor
35th	1923	Bay Shore
36th	1924	Bay Shore
37th	1925	Amityville
38th	1926	Riverhead
39th	1927	Bellport
40th	1928	Huntington
41st	1929	Riverhead
42nd	1930	East Northport

43rd	1931	Southold
44th	1932	Bayport
45th	1933	Huntington
46th	1934	Sayville
47th	1935	Lindenhurst
48th	1936	Riverhead
49th	1937	Port Jefferson
50th	1938	Smithtown Branch
51st	1939	Bayport
52nd	1940	Sayville
53rd	1941	Port Jefferson
	1942	(No Tournament Held)
	1943	(No Tournament Held)
	1944	(No Tournament Held)
	1945	(No Tournament Held)
	1946	(No Tournament Held)
54th	1947	West Sayville
55th	1948	Huntington
56th	1949	St. James
57th	1950	Sayville
58th	1951	Islip
59th	1952	Lindenhurst
60th	1953	Riverhead
61st	1954	Brentwood

62nd	1955	Riverhead
63rd	1956	St. James
64th	1957	Bohemia
65th	1958	Bohemia
66th	1959	Brentwood
67th	1960	Islip Terrace
68th	1961	West Sayville
69th	1962	Sayville
70th	1963	Bayport
71st	1964	Brentwood
72nd	1965	Central Islip
73rd	1966	Central Islip
74th	1967	Islip
75th	1968	Brentwood
76th	1969	West Islip
77th	1970	Islip Terrace
78th	1971	Bay Shore
79th	1972	Lindenhurst
80th	1973	Brentwood
81st	1974	Sayville
82nd	1975	Bayport
83rd	1976	West Sayville
84th	1977	Central Islip
85th	1978	Copiague (Tanner Park)

86th	1979	Lindenhurst
87th	1980	Sayville
88th	1981	Riverhead
89th	1982	Deer Park
90th	1983	Patchogue
91st	1984	Patchogue
92nd	1985	Patchogue
93rd	1986	Riverhead
94th	1987	Patchogue
95th	1988	Ridge
96th	1989	Ridge
97th	1990	Bayport
98th	1991	Central Islip
99th	1992	Central Islip
100th	1993	Copiague (Tanner Park)
101st	1994	Ridge
102nd	1995	Hagerman
103rd	1996	Central Islip
104th	1997	Ridge
105th	1998	Hagerman
106th	1999	Ridge
107th	2000	Copiague (Tanner Park)
108th	2001	Central Islip
109th	2002	Ridge

110th	2003	Hagerman
111th	2004	Copiague (Tanner Park)
112th	2005	Ridge
113th	2006	Central Islip
114th	2007	Ridge
115th	2008	Central Islip

APPENDIX VII: NCFA Tournament Locations

The author regrets that this table is incomplete, as much of this information has been lost to time, and asks that the reader consider this to still be a "work in progress"

1947	Mineola	
1959	East Meadow	
1960	Albertson	
1961	Garden City Park	
1963	Floral Park	
1971	New Hyde Park	(Albertson Host)
1972	Merrick	(North Bellmore Host)
1973	New Hyde Park	(Roslyn Host)
1974	Lynbrook	(Malvern Host)
1975	New Hyde Park	(Manhassett Host)
1976	Merrick	(Freeport Host)
1977	Locust Valley	
1978	Rockville Centre	
1979	New Hyde Park	(Williston Park Host)
1982	North Bellmore	
1983	New Hyde Park	(Albertson Host)

1985	Oyster Bay	
1986	Levittown	
1987	New Hyde Park	(Great Neck "Alerts" Host)
1988	Sea Cliff	
1989	Malverne	
1990	North Massapequa	
1991	Merrick	(Massapequa Host)
1992	Oyster Bay	
1993	Merrick	(Bethpage Host)
1994	New Hyde Park	(Port Washington Host)
1995	Rockville Centre	
1997	New Hyde Park	(Roslyn Rescue Host)
1999	Woodmere	
2000	Hempstead	
2002	Oceanside	
2003	Merrick	
2004	Hempstead	
2005	Oceanside	
2006	New Hyde Park	(Mineola Host)
2007	Oyster Bay	(Bayville Host)
2008	Merrick	(North Bellmore Host)

APPENDIX VIII: NNYVFA Tournament Locations

1st	1895	Adams
2nd	1896	Carthage
3rd	1897	Sackets Harbor
4th	1898	Alexandria Bay
5th	1899	Watertown
6th	1900	Clayton
7th	1901	Theresa
8th	1902	Gouverneur
	1903	(No Tournament Held)
	1904	(No Tournament Held)
	1905	(No Tournament Held)
9th	1906	Dexter
10th	1907	Watertown
11th	1908	Alexandria Bay
12th	1909	Potsdam
13th	1910	Watertown
14th	1911	Canton
15th	1912	Carthage
16th	1913	Watertown
17th	1914	Potsdam
18th	1915	Clayton

19th	1916	Massena
20th	1917	Alexandria Bay
21st	1918	Ogdensburg
22nd	1919	Alexandria Bay
23rd	1920	Potsdam
24th	1921	Clayton
25th	1922	Massena
26th	1923	Carthage
27th	1924	Potsdam
28th	1925	Gouverneur
29th	1926	Alexandria Bay
30th	1927	Massena
31st	1928	Clayton
32nd	1929	Massena
33rd	1930	Carthage
34th	1931	Tupper Lake
35th	1932	Alexandria Bay
36th	1933	Lowville
37th	1934	Gouverneur
38th	1935	Clayton
39th	1936	Carthage
40th	1937	Tupper Lake
41st	1938	Alexandria Bay
42nd	1939	Beaver Falls

43rd	1940	Black River
44th	1941	Lake Placid
45th	1942	Alexandria Bay
46th	1943	Carthage
47th	1944	Massena
48th	1945	Lowville
49th	1946	Potsdam
50th	1947	Clayton
51st	1948	West Carthage
52nd	1949	Saranac Lake
53rd	1950	Massena
54th	1951	Tupper Lake
55th	1952	Alexandria Bay
56th	1953	Lake Placid
57th	1954	Carthage
58th	1955	Clayton
59th	1956	Lowville
60th	1957	Potsdam
61st	1958	West Carthage
62nd	1959	Tupper Lake
63rd	1960	Saranac Lake
64th	1961	Massena
65th	1962	Alexandria Bay
66th	1963	Wilmington

67th	1964	Carthage
68th	1965	Clayton
69th	1966	Lake Placid
70th	1967	Lowville
71st	1968	Tupper Lake
72nd	1969	Saranac Lake
73rd	1970	Massena
74th	1971	Clayton
75th	1972	Alexandria Bay
76th	1973	Tupper Lake
77th	1974	Saranac Lake
78th	1975	Clayton
79th	1976	Potsdam
80th	1977	Alexandria Bay
81st	1978	Saranac Lake
82nd	1979	Tupper Lake
83rd	1980	Clayton
84th	1981	Saranac Lake
85th	1982	Tupper Lake
86th	1983	Lake Placid
87th	1984	Clayton
88th	1985	Massena
89th	1986	Saranac Lake
90th	1987	Clayton

91st	1988	Tupper Lake
92nd	1989	Deerfield
93rd	1990	Massena
94th	1991	Saranac Lake
95th	1992	Saranac Lake
96th	1993	Lake Placid
97th	1994	Clayton
98th	1995	Tupper Lake
99th	1996	Massena
100th	1997	Lake Placid
101st	1998	Tupper Lake
102nd	1999	Deerfield
103rd	2000	Clayton
104th	2001	Lake Placid
105th	2002	Tupper Lake
106th	2003	Lake Placid
107th	2004	Saranac Lake
108th	2005	Lake Placid
109th	2006	Clayton
110th	2007	Massena
111th	2008	Lake Placid

APPENDIX IX: WNYVFA/DTCA Tournament Locations

1st	1900	Lockport
2nd	1901	Albion
3rd	1902	Tonawanda
4th	1903	Medina
5th	1904	Gowanda
6th	1905	Niagara Falls
7th	1906	Brockport
8th	1907	Batavia
9th	1908	Albion
10th	1909	Dansville
11th	1910	Lockport
12th	1911	Springville
13th	1912	Rochester
14th	1913	Lancaster
15th	1914	Bath
16th	1915	Niagara Falls
17th	1916	Middleport
18th	1917	Dansville
19th	1918	LaSalle
20th	1919	North Tonawanda
21st	1920	Batavia

22nd	1921	Albion
23rd	1922	Lancaster
24th	1923	Holley
25th	1924	Williamsville
26th	1925	Warsaw
27th	1926	Kenmore
28th	1927	Niagara Falls
29th	1928	Depew
30th	1929	Tonawanda
31st	1930	Rochester
32nd	1931	Dunkirk
33rd	1932	Arcade
34th	1933	Holley
35th	1934	Tonawanda
36th	1935	Albion
37th	1936	Rochester
38th	1937	Williamsville
39th	1938	Cheektowaga
40th	1939	Fredonia
41st	1940	Brockport
42nd	1941	Dansville
43rd	1942	Niagara Falls
44th	1943	Williamsville
45th	1944	Holley

46th	1945	Batavia
47th	1946	Fredonia
48th	1947	Williamsville
49th	1948	Warsaw
50th	1949	East Rochester
51st	1950	Albion
52nd	1951	Mount Morris
53rd	1952	Williamsville
54th	1953	Greece
55th	1954	Tonawanda
56th	1955	Dansville
57th	1956	Batavia
58th	1957	Lockport
59th	1958	Holley
60th	1959	Williamsville
61st	1960	Hilton
62nd	1961	Dunkirk
63rd	1962	Warsaw
64th	1963	Caledonia
65th	1964	Randolph
66th	1965	Alexander
67th	1966	Williamsville
68th	1967	North Tonawanda
69th	1968	Scottsville

70th	1969	Dansville
71st	1970	Perry
72nd	1971	Bowmansville
73rd	1972	Brockport
74th	1973	Alexander
75th	1974	Albion
76th	1975	Warsaw
77th	1976	Wrights Corners
78th	1977	Williamsville
79th	1978	Gowanda
80th	1979	Hamlin
81st	1980	Dansville
82nd	1981	Alexander
83rd	1982	Albion
84th	1983	Wilson
85th	1984	Dunkirk
86th	1985	Hilton
87th	1986	Dansville
88th	1987	Elba
89th	1988	Albion
90th	1989	Warsaw
91st	1990	Henrietta
92nd	1991	Spring Brook
93rd	1992	Perry

94th	1993	Wheatfield
95th	1994	Dansville
96th	1995	Alexander
97th	1996	Dunkirk
98th	1997	Albion
99th	1998	Perry
100th	1999	Shawnee
101st	2000	(No Tournament Held)
102nd	2001	(No Tournament Held)
103rd	2002	(No Tournament Held)
104th	2003	Albion

NOTE: Western teams discontinued holding tournaments at the Western Convention after 2003 due to concerns over track safety at host facilities

APPENDIX X: Central New York State Firematic Drill Team Association Past Champions

1963 West Albany

1964 West Albany

1965 South Schenectady

1966 South Schenectady

1967 South Schenectady

1968 Delmar

1969 Delmar

1970 Delmar

1971 Delmar

1972 Fuller Road

1973 Fuller Road

1974 Fuller Road

1975 Harmony

1976 Boght

1977 Elsmere

1978 Elsmere

1979 Elsmere

1980 Elsmere

1981 Elsmere

APPENDIX XI: Selected Special Parade Prizes From the 1800's

The Tri-County Firemen's Association Convention, 1894

Finest Appearing Company in Line *($75.00)*

Company Having the Largest Number of Men in Line *($25.00)*

Company Coming the Longest Distance *($25.00)*

Hand Engine and Company Coming Over the Greater Number of Railroads *(Case of Bottled Ale)*

Finest Hose Carriage in Line *(Case of Wine)*

Finest Looking Foreman in Line *(Picture)*

Finest Appearing Chief in Line *(Nickel Plated Lantern)*

Smallest Fireman in Line *(Derby Hat)*

Leanest Looking Fireman in Line *(Pair of Silk Suspenders)*

The Most Popular German in Line *(Lemonade Set)*

Tiredest Looking Fireman in Line *(Hammock)*

Fireman with the Longest Moustache in Line *(Mug)*

Fireman with the Smallest Feet *(Pair of shoes)*

Finest Appearing Exempt Firemen's Association in Line *(Keg of Burton Ale)*

Company Having the Most Beer Drinkers *(Keg of beer)*

Company Having the Largest Number of Smokers in Line *(Two boxes "Old Hutch" cigars)*

Sweetest Looking Company in Line *(Candy Cane)*

Fireman Having the Longest Nose *(Pair of Slippers)*

Fireman Having the Largest Feet *(Foot Rest)*

Hungriest Fireman in Line *(One Ham)*

Tallest Fireman in Line *(Sweater)*

Company Having the Finest Band *(Picture)*

Biggest Dude [sic] in Line *(Bottle of Parfumery)*

Blindest Fireman in Line *(Pair of Spectacles)*

Fireman Having the Largest Number of Children *(One Lamb)*

Largest German in Line *(Bottle of Rhine Wine)*

Company Having the Oldest Apparatus in Line *(Keg of Beer)*

Company Staying the Longest in Town During the Convention *(Keg of India Pale Ale)*

Craziest Looking Fireman in Line *(Box of Cigars)*

Most Intelligent Looking Fireman in Line *(One-Year Subscription to the Hudson Republican)*

Company Arriving in Town First *(Keg of Beer)*

Sleepiest Looking Fireman in Line *(Sofa Pillow)*

Oldest Chief in Line *(Bottle of Brandy)*

Largest Fireman in Line *(Barrel of Apples)*

Youngest Chief in Line *(Cigar Case)*

Fireman Wearing the Poorest Shoes *(Pair of Shoes)*

Most Ignorant Looking Fireman in Line *(One-Year Subscription to the Hudson Sunday Journal)*

Fireman with the Longest Hair *(Box of Cigars)*

Fireman Wearing the Largest Number of Badges *(Hand-Painted Badge)*

Chief Coming the Longest Distance *(Briarwood Pipe)*

Largest Drum Major in Line *(Barrel of Apples)*

Fireman in Line Having the Youngest Baby *(5-lb Box of Candy)*

Ugliest Looking Fireman *(Milk Can Full of Refreshments)*

Fireman Having the Largest Hands *(Pair of Gloves)*

Fireman Having the Largest Head *(Box of Cigars)*

Laziest Fireman in Line *(Two Sets of Underwear)*

Homeliest Looking Assistant Chief in Line *(Alligator Card Case)*

Fireman Measuring the Most Around the Waist *(Carving Set)*

Company Having the Poorest Uniforms in Line *(One Hundred Cigars)*

Company Having the 10 Heaviest Active Members in Line *(Case of Whisky)*

Finest Hook and Ladder Truck in Line *(One Hundred Cigars)*

Baldest Fireman in Line *(Box of Cigars)*

Noisiest Fireman in Line *(Large Tin Horn)*

Fireman in Line who is the Tallest Carpenter *(Carpenter's Pocket Knife)*

Company Having the Largest Number of Grand Army Men in Line *(United States 4x6 Bunting Flag)*

Jefferson County Firemen's 1st Annual Convention, 1895

Leanest Looking Fireman in Line *(Sack of Flour)*

Fireman with the Largest Feet *(Box of Cigars)*

Fireman with the Largest Nose *(10 lbs. Granulated Sugar)*

Fireman with the Most Wrinkles in his Face *(Free Shave)*

Homeliest Chief *(Box of Cigars)*

Handsomest Chief *(Derby Hat)*

Fireman with the Largest Mouth *(Box of Cigars)*

Fireman with the Least Number of Teeth *(Plate of Oysters)*

Fireman with the Largest Number of Children *(A Horsewhip)(!)*

Fireman Having the Youngest Boy Baby *(Pair of Vases)*

Fireman Having the Youngest Girl Baby *(Pair of Shoes)*

Fireman Having the Youngest Twin Babies *(1 Box Home Medicine Remedy)*

Poorest Fireman *(10 lbs. Granulated Sugar)*

Fattest Fireman *(Sack of Flour)*

Tallest Fireman *(Plated Teapot)*

Shortest Fireman *(1 lb. Tea)*

Youngest Fireman *(Fancy Cup & Saucer)*

Oldest Fireman *(10 lbs. Granulated Sugar)*

Fireman Who Could-But-Never-Did Tell a Lie *(A Hatchet)*

Fireman Telling the Biggest Fish Story *(Piece of Music)*

Most Idiotic-Looking Fireman *(Pair of Rubbers)*

Darkest Fireman *(A Bath)*

Fireman with the Dirtiest Teeth *(Bottle of Dentifrice & Toothbrush)*

Heaviest Fireman *(1 Yr. Subscription to the Journal)*

Fireman Having the Largest Ears *(Nickel-Plated Tea Kettle)*

Fireman Taking 1st Prize in Drill *(Silk Umbrella)*

Fireman Having the Smallest Feet *(Souvenir Spoon)*

Fireman Having the Smallest Head *(A Cap)*

Handsomest Fireman *(Pair of Kid Gloves)*

Fireman with the Largest Head *(Shampoo)*

Tallest Chief Engineer *(Cigar Case)*

Foreman Having Most Men in Line *(Sewing Table)*

Tallest Fireman *(1 lb. Coffee)*

Youngest Married Fireman *(Carpet Sweeper)*

Scaliest-Looking Fireman *(Fish Pole)*

Sickliest-Looking Fireman *(One Box Home Remedy)*

ILLUSTRATION CREDITS

Unless otherwise specified below, all photos and illustrations were provided from the personal collections of the various individuals and teams from throughout all of New York State that make up the sport of firematic drill team competition. The author is indebted to them for their willingness to share them for this book.

Page	Credit
12-16	Michael Heller/911 Pictures
22	The Metropolitan Museum of Art, The Edward W.C. Arnold Collection of New York Prints, Maps and Pictures, Bequest of W.C. Arnold, 1954 (54.90.496). Image © The Metropolitan Museum of Art
24	I.N. Phelps Stokes Collection, Miriam and Ira D. Wallach Division of Art, Prints and Photographs, The New York Public Library, Astor, Lenox and Tilden Foundation
28	(Provided by Author)
31	Engraving Depicting Firemen in a Brawl as Buildings Burn by Frederic Theodore Lix Source: Corbis
39	Courtesy of Steve Thomas of the Bath, Maine Fire Department
41	(Provided by Author)
42	The Race, Currier & Ives PR 068, Negative #2416. Collection of the New York Historical Society
49	(Provided by Author)
50	(Provided by Author)
58	Courtesy of the New York State Museum of Firefighting, Hudson, N.Y. (Photo by Author)
62	Courtesy of the Lockport Public Library
63	(Provided by Author)
66	(Provided by Author)
67	(Provided by Author)
70	From The Suffolk County Volunteer Firemen's Blue Book, J. Fred Flugrath, © 1933
71	Courtesy of the New York State Museum of Firefighting, Hudson, N.Y. (Photo by Author)
72	Courtesy of the Brookhaven Town Volunteer Firefighter's Museum
76	Courtesy of the New York State Museum of Firefighting, Hudson, N.Y. (Photo by Author)
77	(Provided by Author)

80	Courtesy of the New York State Museum of Firefighting, Hudson, N.Y. (Photo by Author)
81	Courtesy of the New York State Museum of Firefighting, Hudson, N.Y. (Photo by Author)
83	Courtesy of the Long Island Collection of the East Hampton Public Library
84-86	Courtesy of the New York State Museum of Firefighting, Hudson, N.Y. (Photo by Author)
88	Courtesy of the Greenport Fire Department
89	Courtesy of the Riverhead Fire Department
92	Courtesy of the New York Historical Society (Photo by Author)
93	From The Suffolk County Volunteer Firemen's Blue Book, J. Fred Flugrath, © 1933
94	Courtesy of the New York Historical Society (Photo by Author)
96	Courtesy of the Theresa Fire Department
99	The Brooklyn Daily Eagle Online, The Brooklyn Public Library-Brooklyn Collection
101	Courtesy of the New York State Museum of Firefighting, Hudson, N.Y. (Photo by Author)
103	The Brooklyn Daily Eagle Online, The Brooklyn Public Library-Brooklyn Collection
104	Courtesy of the Thousand Island Museum, Clayton, N.Y.
105	Courtesy of the New York Historical Society (Photo by Author)
106	The Queens Borough Public Library, Long Island Division, Howard Conklin Photographs
112	Courtesy of the Sag Harbor Fire Museum
113	The Queens Borough Public Library, Long Island Division, Howard Conklin Photographs
114	Courtesy of the Babylon Historical Society (both images)
116	Courtesy of the Greenport Fire Department
117	Courtesy of the Brookhaven Town Volunteer Firefighter's Museum (Photo by Author)
118, 119	Courtesy of the New York State Museum of Firefighting, Hudson, N.Y. (Photo by Author)
120	Courtesy of the Lockport Public Library
121, 122	The Queens Borough Public Library, Long Island Division, Joseph Burt Photographs
123	Courtesy of the West Sayville Fire Department
124	Courtesy of the Greenport Fire Department
125	Courtesy of the New York State Museum of Firefighting, Hudson, N.Y. (Photo by Author)
126	Courtesy of the Thousand Island Museum, Clayton, N.Y. (Photo by Author)

127	Courtesy of the Sag Harbor Fire Museum
129	Courtesy of the Greenport Fire Department
132	Courtesy of the Greenport Fire Department (Photo by Author)
133	From The Suffolk County Volunteer Firemen's Blue Book, J. Fred Flugrath, © 1933
135	(Provided by Author)
136	Courtesy of the Riverhead Fire Department
137	Courtesy of the Huntington Fire Department
139	County of Nassau, Department of Parks, Recreation and Museums, Division of Museum Services, Photo Archives Center
143	Courtesy of the New York State Museum of Firefighting, Hudson, N.Y. (Photo by Author)
147	The Queens Borough Public Library, Long Island Division, Joseph Burt Photographs (both images)
148	Courtesy of the Potsdam Fire Department (top photo)
149	Courtesy of the Thousand Island Museum, Clayton, N.Y.
150	Courtesy of the Greenport Fire Department (top image)(Photo by Author)
152	Courtesy of the West Sayville Fire Department
154	Courtesy of the North Tonawanda Public Library
155	Courtesy of the New York State Museum of Firefighting, Hudson, N.Y. (Photo by Author)
156	Courtesy of the Riverhead Fire Department
157	Courtesy of the Historical Association of South Jefferson County (Photo by Author)
162	Courtesy of the New York State Museum of Firefighting, Hudson, N.Y. (Photo by Author)
165	Courtesy of the Thousand Island Museum, Clayton, N.Y.
167	Courtesy of the New York State Museum of Firefighting, Hudson, N.Y. (Bottom image) (Photo by Author)
174	Courtesy of the New York State Museum of Firefighting, Hudson, N.Y. (Photo by Author)
175	Courtesy of the Brookhaven Town Volunteer Firefighter's Museum (Photo by Author)
178	Courtesy of the West Sayville Fire Department
179	Courtesy of the Jervis Public Library, Rome, N.Y.
183	Courtesy of the Thousand Island Museum, Clayton, N.Y.
187	Courtesy of the Bay Shore Fire Department

190	Courtesy of the Thousand Island Museum, Clayton, N.Y.
193	Courtesy of the Firemen's Memorial Exhibit Center of Western New York
198	Courtesy of the West Sayville Fire Department
207	Courtesy of the West Sayville Fire Department
208	Courtesy of the New York State Museum of Firefighting, Hudson, N.Y. (Photo by Author)
212	Courtesy of the New York State Museum of Firefighting, Hudson, N.Y. (bottom image) (Photo by Author)
214	Courtesy of the Lindenhurst Fire Department
218, 219	Courtesy of the West Sayville Fire Department
222	Courtesy of the New York State Museum of Firefighting, Hudson, N.Y. (Photo by Author)
231	Courtesy of the New York State Museum of Firefighting, Hudson, N.Y. (both images) (Photos by Author)
232	Courtesy of the Brookhaven Town Volunteer Firefighter's Museum (bottom image) (Photo by Author)
240	Courtesy of the West Sayville Fire Department
243	Michael Heller/911 Pictures
257	Courtesy of Jay Zinger/Zingphoto
258	Michael Heller/911 Pictures
259	Courtesy of the New York State Museum of Firefighting, Hudson, N.Y. (Photo by Author)
260	Michael Heller/911 Pictures

REFERENCES

[i] *America's Own & Firemen's Journal*, December 29th, 1849, p.3
[ii] So Others Might Live, Terry Golway, © 2002 Basic Books, p.11
[iii] Outline of the History of Long Island, compiled by Harriett Stryker-Rodder, published by the Long Island Historical Society, 1955
[iv] So Others Might Live, Terry Golway, © 2002 Basic Books, p.16
[v] The Romance of Firefighting, Robert S. Holzman © 1956 Bonanza Press, p.3
[vi] The Story of the Volunteer Fire Department of the City of New York, George W. Sheldon, 1882
[vii] The Romance of Firefighting, Robert S. Holzman © 1956 Bonanza Press, p.61
[viii] As You Pass By, Kenneth Holcomb Dunshee, ©1952, Hastings House, p.114
[ix] The Romance of Firefighting, Robert S. Holzman © 1956 Bonanza Press, p.62
[x] Our Firemen: The History of the New York Fire Departments, A.E. Costello, Chapter 13
[xi] The Romance of Firefighting, Robert S. Holzman © 1956 Bonanza Press, p.62
[xii] The Old New York Volunteer Fire Department, George William Sheldon, *Harpers New Monthly Magazine*, January, 1881
[xiii] Fire Engines, Firefighters, Paul C. Ditzel, © 1984 Bonanza Books
[xiv] The Story of the Volunteer Fire Department of the City of New York, George W. Sheldon, 1882
[xv] ibid
[xvi] As You Pass By, Kenneth Holcomb Dunshee, ©1952, Hastings House
[xvii] The Story of the Volunteer Fire Department of the City of New York, George W. Sheldon, 1882
[xviii] The Romance of Firefighting, Robert S. Holzman © 1956 Bonanza Press, p.63
[xix] "Enjine! Enjine!", Kenneth Holcomb Dunshee, published by Harold Vincent Smith © 1939 for the Home Insurance Co., p.56
[xx] Outline of the History of Long Island, compiled by Harriett Stryker-Rodder, published by the Long Island Historical Society, 1955
[xxi] *America's Own & Firemen's Journal*, August 18th, 1849
[xxii] As You Pass By, Kenneth Holcomb Dunshee, ©1952, Hastings House
[xxiii] The Old New York Volunteer Fire Department by George William Sheldon, in *Harpers New Monthly Magazine*, February 1881
[xxiv] Our Firemen: The History of the New York Fire Departments, A.E. Costello, Chapter 13
[xxv] The Old New York Volunteer Fire Department by George William Sheldon, in *Harpers New Monthly Magazine*, February 1881
[xxvi] Our Firemen: The History of the New York Fire Departments, A.E. Costello, Chapter 13
[xxvii] The Old New York Volunteer Fire Department by George William Sheldon, in *Harpers New Monthly Magazine*, February 1881
[xxviii] Our Firemen: The History of the New York Fire Departments, A.E. Costello, Chapter 12
[xxix] [xxix] So Others Might Live, Terry Golway, © 2002 Basic Books, p.72
[xxx] Our Firemen: The History of the New York Fire Departments, A.E. Costello, Chapter 12
[xxxi] ibid
[xxxii] *The Long Islander*, May 22nd, 1840
[xxxiii] As You Pass By, Kenneth Holcomb Dunshee, ©1952, Hastings House, p.117
[xxxiv] Our Firemen: The History of the New York Fire Departments, A.E. Costello, Chapter 12
[xxxv] *America's Own and Fireman's Journal*, May 5th, 1849, p.1
[xxxvi] So Others Might Live, Terry Golway, © 2002 Basic Books, p.121
[xxxvii] ibid, p.125
[xxxviii] Our Firemen: The History of the New York Fire Departments, A.E. Costello, Chapter 47
[xxxix] Per their website, http://www.hudsonfire.com/content/beginning/
[xl] The Centennial History of the Utica Fire Department, p.17
[xli] The Firemen and Fire Departments of Troy, N.Y., Arthur James Weise, ©1895 Weed-Parsons Printing Co., p.103
[xlii] ibid, p.109-110
[xliii] ibid
[xliv] The *Utica Daily Observer*, June 19th, 1849
[xlv] The Romance of Firefighting, Robert S. Holzman © 1956 Bonanza Press, p.49
[xlvi] *America's Own and Firemen's Journal*, August 13th, 1853, p.3
[xlvii] The *Niagara Falls Gazette*, June 9th, 1858, p.3
[xlviii] The Firemen and Fire Departments of Troy, N.Y., Arthur James Weise, ©1895 Weed-Parsons Printing Co., p.149
[xlix] The *Albany Evening Journal*, September 29th, 1858

[l] The Albany *Atlas & Argus*, September 30th, 1858
[li] ibid
[lii] The *Albany Evening Journal*, September 30th, 1858
[liii] The Albany *Atlas & Argus*, October 1st, 1858
[liv] The *Albany Evening Journal*, October 1st, 1858
[lv] The Albany *Atlas & Argus*, October 1st, 1858
[lvi] The *Brooklyn Daily Eagle*, "City News & Gossip", October 1st, 1858, p.3
[lvii] As You Pass By, Kenneth Holcomb Dunshee, ©1952, Hastings House, p.117
[lviii] The *Brooklyn Daily Eagle*, "City News & Gossip", February 4th, 1854, p.3
[lix] The Firemen's Muster – America's Sport, Stan Dixon
[lx] ibid
[lxi] The Firemen and Fire Departments of Troy, N.Y., Arthur James Weise, ©1895 Weed-Parsons Printing Co., p.150-151
[lxii] ibid
[lxiii] ibid, p.152
[lxiv] The Firemen's Muster – America's Sport, Stan Dixon
[lxv] ibid
[lxvi] As You Pass By, Kenneth Holcomb Dunshee, ©1952, Hastings House
[lxvii] The *Brooklyn Daily Eagle*, September 28th, 1863
[lxviii] The *Lockport Journal*, June 14th, 1867
[lxix] *The Long Islander*, September 14th, 1860
[lxx] The Romance of Firefighting, Robert S. Holzman © 1956 Bonanza Press, p.129
[lxxi] The Firemen's Muster – America's Sport, Stan Dixon
[lxxii] The Romance of Firefighting, Robert S. Holzman © 1956 Bonanza Press, p.79
[lxxiii] Our Firemen: The History of the New York Fire Departments, A.E. Costello, Chapter 47
[lxxiv] The Firemen's Muster – America's Sport, Stan Dixon
[lxxv] 100 Years of Service, 1872-1972, compiled and edited by Charles L. Radzinsky, ©1972 Firemen's Association of the State of New York, p.14
[lxxvi] ibid, p.16
[lxxvii] ibid
[lxxviii] ibid
[lxxix] ibid, p.26
[lxxx] FASNY Secretary's notes, 1874
[lxxxi] ibid, 1875
[lxxxii] ibid
[lxxxiii] ibid, 1876
[lxxxiv] ibid, 1880
[lxxxv] ibid, 1882
[lxxxvi] ibid, 1881
[lxxxvii] The Firemen's Muster – America's Sport, Stan Dixon
[lxxxviii] The *Suspension Bridge Journal*, August 18th, 1883, p.3
[lxxxix] ibid
[xc] *The New York Times*, May 31st, 1882
[xci] *The Budget*, June 3rd, 1882, p.3
[xcii] The *South Side Signal*, June 3rd, 1882
[xciii] *The Budget*, June 3rd, 1882, p.1
[xciv] *The Budget*, September 2nd, 1882
[xcv] FASNY Secretary's notes, 1882
[xcvi] The *South Side Signal*, September 29th, 1883
[xcvii] ibid
[xcviii] FASNY Secretary's notes, 1884
[xcix] ibid
[c] *The Sag Harbor Corrector*, September 20th, 1884
[ci] *The Long Island Traveler*, September 26th, 1884, p.1
[cii] ibid
[ciii] ibid

[civ] *The Sag Harbor Corrector,* September 20th, 1884
[cv] *The Brooklyn Daily Eagle,* September 16th, 1897, p.3
[cvi] *The Hempstead Inquirer,* September 11th, 1885
[cvii] *The New York Times,* September 10th, 1885, p.2
[cviii] The Suffolk County Volunteer Firemen's Blue Book, J. Fred Flugrath, © 1933, p.12
[cix] *The Hempstead Inquirer,* September 11th, 1885
[cx] *The Argus,* Patchogue, N.Y., 1898, Vol. XIV, No.43
[cxi] *The Brooklyn Daily Eagle,* September 16th, 1902
[cxii] The Suffolk County Volunteer Firemen's Blue Book, J. Fred Flugrath, © 1933, p.12
[cxiii] *The Patchogue Advance,* September 11th, 1886
[cxiv] ibid
[cxv] The Suffolk County Volunteer Firemen's Blue Book, J. Fred Flugrath, © 1933, p.24
[cxvi] *The New York Sun,* September 9th, 1886
[cxvii] *The Patchogue Advance,* September 11th, 1886
[cxviii] *Shore Lines,* March, 1987
[cxix] ibid
[cxx] ibid
[cxxi] ibid
[cxxii] ibid
[cxxiii] *The New York Sun,* September 9th, 1886
[cxxiv] The Suffolk County Volunteer Firemen's Red Book, J. Fred Flugrath, ©1907, p.12
[cxxv] The New York State Volunteer Fireman, June 1937
[cxxvi] The Firemen's Muster – America's Sport, Stan Dixon
[cxxvii] *The Brooklyn Daily Eagle,* September 30th, 1886, p.4
[cxxviii] *The Brooklyn Daily Eagle,* September 16th, 1886
[cxxix] *The Sag Harbor Express,* September 22nd, 1887
[cxxx] ibid
[cxxxi] ibid
[cxxxii] *The Long Island Traveler,* September 23rd, 1887
[cxxxiii] From Minutes of the Babylon Fire Department
[cxxxiv] *The Sag Harbor Express,* September 22nd, 1887
[cxxxv] ibid
[cxxxvi] ibid
[cxxxvii] *The East Hampton Star,* September 9th, 1888
[cxxxviii] ibid
[cxxxix] The Suffolk County Volunteer Firemen's Blue Book, J. Fred Flugrath, © 1933, p.25
[cxl] From the World Wide Web
[cxli] *The Brooklyn Daily Eagle,* June 18th, 1890, p.4
[cxlii] From Wikipedia.org
[cxliii] *The Brooklyn Daily Eagle,* September 5th, 1889, p.4
[cxliv] *The Brooklyn Daily Eagle,* September 4th, 1890, p.4
[cxlv] ibid
[cxlvi] ibid
[cxlvii] ibid
[cxlviii] The Firemen's Muster – America's Sport, Stan Dixon
[cxlix] From http://www.handtubs.com/nesvfl.htm
[cl] The Suffolk County Volunteer Firemen's Blue Book, J. Fred Flugrath, © 1933, p.25
[cli] *The Brooklyn Daily Eagle,* September 5th, 1894
[clii] The Suffolk County Volunteer Firemen's Blue Book, J. Fred Flugrath, © 1933, p.33
[cliii] From The History of the Bayport Fire Department – 100 Years of Dedicated Service, 1891 - 1991
[cliv] The *Niagara Falls Gazette,* August 17th, 1892, p.1
[clv] ibid
[clvi] FASNY Secretary's notes, 1893
[clvii] *The New York Times,* Sunday, August 13th, 1893, p.2
[clviii] *The Brooklyn Daily Eagle,* September 16th, 1893, p.5

[clix] ibid
[clx] ibid
[clxi] ibid
[clxii] *The Sag Harbor Express*, September 6th, 1894
[clxiii] The History of the Bayport Fire Department – 100 Years of Dedicated Service, 1891 - 1991
[clxiv] *The Brooklyn Daily Eagle*, April 23rd, 1894, p.12
[clxv] *The Brooklyn Daily Eagle*
[clxvi] *The Brooklyn Daily Eagle*, June 16th, 1897, p.3
[clxvii] The *Jefferson County Journal*, June 4th, 1895, p.5
[clxviii] The *Jefferson County Journal*, September 24th, 1895
[clxix] ibid
[clxx] Courtesy Charles Zigrosser
[clxxi] From History of the Sag Harbor Fire Department 1819 – 1853 by Russella J. Hazard, p.10
[clxxii] From the West Sayville FD website, www.wsfd.com
[clxxiii] From an article by Darrick Hart, at http://members.aol.com/fireriter/
[clxxiv] The Suffolk County Volunteer Firemen's Blue Book, J. Fred Flugrath, © 1933, p.29
[clxxv] *The Riverhead News*, September 6th, 1897
[clxxvi] *The Riverhead News*, September 11th, 1897
[clxxvii] ibid
[clxxviii] *The Brooklyn Daily Eagle*, September 16th, 1897, p.3
[clxxix] The Romance of Firefighting, Robert S. Holzman © 1956 Bonanza Press, p.104
[clxxx] *The Brooklyn Daily Eagle*, June 9th, 1898
[clxxxi] *The Watertown Daily Times*, June 17th, 1898, p.5
[clxxxii] *The Watertown Daily Times*, June 9th, 1899, p.5
[clxxxiii] *The Lockport Union Sun*, June 14th, 1900
[clxxxiv] ibid
[clxxxv] *The Brooklyn Daily Eagle*, August 28th, 1899, p.6
[clxxxvi] *The Brooklyn Daily Eagle*, September 13th, 1899, p.8
[clxxxvii] *The Brooklyn Daily Eagle*, September 8th, 1900, p.4
[clxxxviii] ibid
[clxxxix] ibid
[cxc] The Firemen's Muster – America's Sport, Stan Dixon
[cxci] both ibid
[cxcii] *The Brooklyn Daily Eagle*, July 26th, 1900, p.16
[cxciii] The Romance of Firefighting, Robert S. Holzman © 1956 Bonanza Press, p.123
[cxciv] ibid
[cxcv] *The Brooklyn Daily Eagle*, August 21st, 1901, p.1
[cxcvi] *The Brooklyn Daily Eagle*, September 11th, 1901, p.8
[cxcvii] *The Port Jefferson Echo*, September 14th, 1901, p.2
[cxcviii] *The Orleans Republican*, August 7th, 1901, p.2
[cxcix] *The Orleans American*, August 1st, 1901, p.3
[cc] *The Orleans Republican*, August 7th, 1901, p.2
[cci] *The Brooklyn Daily Eagle*, September 16th, 1902, p.7
[ccii] *The Riverhead News*, September 20th, 1902
[cciii] The Suffolk County Volunteer Firemen's Red Book, J. Fred Flugrath, ©1907, p.43
[cciv] *The Brooklyn Daily Eagle*, September 16th, 1902
[ccv] *The Brooklyn Daily Eagle*, August 13th, 1902
[ccvi] *The Riverhead News*, September 20th, 1902
[ccvii] *The Flushing Daily Times*, September 8th, 1903
[ccviii] The Suffolk County Volunteer Firemen's Blue Book, J. Fred Flugrath, © 1933, p.116
[ccix] ibid
[ccx] Fire Engines, Firefighters, Paul C. Ditzel, © 1984 Bonanza Books
[ccxi] The Romance of Firefighting, Robert S. Holzman © 1956 Bonanza Press, p.193
[ccxii] The *Niagara Falls Gazette*, July 27th, 1905, p.1
[ccxiii] The Firemen's Muster – America's Sport, Stan Dixon

[ccxiv] *The Orleans American,* July 20th, 1908, p.1
[ccxv] *The Dansville Express,* July 29th, 1909, p.2
[ccxvi] *The Potsdam Herald-Recorder,* July 2nd, `1909, p.1
[ccxvii] ibid
[ccxviii] The Suffolk County Volunteer Firemen's Blue Book, J. Fred Flugrath, © 1933, p.40
[ccxix] *The Dansville Express,* July 29th, 1909, p.2
[ccxx] ibid
[ccxxi] ibid
[ccxxii] The Suffolk County Volunteer Firemen's Blue Book, J. Fred Flugrath, © 1933, p.27
[ccxxiii] The Firemen's Muster – America's Sport, Stan Dixon
[ccxxiv] *The Lockport Union-Sun,* July 28th, 1910, p.1
[ccxxv] The Suffolk County Volunteer Firemen's Blue Book, J. Fred Flugrath, © 1933, p.123
[ccxxvi] ibid, p.29
[ccxxvii] *The Lockport Union-Sun,* July 27th, 1910, p.1
[ccxxviii] *The Carthage Republican,* June 6th, 1912
[ccxxix] ibid
[ccxxx] *The Lockport Union-Sun,* July 28th, 1910, p.1
[ccxxxi] *The Watertown Daily Times,* June 7th, 1913
[ccxxxii] The Suffolk County Volunteer Firemen's Blue Book, J. Fred Flugrath, © 1933, p.40
[ccxxxiii] *The South Side Signal,* August 29th, 1913, p.1
[ccxxxiv] ibid
[ccxxxv] *The Riverhead News,* September 4th, 1914
[ccxxxvi] *The Potsdam Courier-Freeman,* June 3rd, 1914
[ccxxxvii] *The Patchogue Advance,* September 3rd, 1915
[ccxxxviii] *The South Side Signal,* September 3rd, 1915, p.1
[ccxxxix] ibid
[ccxl] *The Suffolk County News,* September 3rd, 1915, p.1
[ccxli] *The South Side Signal,* September 3rd, 1915, p.1
[ccxlii] ibid
[ccxliii] The Suffolk County Volunteer Firemen's Blue Book, J. Fred Flugrath, © 1933, p.29
[ccxliv] The History of the Sag Harbor Fire Department 1819-1853, Russella J. Hazard, p.18
[ccxlv] *The Watertown Daily Times,* June 3rd, 1915, p.7
[ccxlvi] ibid, p.6
[ccxlvii] The Suffolk County Volunteer Firemen's Blue Book, J. Fred Flugrath, © 1933, p.29
[ccxlviii] *The Watertown Daily Times,* June 3rd, 1915, p.6
[ccxlix] The Suffolk County Volunteer Firemen's Blue Book, J. Fred Flugrath, © 1933, p.18
[ccl] 100 Years of Service, 1872-1972, compiled and edited by Charles L. Radzinsky, ©1972 Firemen's Association of the State of New York, p.119
[ccli] *The Riverhead News,* September 1st, 1916
[cclii] *The Ogdensburg Advance,* June 20th, 1918, p.5
[ccliii] *The North Tonawanda Evening News,* July 31st, 1919, p.1
[ccliv] The Suffolk County Volunteer Firemen's Blue Book, J. Fred Flugrath, © 1933, p.24
[cclv] ibid
[cclvi] ibid
[cclvii] *The Orleans Republican,* August 31st, 1921, p.3
[cclviii] *The Riverhead News,* September 9th, 1921
[cclix] The Firemen's Muster – America's Sport, Stan Dixon
[cclx] ibid, p.40
[cclxi] From The History of the Bayport Fire Department – 100 Years of Dedicated Service, 1891 - 1991
[cclxii] *The New York Times,* September 21st, 1942, p.22
[cclxiii] *The Western New Yorker,* July 20th, 1925, p.1
[cclxiv] From Wikipedia.com
[cclxv] ibid
[cclxvi] The Suffolk County Volunteer Firemen's Blue Book, J. Fred Flugrath, © 1933, p.124
[cclxvii] ibid, p.42

[cclxviii] ibid, p.43
[cclxix] *The Riverhead News*, September 3rd, 1926
[cclxx] ibid
[cclxxi] ibid
[cclxxii] The Suffolk County Volunteer Firemen's Blue Book, J. Fred Flugrath, © 1933, p.108
[cclxxiii] ibid
[cclxxiv] *The New York Times*, August 12th, 1928, p.1
[cclxxv] *The Watertown Daily Times*, June 14th, 1928, p.17
[cclxxvi] *The Watertown Daily Times*, June 21st, 1929, p.2
[cclxxvii] ibid
[cclxxviii] 100 Years of Service, 1872-1972, compiled and edited by Charles L. Radzinsky, ©1972 Firemen's Association of the State of New York, p.148-149
[cclxxix] ibid
[cclxxx] ibid
[cclxxxi] The Suffolk County Volunteer Firemen's Blue Book, J. Fred Flugrath, © 1933, p.108
[cclxxxii] ibid
[cclxxxiii] ibid
[cclxxxiv] ibid
[cclxxxv] From FASNY Secretary's Notes, 1930
[cclxxxvi] 100 Years of Service, 1872-1972, compiled and edited by Charles L. Radzinsky, ©1972 Firemen's Association of the State of New York, p.149
[cclxxxvii] *The Riverhead News*, August 28th, 1931
[cclxxxviii] The Firemen's Muster – America's Sport, Stan Dixon
[cclxxxix] ibid
[ccxc] The History of the Bayport Fire Department - 100 Years of Dedicated Service, 1891-1991
[ccxci] ibid
[ccxcii] ibid
[ccxciii] *The Ogdensburg Advance-News*, June 9th, 1934, p.6
[ccxciv] *The New York Times*, October 16th, 1932, p.3
[ccxcv] *The Tonawanda Evening News*, July 25th, 1934
[ccxcvi] ibid, p.2
[ccxcvii] ibid
[ccxcviii] The Suffolk County Volunteer Firemen's Blue Book, J. Fred Flugrath, © 1933, p.21
[ccxcix] ibid, p.41
[ccc] *The Riverhead News-Review*, July 16th, 1937, p.1
[ccci] *The Watertown Daily Times*, June 12th, 1936, p.8
[cccii] ibid
[ccciii] *The Rochester Democrat and Chronicle*, July 30th, 1936, p.16
[ccciv] ibid
[cccv] ibid, p.15
[cccvi] ibid, p.16
[cccvii] *The Suffolk County News*, July 15th, 1938. p.1
[cccviii] *The New York State Volunteer Fireman*, Vol. 5, No. 2, p.1
[cccix] 100 Years of Service, 1872-1972, compiled and edited by Charles L. Radzinsky, ©1972 Firemen's Association of the State of New York, p.15
[cccx] *The Long Island Daily Press*, August 18th, 1939
[cccxi] *The Long Island Star-Journal*, August 18th, 1939, p.8
[cccxii] *The Long Island Daily Press*, August 18th, 1939
[cccxiii] *The Long Island Star-Journal*, August 18th, 1939, p.1
[cccxiv] From the West Sayville FD website, www.wsfd.com
[cccxv] *The Brockport Republic-Democrat*, July 25th, 1940, p.1
[cccxvi] *The Genesee County Express and Advertiser*, July 24th, 1941, p.5
[cccxvii] *The Riverhead News*, July 11th, 1941, p.1
[cccxviii] The Firemen's Muster – America's Sport, Stan Dixon
[cccxix] FASNY Secretary's Notes, 1941, p.198

[cccxx] *The Tonawanda Evening News*, July 25th, 1934
[cccxxi] *The Watertown Daily Times*, June 7th, 1943, p.8
[cccxxii] *The Amherst Bee*, July 29th, 1943, p.1
[cccxxiii] 100 Years of Service, 1872-1972, compiled and edited by Charles L. Radzinsky, ©1972 Firemen's Association of the State of New York, p.160
[cccxxiv] *The Riverhead News*, April 5th, 1946, p.2
[cccxxv] FASNY Secretary's Notes, 1944, p.139
[cccxxvi] ibid
[cccxxvii] 100 Years of Service, 1872-1972, compiled and edited by Charles L. Radzinsky, ©1972 Firemen's Association of the State of New York, p.172
[cccxxviii] From American Decades, 1940 – 1949, by Victor Bondi, © 1995 by Gale Research, p.331
[cccxxix] ibid, p.337
[cccxxx] ibid, p.340
[cccxxxi] *The Batavia Daily News*, July 25th, 1945, p.1
[cccxxxii] From the West Sayville FD website, www.wsfd.com
[cccxxxiii] The Riverhead *County Review*, August 29th, 1946
[cccxxxiv] *The Riverhead News*, August 23rd, 1946, p.1
[cccxxxv] The Riverhead *County Review*, August 29th, 1946
[cccxxxvi] ibid
[cccxxxvii] *The Riverhead News*, August 23rd, 1946, p.1
[cccxxxviii] ibid
[cccxxxix] ibid
[cccxl] ibid
[cccxli] ibid
[cccxlii] ibid
[cccxliii] *The Amherst Bee*, July 24th, 1947, p.1
[cccxliv] 100 Years of Service, 1872-1972, compiled and edited by Charles L. Radzinsky, ©1972 Firemen's Association of the State of New York, p.175
[cccxlv] ibid
[cccxlvi] *The Riverhead News*, August 27th, 1948, p.1
[cccxlvii] *Newsday*, Suffolk Edition, August 20th, 1948
[cccxlviii] *The Rome Daily Sentinel*, August 20th, 1948, p.1
[cccxlix] *The Rome Daily Sentinel*, August 21st, 1948, p.3
[cccl] *The Riverhead News*, August 27th, 1948, p.1
[cccli] American Decades, 1940 – 1949, by Victor Bondi, © 1995 by Gale Research, p.330
[ccclii] So Others Might Live by Terry Golway, Basic Books, 2002, p.207
[cccliii] The Firemen's Muster – America's Sport, Stan Dixon
[cccliv] From the West Sayville FD website, www.wsfd.com
[ccclv] The Riverhead *County Review*, September 1st, 1949
[ccclvi] *The Syracuse Herald-Journal*, August 18th, 1950, p.4
[ccclvii] *The Syracuse Herald Journal*, August 17th, 1950, p.1
[ccclviii] ibid
[ccclix] ibid, p.20
[ccclx] From FASNY Secretary's notes, 1951
[ccclxi] 100 Years of Service, 1872-1972, compiled and edited by Charles L. Radzinsky, ©1972 Firemen's Association of the State of New York, p.182
[ccclxii] *The Amherst Bee*, July 31st, 1952, p.1
[ccclxiii] From FASNY Secretary's notes, 1954
[ccclxiv] *The Tonawanda Evening News*, July 27th, 1954, p.1
[ccclxv] ibid
[ccclxvi] *The Tonawanda Evening News*, July 28th, 1954, p.1
[ccclxvii] ibid
[ccclxviii] *The News of the Tonawandas*, July 29th, 1954, p.11
[ccclxix] ibid
[ccclxx] *The Tonawanda Evening News*, July 29th, 1954, p.1

[cccclxxi] ibid
[ccclxxii] ibid
[ccclxxiii] ibid
[ccclxxiv] *The Riverhead News-Review*, August 7th, 1952, p.3
[ccclxxv] ibid
[ccclxxvi] *The Riverhead News-Review*, July 23rd, 1953, pp.1-3
[ccclxxvii] ibid, p.3
[ccclxxviii] *The Syracuse Herald-Journal*, August 18th, 1950, p.4
[ccclxxix] From FASNY Secretary's Notes, 1954
[ccclxxx] ibid
[ccclxxxi] *The Volunteer Fireman* (FASNY), September, 1956
[ccclxxxii] *The New York Times*, September 6th, 1955, p.27
[ccclxxxiii] From FASNY Secretary's Notes, 1955
[ccclxxxiv] [ccclxxxiv] 100 Years of Service, 1872-1972, compiled and edited by Charles L. Radzinsky, ©1972 Firemen's Association of the State of New York, p.196
[ccclxxxv] *The New York Times*, August 31st, 1956, p.19
[ccclxxxvi] *The Volunteer Fireman* (FASNY), September, 1956
[ccclxxxvii] *The Genesee Country Express*, July 28th, 1955, p.1
[ccclxxxviii] ibid
[ccclxxxix] ibid
[cccxc] *The Genesee Country Express*, July 28th, 1955, p.5
[cccxci] *The Batavia Daily News*, July 24th, 1956, p.1
[cccxcii] *The Batavia Daily News*, July 25th, 1956, p.1
[cccxciii] *The Wyoming County Times*, July 26th, 1962, p.1
[cccxciv] Weiss vs. Franklin Square & Munson Fire District of Town of Hempstead, 128 N.Y.S.2d 702
[cccxcv] ibid
[cccxcvi] Weiss vs. Franklin Square & Munson Fire District of Town of Hempstead, 283 A.D. 1077, 131 N.Y.S.2d 318
[cccxcvii] McKinney's Volunteer Firefighter's Benefit Law, Section 5, ©1988 by West Publishing, p.16-17
[cccxcviii] ibid, p.13
[cccxcix] *Newsday*, August 24th, 1958
[cd] From www.nhra.com
[cdi] The Firemen's Muster – America's Sport, Stan Dixon
[cdii] From http://digger74.proboards1.com/, October 28th, 2005
[cdiii] *The New York Times*, August 19th, 1964, p.39
[cdiv] 100 Years of Service, 1872-1972, compiled and edited by Charles L. Radzinsky, ©1972 Firemen's Association of the State of New York, p.196
[cdv] ibid, p.222
[cdvi] The New York Times, June 19th, 1966, p.84
[cdvii] The *New York Sunday News*, August 10, 1969
[cdviii] Unknown Long Island newspaper, February 2nd, 1975
[cdix] The *New York Sunday News*, August 10, 1969
[cdx] *Newsday*, September 8th, 1973, p.74
[cdxi] From http://digger74.proboards1.com/, December 28th, 2008
[cdxii] From the West Sayville FD website, www.wsfd.com
[cdxiii] *Newsday*, September 2nd, 1986, p.3
[cdxiv] *Newsday*, July 7th, 1988, p.2
[cdxv] From http://digger74.proboards1.com/
[cdxvi] *Newsday*, July 8th, 1994, p. B-23
[cdxvii] *Newsday*, November 17th, 2005
[cdxviii] From http://www.newsday.com/news/nyf-fire-emailform,0,6371900.customform
[cdxix] ibid
[cdxx] *Newsday*, October 6th, 2006
[cdxxi] Administrative Expenses of Fire Districts, 2006-MS-4, Office of New York State Comptroller, p.4
[cdxxii] ibid
[cdxxiii] ibid, p.5

[cdxxiv] ibid, p.23
[cdxxv] ibid, p.24
[cdxxvi] ibid, p.28
[cdxxvii] ibid, p.27
[cdxxviii] The Firemen's Muster – America's Sport, Stan Dixon
[cdxxix] ibid